HERTFORDSHIRE

HERTFORDSHIRE
a landscape history

Anne Rowe & Tom Williamson

HERTFORDSHIRE PUBLICATIONS
an imprint of
University of Hertfordshire Press

First published in Great Britain in 2013 by
Hertfordshire Publications
an imprint of
University of Hertfordshire Press
College Lane
Hatfield
Hertfordshire
AL10 9AB

British Library Cataloguing in Publication Data
A catalogue record for this book is available from the British Library

Front cover images l–r
Much Hadham church © Nigel Otter
Fields near Therfield © Bill Martin
Veteran oak pollard, Knights Hill, Westmill
Pargeting, Furneux Pelham
Hatfield House
Lavender fields, Cadwell Farm, Hitchin
River Mimram near Panshanger © Tom Crowley
Buttercups at Moor Green © Stephen Wooderson
Aerial view of Welwyn Garden City. Courtesy English Heritage

ISBN 978-1-909291-00-3

Design by Arthouse Publishing Solutions Ltd

Printed in Great Britain by Hobbs the Printers Ltd

Contents

Figures

Table

Abbreviations

BL	British Library
HALS	Hertfordshire Archives and Local Studies
HHA	Hatfield House Archive
HHER	Hertfordshire Historic Environment Record
OD	Ordnance Datum
OE	Old English
TNA:PRO	The National Archives:Public Record Office
VCH	*Victoria County History*

Units of measurement and money

1 mile = 1.6 kilometres

1 acre = 0.4 hectares

A pound (£) contained twenty shillings; one shilling (s) contained 12 old pennies (d) = 5p

Acknowledgements

Many people have helped us with this book, providing ideas, encouragement or information, or allowing access to archives or to the landscape itself. We would like to thank, in particular, Sue Flood and her team at Hertfordshire Archives and Local Studies, for all their patience and cheerful assistance; Stewart Bryant and Isobel Thompson at the Hertfordshire County Council Historic Environment Unit; the members of the Hertfordshire Gardens Trust Research Group; Bushey Museum; Mike Stanyon of the Apsley Paper Trail; David Short; Mick Thompson of Ashridge College; Sarah Spooner of the University of East Anglia; Gerry Barnes; and John Cherry, Rosemary Pateman, Julie and Bob Williams, Peter Handy and Marcus and Debbie Taverner. We would also like to thank the many people who, over the decades, have inspired us to research Hertfordshire's past, including the late Grant Longman, the late Lionel Munby, Mark Bailey and Christopher Taylor. Our special thanks go to Liz and Charlie, without whose unstinting support this book could not have been written.

The diagrams and illustrations are by the authors, with the exception of 2.2–3, 3.11, 6.6, 6.8, 8.3, 8.5–6, 8.11, 9.3–4, 10.5 and 10.7, which are reproduced by kind permission of Hertfordshire Archives and Local Studies; 1.5, 3.2 and 10.6, which are reproduced courtesy of English Heritage; the parish map (p. xii) which is reproduced with the kind permission of David Short; 6.5 which is reproduced with the kind permission of Harpenden & District Local History Society; 6.7 which is reproduced with the kind permission of Herts and Middlesex Wildlife Trust; 8.10, which is by Chris Beddall; and 9.10, which is reproduced with the kind permission of Apsley Paper Trail.

↘ Detached parts of a parish

Parish map of Hertfordshire

1. Royston was created from Therfield, Reed and Barkway in 1540
2. Baldock was created from Weston in the mid-12th century
3. Broadfield had become attached to the parish of Cottered by the 18th century
4. Nuthampstead was created from Barkway in the late 19th century
5. Hoddesdon was created from Broxbourne and Great Amwell in 1844
6. Markyate was created from Flamstead, Caddington and parts of neighbouring Bedfordshire parishes in 1888
7. Studham was moved into Bedfordshire in 1897
8. Kensworth was moved into Bedfordshire in 1897
9. Caddington was moved into Bedfordshire in 1897
10. Whipsnade, a detached part of the Bedfordshire parish which became part of Hertfordshire in 1897
11. Bovingdon and Flaunden were created from Hemel Hempstead in 1841
12. Stagenhoe was in St Paul's Walden parish but in Hitchin hundred while the rest of St Paul's Walden was in Cashio hundred

– 1 –

A county in context

Introduction

This book is about the landscape of the county of Hertfordshire. It explains the historical processes that created the modern physical environment, concentrating on such matters as the form and location of villages, farms and hamlets, the character of fields, woods and commons, and the varied forms of churches, vernacular houses, and great houses with their associated parks and gardens. But we also use these features, in turn, as forms of historical evidence in their own right, to throw important new light on key debates in social, economic and environmental history. Our focus is not entirely on the rural landscape. Most Hertfordshire people, like the majority of their fellows elsewhere in the country, live in towns and suburbs, and these too – although often created relatively recently – are a part of the county's historic landscape and have a story to tell. The purpose of this opening chapter is to set the scene, explaining some of the physical contexts and broad patterns of historical development which form the essential background to the more detailed studies presented in the chapters that follow.

Covering a mere 632 square miles (1,638 square kilometres), Hertfordshire is one of the smallest counties in England and – with a population in 2011 of over 1.1 million – among the most densely populated. In some ways it is one of the less remarkable, with no coastline, no very dramatic ranges of hills, no extensive heaths or wetlands. Lionel Munby, writing his seminal *The Hertfordshire Landscape* in the 1970s, suggested – perhaps a little unfairly – that 'no stranger would think of holidaying here'.[1] In fact, as Munby's text itself makes clear, the county has much to detain the student of history, archaeology and, above all, landscape history, not least because, in landscape terms, there is so much variety in a small compass: for Hertfordshire is a county of remarkable contrasts. Today much of it is urbanised, or suburbanised, and substantial areas of the south now form, in effect, a continuation of London. But the west, and especially the east, can still boast extensive stretches of 'unspoilt' countryside which display a rich variety, ranging from the beech woods of the Chilterns through the intimate, almost secretive clayland countryside around

Figure 1.1. View of the beech woodland of Berkhamsted Frith in the rolling landscape of the Chiltern Hills north of Berkhamsted.

Figure 1.2. The intimate boulder clay landscape of east Hertfordshire south of Braughing: the village of Standon nestles in the valley of the river Rib, surrounded by wooded hills.

Braughing and the Hadhams and the ancient coppiced hornbeam woodlands west of the Lea valley, to the sweeping panoramas of the chalklands near Royston, Baldock, Hexton and Tring (Figures 1.1 and 1.2). None of these fine landscapes is peculiar to the county itself, however. In all directions different arrangements of fields and settlements, woods and commons flow without interruption into neighbouring counties. The phrase 'the Hertfordshire landscape' is, to a large extent, meaningless.

In part this circumstance reflects the fact that the county's boundaries are strangely arbitrary and largely unrelated to natural topography. To the east Hertfordshire is separated from Essex, along most of its boundary, by the rivers Lea and Stort, but the configuration of soils and landforms to either side is virtually identical in character. The county's long northern boundary very roughly follows the line of the Chiltern escarpment and, until the end of the nineteenth century, included the parishes of Kensworth and Caddington and part of Studham (all now mainly in Bedfordshire).[2] But two marked 'salients' reach out onto the level plain of the Midlands to the north, one containing the parishes of Tring and Puttenham and the other, to the north of the town of Baldock, made up of Hinxworth, Ashwell, Radwell, Bygrave, Caldecote and Newnham.[3] The other boundaries follow no natural feature at all, seeming to pick their way through the landscape in an arbitrary manner. Until boundary changes in 1965, they included another marked 'peninsula' extending out into Middlesex to the south, embracing Totteridge, East Barnet and Chipping Barnet.[4] As the authors of the *English Place-Name Society* put it in their volume for the county, published in 1938, 'There can hardly be a county in southern England which is more obviously artificial than Hertfordshire'.[5] But such things are not so much a problem as part of Hertfordshire's interest. Embracing as it does varied countrysides which extend beyond its boundaries, it provides a particularly good opportunity to study the kinds of factors which have shaped neighbouring but contrasting landscapes.

A focus on this essential diversity of Hertfordshire's landscapes structures much of what follows and serves to some extent to distinguish our book from its great predecessor, Munby's *The Hertfordshire Landscape*, published in 1977. That book, while certainly recognising the complex variety of Hertfordshire's landscapes, nevertheless adopted a more thematic and chronological approach, discussing the physical environment of the county as a whole as this developed through successive periods, rather than concentrating on the different experience of its constituent parts. Munby's book is now more than 35 years old, moreover, and much research has taken place – into Hertfordshire specifically, and landscape history and archaeology more generally – which has served to challenge or at least modify some of its main conclusions. Some of the most important new work has appeared over the last decade: notable examples include Ros Niblett and Isobel Thompson's remarkable synthesis of the archaeology and history of St Albans, Julia Crick's publication of the charters of St Albans Abbey, the new *Historical Atlas* edited by David Short, and the impressive synthesis of the county's geology recently edited by John Catt.[6] In the chapters that follow we draw extensively on these works and on other research which has been produced over the last few decades by a wide variety of local historians, archaeologists and others. But, while presenting and interpreting the very latest work, this book does not purport to be a final and definitive statement on the county's landscape history. Such a thing does not and can never exist. New discoveries and new approaches will always undermine or modify old orthodoxies and, if this volume remains as relevant in 35 years' time as Munby's does today, we would both be very pleasantly surprised.

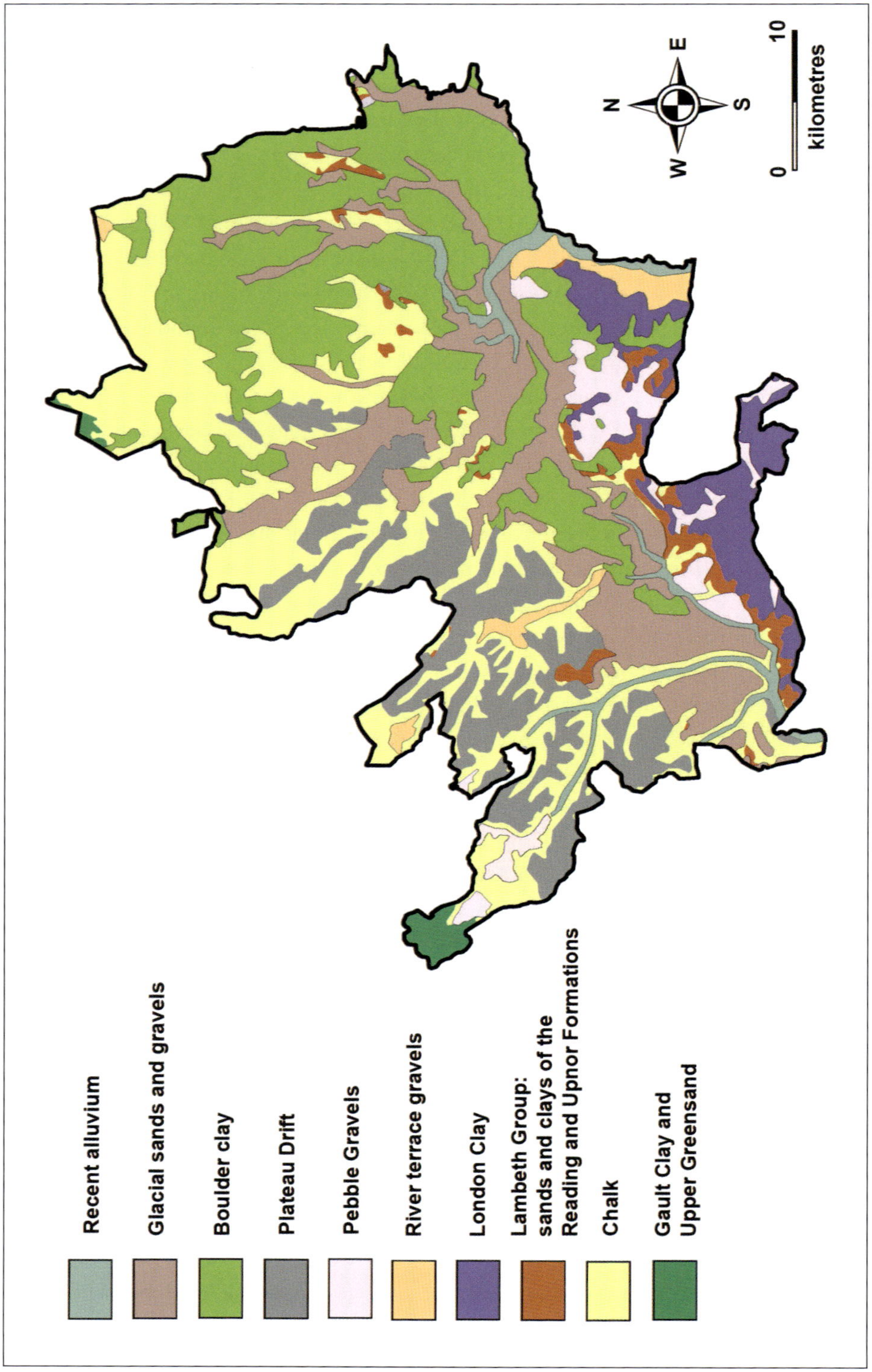

Figure 1.3. Hertfordshire: simplified surface geology.

Contexts: geology and topography

As Munby recognised, any understanding of Hertfordshire's history must begin with the natural environment and, in particular, with patterns of topography, soils, drainage and geology. The gently rolling Hertfordshire landscape has mainly been carved from Chalk, and from the relatively soft and recent sediments deposited in the last sixty million years above it. Only in the far north, on the very fringes of the county, is the underlying Gault formation, comprising sticky and poorly draining mudstones, exposed, together with small areas of the Upper and Lower Greensand. The majority of the county overlies the northern section of a great syncline or trough of Chalk, with London and the Thames at its centre, the North Downs forming its southern edge and the escarpment of the Chiltern Hills and their north-easterly continuation, the East Anglian Heights, its northern boundary. This escarpment is higher and steeper towards the south-west, in the Chiltern Hills *sensu stricto* to the south-west of Hitchin, reaching a height of 244 metres OD at Hastoe Farm in Tring, and becomes more muted and diffuse as it heads north-east. To the south and east of this strip of relatively high ground the land slopes away as a gentle dipslope which is covered by a wide variety of later deposits, and dissected to varying degrees by valleys within which the underlying Chalk is often exposed (Figure 1.3).

The Chalk, which was laid down on the bed of a warm sea during the Cretaceous Period, between 145 and 65.5 million years ago, comprises a very fine-grained soft white limestone derived from coccoliths secreted by marine algae, together with the skeletons of other marine creatures.[7] Originally forming horizontal beds, it was only later folded into the great syncline as a result of tectonic activity. The Chalk includes bands of irregular nodules of flint, a form of chert (silica dioxide) made up of tiny quartz crystals: a substance which, like glass, breaks with a marked 'conchoidal' (or shell-like) fracture. Early man skilfully fashioned it to make a wide variety of cutting, chopping and scraping tools and weapons; later societies, as we shall see, employed it as a major building material. The walls of the Roman town at Verulamium, and most of the county's medieval churches, are built from it.

Following the end of the Cretaceous, in the Tertiary period, much of south-east England, including Hertfordshire, was periodically inundated and eroded by the sea, and a series of formations was deposited on the eroded surface of the Chalk.[8] The first of these Tertiary deposits was the Thanet Sand Formation, which is largely restricted to the far east of the county. A little later (after another period of uplift and erosion) a brief marine invasion deposited the Upnor Formation more widely across the county. This was followed in turn by the deposition of the Reading, Harwich and London Clay Formations. The Upnor Formation comprises a narrow (usually less than three metres thick) band of sands or clayey sands containing black-coated flint pebbles deposited in shallow water.[9] More important in its effects on the landscape is the Reading Formation, largely comprising multi-coloured non-marine clays and sands, which have been widely exploited for brick-making in the county, with sporadic beds of pebbles. In some areas the coarser sands and pebbles have been cemented together by silica deposited from groundwater to form the only rock for which the county is famous, the so-called 'Hertfordshire Puddingstone'.[10] The most important of these Tertiary deposits, however, is the heavy, impermeable London Clay, which covers much of the south of the county, approaching a thickness of 90 metres on the boundary with Middlesex near Bushey Heath, Elstree and Arkley.[11]

Following this period of deposition, which ended some 51 million years ago, the land was subject to extensive fluvial erosion which gradually moulded its topography into a form approximating to that of today. It also gradually removed more than 140 metres of the sediments overlying the Chalk – the Upnor, Reading, Harwich and London Clay formations – across much of the north and west of the county. Once their thickness had been reduced to about 15 metres, they became more permeable, so that water percolated through them into the Chalk below, rather than flowing as surface streams, and erosion largely came to an end. The basal layers of the Upnor and lower Reading Formations thus survived but were largely transformed by weathering and disturbance during the Quaternary period (the last 2.5 million years) into the clay-rich Plateau Drift which now covers much of the Chiltern dipslope in the west of the county, and which formerly covered much of north-east Hertfordshire as well.[12] This drift, which varies in character but essentially comprises varying mixtures of pebbles and sandy clay (clay-with-flints and pebbly clay), is thickest (up to 15 metres) in the middle of the dipslope interfluves, furthest from the valleys. Only occasionally do outliers of the Reading and Upnor Formations survive unmodified on the Chiltern dipslope, as around Sarratt and Micklefield Green.[13] They give rise to areas of acid soil which are still frequently characterised by extensive tracts of woodland and commons.

The Quaternary period was marked by a significant deterioration in the earth's climate, producing about fifty glacial–interglacial cycles which are commonly (although inaccurately) described simply as the 'Ice Age'. In the earlier stages of the Quaternary the county was crossed from west to east by the precursor of the river Thames (the 'proto-Thames') which flowed far to the north of that river's present course: north-eastwards from Watford to Roestock, then northwards towards Stanborough before swinging eastwards towards Brickendonbury, northwards towards Bengeo, eastwards again to Mardocks Farm, then south-east to Eastwick, entering what is now Essex between Sawbridgeworth and Harlow.[14] The greatest Quaternary influence on the landscape was, however, the Anglian ice sheet, which covered much of Hertfordshire between 474,000 and 427,000 years ago, eroding and in places degrading the Chalk escarpment on its northern fringes and depositing river gravels, glacial lake sediments and above all the chalky boulder clay which covers the eastern side of the county, obscuring whatever remained of the Plateau Drift after this had been eroded by ice and subglacial meltwater streams.[15] The great ice sheet blocked the easterly flow of the 'proto-Thames' at a point to the east of Ware, creating a large lake (known to geologists as 'Lake Hertford') which extended westwards along the Vale of St Albans.[16] This may have existed for several centuries before the ice advanced over it, impounding the river further to the west and forming another lake to the south of St

Albans (the 'Moor Mill Lake'). It eventually overflowed southwards, forming the first part of a new course of the river Thames through London.[17]

By the end of the first phase of the Anglian glaciation boulder clay, known as the Ware till, had been deposited over much of central, northern and eastern Hertfordshire. As the ice sheet stagnated and started to retreat, meltwaters flowed away from it in two main directions: south-westwards along the Vale of St Albans and southwards down the lower Lea valley, depositing the extensive spread of gravels which are now a distinctive feature of both. The ice subsequently readvanced on three occasions, depositing further layers of boulder clay known to geologists as the Stortford till, the Ugley till and the Westmill till, the last of these a layer of calcareous clay up to eight metres thick which generally forms the surface of the interfluves in east Hertfordshire today. Lobes of ice extended down the Vale of St Albans as far as Hatfield and along the lower Lea valley as far as Hoddesdon, but they did not encroach upon the main area of the London Clay uplands to the south-west of Hertford Heath.[18] Nor, during any of the glacial stages, did the ice sheets approaching the county from the north-east manage to rise over the Chalk escarpment west of Luton, consequently leaving much of the Chiltern dipslope unglaciated.

The geological history of Hertfordshire is thus highly complex but the county may, for convenience, be divided into four or five main landscape areas which, very broadly, correlate with the character of the surface deposits just described (Figure 1.4). In the north of the county, on and below the Chiltern escarpment, the Chalk is exposed in a wide, sweeping landscape, while on the level plains to the north – on the very fringes of the county – the underlying Gault formation, comprising poorly draining mudstones, is found on the surface, together with small areas of the Upper and Lower Greensand. Both the escarpment and the plains below were characterised in the Middle Ages by clustered villages farming extensive 'open fields' of the kind familiar from school textbooks, in which the holdings of individual farmers comprised numerous small, unhedged strips which were intermingled with those of neighbours and scattered evenly through extensive 'fields', over which communal routines and rotations were imposed. This 'champion' countryside, as it was called by sixteenth-century topographers, generally survived into the eighteenth or nineteenth centuries. Open fields were then removed by large-scale, planned enclosure, usually effected by parliamentary acts, and as a result the district is today characterised by straight-sided fields defined by species-poor hawthorn hedges.

Countryside like this extends northwards, far into the Midlands, but the Chiltern escarpment forms a major 'frontier' in southern Britain and to the south of it rather different kinds of landscape can be found, sometimes described as 'ancient countryside' by modern academics and as 'woodland' by early topographers, in part because they contained (and often still contain) large areas of managed woodland, but also because they boasted an abundance of hedges and hedgerow trees.[19] Open fields of a kind could be found in these districts but they usually disappeared at an early date and were everywhere accompanied by land farmed 'in severalty', as hedged fields in individual occupancy. The settlement

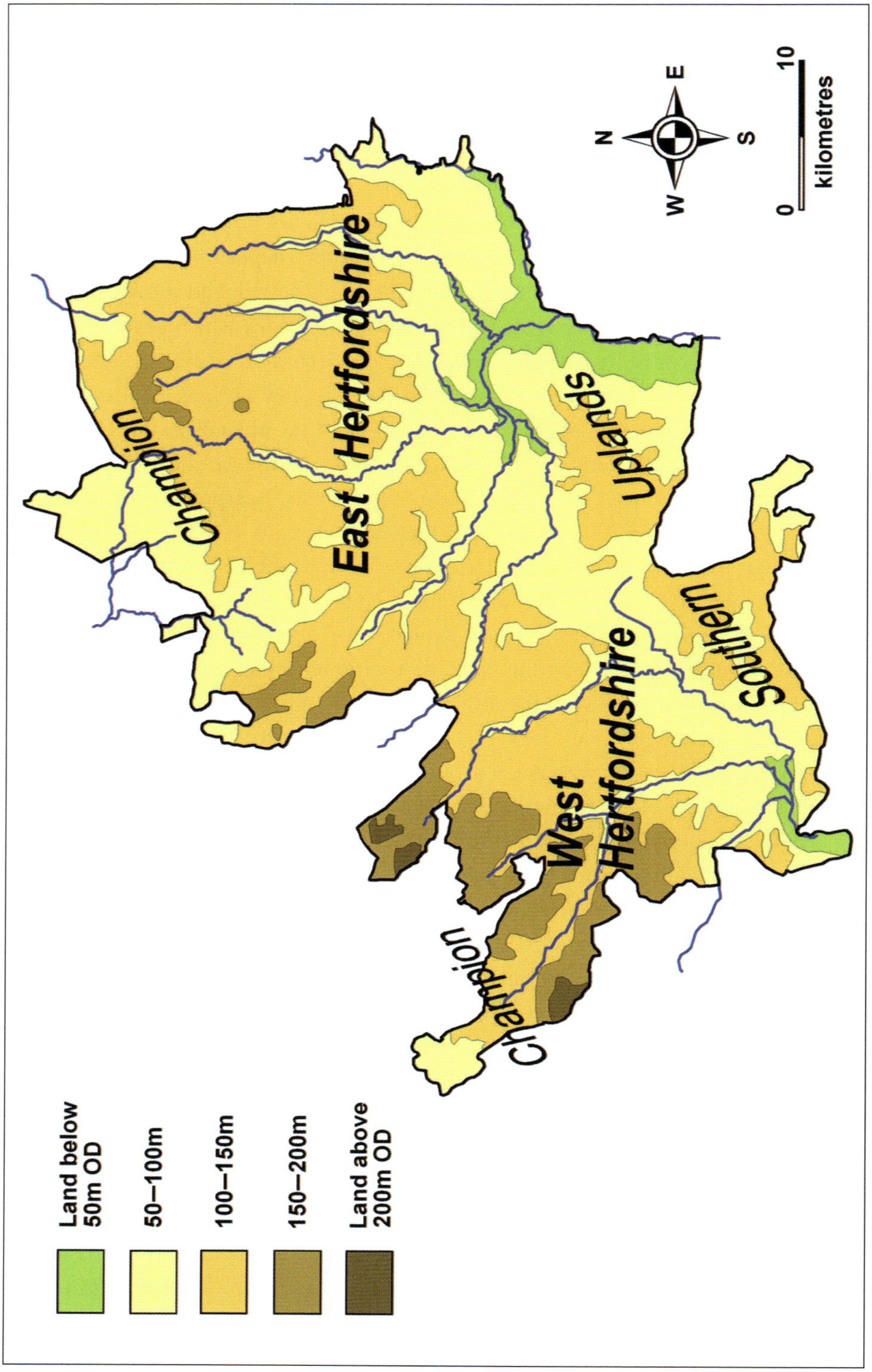

Figure 1.4. Hertfordshire: topography and landscape regions.

pattern, instead of consisting solely of nucleated villages, also included numerous isolated farms and small hamlets. Landscapes of this type took a variety of forms, corresponding to the main topographical districts within the county.

In the west, on the Chiltern dipslope, the Chalk is exposed in the main valleys but is mainly buried beneath the Plateau Drift – the 'clay-with-flints' (a fairly permeable clay with flint pebbles) and the 'pebbly clay' (a more complex mixture of stony clays, sands and gravels), which give rise to relatively infertile and acidic soils. These are interspersed with surviving fragments of the Upnor and Reading beds, associated with pockets of even more acidic soil. For much of its history this remained a sparsely populated district, and even today large areas of woodland survive on the highest ground, above the escarpment, often occupying tracts of former common land, while other commons – Chipperfield Common, Nomansland Common in Wheathampstead and Harpenden Common – still remain on the poorest soils of the dipslope. The largest settlements were always in the principal river valleys – those of the Chess, Ver, Gade, Bulbourne and upper Lea – and here extensive areas of open-field land often existed in medieval times. The wide interfluves between, in contrast, were characterised by more scattered settlement – small hamlets, isolated farms, loose girdles of dwellings around the margins of commons – and by a mixture of enclosed fields and smaller areas of common arable. It should perhaps be emphasised at this point that both in this part of Hertfordshire and elsewhere most areas of common land were probably wooded in the early Middle Ages. Some continued to be occupied by 'wood-pastures' into post-medieval times, but most gradually became more open land as individual trees died or were felled, and could not be replaced because of the intensity of grazing: commoners' livestock simply consumed the young trees. Over the last century or so, however, grazing on most commons has come to an end, so that they have gradually become covered in scrub and trees once more.

In the east of the county the gently tilted plateau between the principal valleys is covered with boulder clay. This clay contains a variety of rock debris, including large amounts of chalk, ensuring that where it occurs on sloping ground, or towards the edges of the plateau and lies thinly over the solid Chalk, it gives rise to fertile, calcareous and tractable soils. Here, once again, villages farming extensive areas of open-field land existed in the Middle Ages. The level areas away from the valleys are more poorly draining and were mainly cultivated from the start in hedged fields, with deep perimeter ditches to help drain the sodden land, the water emptying into numerous surface streams. Even today many small greens and commons survive here, around which groups of farms and cottages are clustered; some entirely isolated farms are also found, standing alone within their own fields. This remains for the most part attractive, well-wooded countryside, pleasantly undulating, with an abundance of ancient farms and cottages, a maze of minor lanes and footpaths, and wide fields almost exclusively under arable cultivation. In many ways it is reminiscent of the landscapes of Essex, or even Suffolk, to the east and north-east. As we shall see, it shares much in common with these districts, especially in terms of styles of vernacular architecture.

To the south of these two dipslope landscapes lies the broad area of low-lying countryside traditionally called the Vale of St Albans, running from Watford in the south-west to Hertford in the east – a wide depression containing a complex collection of sands and gravels laid down partly by the 'proto-Thames' and partly by glacial meltwaters. Today characterised by worked-out (and in some places still active) gravel pits, in medieval and early post-medieval times the band of sand and gravel was mirrored by the distribution of rabbit warrens.[20] To the south of this the land rises again towards the Middlesex border. The sands and gravels of the Lambeth Group – of the Reading and Upnor Formations – are exposed on the edges of these Southern Uplands, but most of the interior is occupied by the stiff, impervious London Clay, which gives rise to soils both poorly draining and – owing to the oxidation of the iron in the clay – acidic in character. Poor though these soils are, and unappealing to early farmers, the Pebble Gravels (of debated origin) which cap the highest ground give rise to soils even less appealing, and which mainly carried extensive areas of heathland until enclosure in modern times. This is a district of abundant surface water; there are numerous ponds feeding many streams, which sometimes disappear suddenly where the clay has been dissolved and washed down into the underlying Chalk, forming a 'swallow hole'. This combination of acid, pebbly gravels and waterlogged acid clays ensured that this district, much of it now extensively built up and suburbanised, was in all periods before the nineteenth century the most sparsely settled part of the county. It was a third variant of 'ancient countryside', with scattered farms, small hamlets and fields carved, for the most part, directly from the woodland and waste.

Contexts: pre-medieval settlement

When Munby wrote his book in the 1970s it was already clear that Hertfordshire had been settled and exploited on some scale from the earliest times. Archaeological research over the following thirty years or so has done nothing but strengthen this impression. But our main concern in this book is with landscape – that is, with the upstanding traces of the past – rather than with archaeology *per se*, and comparatively few earthworks or other upstanding remains from early times remain in the county today, most having been levelled by medieval or post-medieval agriculture. Nevertheless, some account of the county's development before the Middle Ages needs to be presented, for reasons which will become apparent, although limitations of space dictate that this must be brief.

Much of the county has been so extensively urbanised or affected by mineral extraction over the last century or so that evidence for early settlement has been lost, unrecorded by modern archaeology, while in many parts heavy soils make aerial photography comparatively ineffective and extensive woods and pasture limit the potential of fieldwalking – the systematic examination of the ploughsoil for the spreads of pottery and other debris indicating the location of settlements. On the other hand, over recent decades a number of important fieldwalking surveys have been carried out in the more rural and arable areas, while changes in planning legislation in the 1990s ensured that all major construction projects since then have been preceded by archaeological reconnaissance, sometimes leading to large-scale excavation. As a result of all this, and largely thanks to a number of dedicated 'amateurs' and the work of the county council's archaeology service, our knowledge of the county's pre-medieval settlement has increased considerably over the last few decades.

Hertfordshire was clearly occupied during the Palaeolithic – the immense period of time leading up to the end of the final glaciation some 12,000 years ago – and in the Mesolithic – the period from the end of the last glaciation up to the arrival of Neolithic farmers in the fourth millennium BC.[21] But early hunter-gatherers have left no above-ground traces of their existence and even for the Neolithic, while important settlements have been excavated at Foxholes Farm near Hertford, Old Parkbury in St Albans, The Grove near Watford and elsewhere,[22] surviving 'monuments' are restricted to the long barrow on Therfield Heath, near Royston, and the probable remains of another on the county boundary with Bedfordshire at Pegsdon Common, although others, and examples of other Neolithic 'ritual' monuments, are known from excavation and aerial photography. These include the probable 'causewayed enclosure' near Sawbridgeworth and the henge currently being excavated at Norton; many others doubtless await discovery.[23] For the subsequent Bronze Age (*c.*2,200–800 BC) we have relatively little evidence for settlements, but large numbers of round barrows, the characteristic burial mounds of the period, are known. They mainly survive only as cropmarks but there are fine upstanding examples at Therfield Heath near Royston, on Chipperfield Common[24] and sporadically elsewhere

Figure 1.5. Aerial view of Therfield Heath barrow cemetery, the largest and best-preserved example in the county. A group of round barrows, erected in the early–middle Bronze Age, cluster around a much earlier Neolithic long barrow. The site evidently maintained a ritual significance for over a thousand years. Courtesy English Heritage.

(Figure 1.5). They are often positioned on major watersheds and were possibly placed on the margins of social territories based on lower ground, as appears to have been the case elsewhere in England.

Neolithic and early Bronze Age settlements were concentrated in districts characterised by light, freely draining soils, presumably because these were the main areas of arable land at a time when only simple ploughs – 'ards', incapable of turning a proper furrow – were in use. The later Bronze and Iron Age seem to have witnessed some expansion in the area under cultivation, with settlement moving beyond the lighter land of the major valleys and the Chalk escarpment up onto the margins of the higher drift-covered ground, as evidenced by the sites at Thorley, Wood End and Raffin Green,[25] although growth was doubtless interrupted by periods of decline and retrenchment, as may have occurred in the Chilterns during the first millennium from *c.*600 BC to *c.*300 BC.[26] A number of the monuments called 'hillforts' – enclosures of varying size defined by substantial banks and ditches – were erected in the county during the early and middle Iron Age: Arbury Banks in Ashwell (an unimpressive example); Wilbury Hill in Letchworth; The Aubreys at Redbourn; Caley Wood in Little Hadham; and Ravensborough Castle in Hexton. Two further examples just beyond the county boundary exist at Wallbury, two kilometres to the south of Bishop's Stortford but within Essex, and at Ivinghoe Beacon, some three kilometres north of the county boundary in Buckinghamshire.[27] Another has been suggested at Gatesbury near Braughing, but this is contested.[28] Others may well have existed which were levelled by the plough long ago. The bounds of an estate at Gaddesden, described in a document of *c.*970, turned west from the river Gade at a point described as 'to the south of the *byrig*' – the Old English term for a fort, often a hillfort.[29] Aerial photographs suggest a ploughed-out oval enclosure, possibly the lost fort, occupying a prominent spur at this point. Excavated Hertfordshire hillforts usually contain limited evidence of settlement, as at The Aubreys, suggesting only 'seasonal, not permanent, occupation'.[30] Often regarded simply as permanently occupied strongholds or as refuges in times of trouble, these enigmatic enclosures probably fulfilled a range of functions, perhaps in part religious or ceremonial in character, although they certainly suggest a society geared to conflict.[31] In addition, in the Chiltern Hills a number of linear earthworks or 'dykes' of Iron Age date occur, most now levelled, but originally comprising single or multiple banks and ditches varying greatly in length and scale.[32] They include the 'Grims Ditch' in the western Chilterns, extending into Buckinghamshire. The purpose of these monuments is, again, unknown, but they suggest that the landscape was filling up with people and that there was an increasing need to demarcate territory and allocate resources between competing social groups.

Indeed, by the later Iron Age settlement was expanding out across the more difficult soils of the clay-covered interfluves at some distance from the margins of the principal valleys. Of particular interest in this context are the results of Angus Wainwright's survey of the Ashridge woods, on the Plateau Drift soils of the Chiltern Hills, where (unusually for

Hertfordshire) extensive areas remained unploughed throughout the medieval and post-medieval period as common land or woodland, allowing evidence of prehistoric activity to survive in the form of upstanding earthworks.[33] Networks of former lanes were discovered that linked settlements set in the midst of groups of embanked fields dating from the early, middle but mainly the later Iron Age and the Roman period. They were separated by areas of undivided ground which were presumably occupied by pasture and woodland, the latter apparently exploited as fuel for a local iron smelting industry. Some late Iron Age settlements could be found even on the inhospitable Southern Uplands, although always on deposits of gravel and pebble rather than on the London Clay, as at Foxholes Farm near Hertford and Hertford Heath. But while the county's clay-covered interfluves were thus occupied by isolated farms, most probably stock ranches, the main areas of settlement, not surprisingly, remained on lighter and more easily worked soils – on or below the escarpment in the north, in the principal river valleys and in the Vale of St Albans – and in some of these locations very large areas of occupation could be found by the very end of the Iron Age, to which archaeologists give the name *oppida*.[34]

Oppida are one of the characteristic features of Hertfordshire's archaeology. They comprised extensive but discontinuous spreads of occupation which were usually defined by, or included within their area, disconnected lengths of substantial bank and ditch. They often display some apparently 'urban' characteristics, in the form of metalworking, pottery manufacture and the use and production of coinage (which now made its first appearance in Britain), as well as evidence for the importation of goods from the Roman world – especially Italian *amphorae* which presumably contained wine and oil. The latter are also found in the wealthy burials which are often located within or close to these places. The most important *oppidum* was at St Albans and was called Verlamion, later to become the Roman Verulamium.[35] It may have developed as early as *c.*10 BC, when a ruler called Tasciovanus was minting coins there. It comprised a loose network of aristocratic households scattered around the margins of a central low-lying area, apparently a focus for ritual activity, coin production and metalworking, beside but separated from the river Ver by a system of linear dykes. Another large example, earlier in date (foreign imports were reaching here by 25 BC) and with less evidence for linear earthworks, was at Braughing, in the east of the county, near to the confluence of the rivers Quin and Rib. Others, smaller in scale, existed in the valley of the Lea around Wheathampstead, in the valley of the Mimram around Welwyn and on and below the Chalk escarpment at Baldock, in the north of the county. It has been argued that, as southern Britain was increasingly drawn into the political and economic orbit of the Roman world, chieftains – in Hertfordshire and Essex especially – grew powerful through the control of the import and distribution of prestige goods from Gaul, including fine pottery and silverware, and that this encouraged the development of progressively larger and more complex political entities, tribal 'kingdoms', centred on the larger *oppida*.[36] In the case of Hertfordshire and surrounding areas, a number of smaller tribal units appear to have coalesced around the start of the first millennium to form the kingdom of the *Catuvellauni*.

There were other major late Iron Ages sites in the county, as at Ware, possibly an *entrepôt* at which imports were transhipped from vessels coming up the Lea from the Thames, and at Cow Roast Inn, in the Bulbourne valley west of Northchurch, apparently a major metalworking centre.[37] Whether these sites might also be classed as *oppida* is unclear. The term itself is poorly defined, and *oppida* usually represent not large and clearly defined settlements but several dense clusters of later Iron Age activity within much wider, more diffuse areas of occupation. Indeed, rather than simply emphasising their role as centres of power and exotic consumption, we might also see them as a manifestation of a landscape filling up with people, and in which particularly dense concentrations of farms developed where soils were most amenable to contemporary agriculture.

Some of the rulers of these emerging tribal kingdoms appear by name in the classical sources which describe the abortive invasion by Julius Caesar in 54 BC, the political relations of Rome and the wider world over the following century and the Claudian invasion of AD 43, which finally brought southern Britain within the Roman empire. The cremated remains of one such leader were buried in an extraordinary tomb discovered by Ros Niblett during an excavation at Folly Lane, St Albans, in 1991/2, within the area of the Verlamion *oppidum*.[38] A pit, originally covered by a turf mound, contained a cremation and grave goods which included metalwork and imported pottery dating to the period *c.* AD 50 – that is, *after* the Roman invasion. The site was later maintained through the Roman period as a temple, and Niblett has speculated that a grateful population may have venerated a leader – possibly Adminus, son of Cunobelin – who negotiated a peaceful transition to Roman rule and served after the Conquest as a client king, thus averting the bloodshed of war. Others have suggested, more generally, that in Hertfordshire, as in neighbouring parts of Essex, Roman rule may have been accepted voluntarily by a tribal elite already familiar with Roman culture. It is noteworthy that there are no certain examples of Roman military forts in the county. The early appearance of Romanised buildings at Verlamion has been plausibly interpreted as evidence of 'a local elite losing no time in associating itself with the dominant power and demonstrating its wealth and loyalty by the construction of showy "Roman" buildings and the consumption of imported goods'.[39]

It is certainly clear that, while in many ways the Conquest brought upheaval and change, the economic and territorial structures of the late Iron Age endured, and moulded in many ways the development of the Romano-British landscape. Verlamion thus became the most important Roman town in the region, a *municipium* which served as a central place for the Catuvellauni, complete with a monumental forum, amphitheatre and temples. It is still an impressive archaeological site, especially its walls and amphitheatre. The *oppida* at Baldock and Braughing, moreover, developed into the two other main urban settlements in the county. Both lacked much in the way of public buildings but may nevertheless have had political and social as well as economic roles. Significant Roman settlements at Ware, Welwyn and around the Cow Roast Inn in Northchurch all appear to have developed, once again, directly from important Iron Age sites. But additional places with commercial or industrial functions also emerged on new sites, at Cheshunt, Bishop's Stortford and elsewhere: there was change as well as continuity.[40] All these towns and other major places lay on roads, some simply tidied-up versions of earlier trackways, but others – Watling Street and Ermine Street – entirely new military routes, carefully constructed in a series of long, ruler-straight alignments.

These urban and industrial centres were surrounded by a rural landscape which was, by the second century, extensively settled, with numerous villas and farmsteads. Hertfordshire has seen fewer systematic fieldwalking surveys than many areas of England and, as already noted, in many parts of the county modern land use or urbanisation precludes the effective deployment of this technique. Such surveys as have been carried out, together with archaeological material discovered by accident or during watching briefs and the evidence of excavations and aerial photography, suggest a density of perhaps one settlement for every one or two square kilometres, except on the infertile and intractable London Clay soils of the Southern Uplands.[41] Even here villas were closely spaced along the valleys of the Colne and the Lea, while a number of Roman sites are known from the clay uplands, some perhaps isolated grazing ranches, others pottery kilns, utilising both the local clays and the reserves of fuel provided by the extensive woodlands. The available evidence thus indicates that Hertfordshire's Roman landscape was a well-exploited one although, as in prehistory, there was a distinction between the major exposures of lighter soils in the main valleys and on and below the Chalk escarpment to the north, where the main towns, largest villas and farms were located; and the clay-covered interfluves where farms were, for the most part, more widely spaced, smaller and poorer. It was a distinction which was to continue into the medieval period, and which is today still deeply etched, in innumerable ways, into the landscape of the county.

The continuity question

When Munby wrote his pioneering book it was widely assumed that little survived in the English landscape, other than a few individual archaeological 'sites' such as barrows or hillforts, from before the late Anglo-Saxon period. Munby himself was aware of Hertfordshire's rich legacy of Iron Age and Roman monuments, and suggested a number of places, including Reed, where the arrangement of fields and roads might have survived from these remote periods.[42] But for the most part he believed that the extent of settlement and agriculture in these periods had been limited, and that the Hertfordshire landscape was essentially created by the clearance of extensive tracts of woodland and 'waste' in the period between the ninth and the fourteenth centuries.[43] Since the 1970s there has been a growing recognition that the Roman countryside was, as we have noted, very extensively tamed and cleared, in Hertfordshire as elsewhere. But debate has raged, and continues to rage, over how much of the basic medieval and modern pattern of fields, roads and settlements, was inherited from this remote period. Some believe that there was a considerable degree of continuity and that many of the 'everyday' features of the rural landscape which we usually assume to be of medieval or later origin were actually established in prehistoric or Roman times. Others, in contrast, believe that the density of late prehistoric and Roman settlement can be exaggerated, and was considerably less than in the Middle Ages. They also argue that the scale of population decline in the immediate post-Roman period was so great that fields were engulfed in woodland to such an extent that in most areas, and especially on the heavier soils, clearance had to begin afresh in the Saxon period, so that the landscape was reconstituted, in effect, from scratch.[44] On balance, when it comes to the details of Hertfordshire's landscape, we would perhaps incline more to this latter view, not least because while the Roman countryside was clearly well exploited, that of the high Middle Ages was positively choked with people. Although not denying, as we shall see, that some elements of the countryside were first forged in the remote past, most of its basic fabric was created by the great expansion of population and settlement which occurred between the tenth and the fourteenth centuries.

In terms of administrative and organisational structures, in contrast, it is possible to argue for a greater degree of continuity, although it is important to emphasise that this was not true at the level of major administrative units: Hertfordshire did not develop from the kingdom of the Catuvellauni, and indeed in general there was little continuity between Roman *civitates* and the kingdoms of middle Saxon England – the East Saxons, the Mercians and the rest. This is because, so far as the evidence goes, when Britain ceased to be a part of the Roman empire and was settled, to an extent which is still vigorously debated, by new peoples from across the North Sea – the Angles and Saxons and their neighbours – the Roman provinces and their subdivisions fragmented into a large number of small, autonomous or semi-autonomous tribal territories, each extending over tens or hundreds of square kilometres rather than embracing larger areas.[45] The names of some – such as the *Hicce*, who occupied north Hertfordshire and gave their name to Hitchin – are mentioned in early documents such as the *Tribal Hidage*, a tribute list probably drawn up in the seventh century for a Mercian king; others are indicated in early place-names.[46] Only in the course of the later sixth and seventh centuries did the kingdoms we read about in the pages of the history written by Bede gradually emerge, as some of these small groups gained ascendancy over their neighbours. Hertfordshire was not among the kingdoms so formed, however, and before the early tenth century it did not exist as a unit at all. Instead its area was divided between two polities, the centre and west forming part of Mercia, the east falling within the kingdom of the East Saxons.[47]

At this level, then, there was little continuity. But locally there was perhaps more. For it is striking that many of the places that we know to have been important in middle and later Saxon times also appear to have been significant in the Roman and, in many cases, the Iron Age periods. The most obvious and dramatic case is St Albans. Here there is evidence that the burial place of the third-century martyr Alban – or what was assumed to be his burial place – within a cemetery lying outside the walls of the Roman town developed directly into a major Saxon shrine, a church which is described by Bede in the early eighth century

Figure 1.6. View of St Albans Abbey from the Roman town of Verulamium. The enduring importance of St Albans, from the late Iron Age into modern times, is the clearest of many examples of long-term settlement continuity in Hertfordshire.

(below, p. 240).[48] In the immediate post-Roman period a tribal group called the *Wæclingas* appear to have established themselves in or around the site of the old *municipium*, and were eventually absorbed into the Mercian kingdom. The Mercian kings then established a defended place, a *burh*, within the walls of the now derelict Roman city – Kingsbury – and endowed the old church as an abbey.[49] The monks, in turn, probably in the eleventh century, established a market place at its gates, the core of the modern town, on the hill overlooking the *municipium* in the valley below. The great medieval city of St Albans thus developed, albeit indirectly, from the Iron Age Verlamion (Figure 1.6). In a similar way, the *oppidum* and Roman town at Braughing appear, as we shall see, to have developed into the focus for the tribal territory of the *Brahingas*, 'the people of Brahha', and when they were absorbed into the East Saxon kingdom the place continued to function as a major royal estate centre, with an important church or *minster*, which served the spiritual needs of an extensive territory. Braughing never developed into a significant medieval town but it was the meeting place for (and gave its name to) one of the late Saxon administrative subdivisions of the shire, known as hundreds.[50] Many other important places in the county in middle and later Saxon times – major royal estate centres or early church sites – appear to be closely associated with large Iron Age or Roman settlements, including Welwyn,

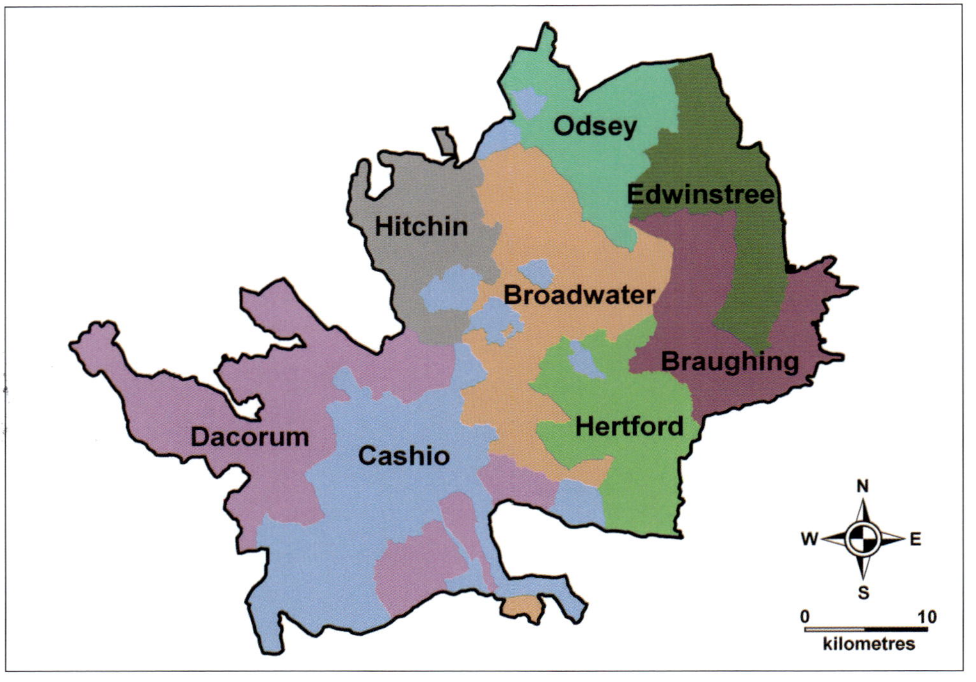

Figure 1.7. Hertfordshire: the pattern of medieval hundreds.

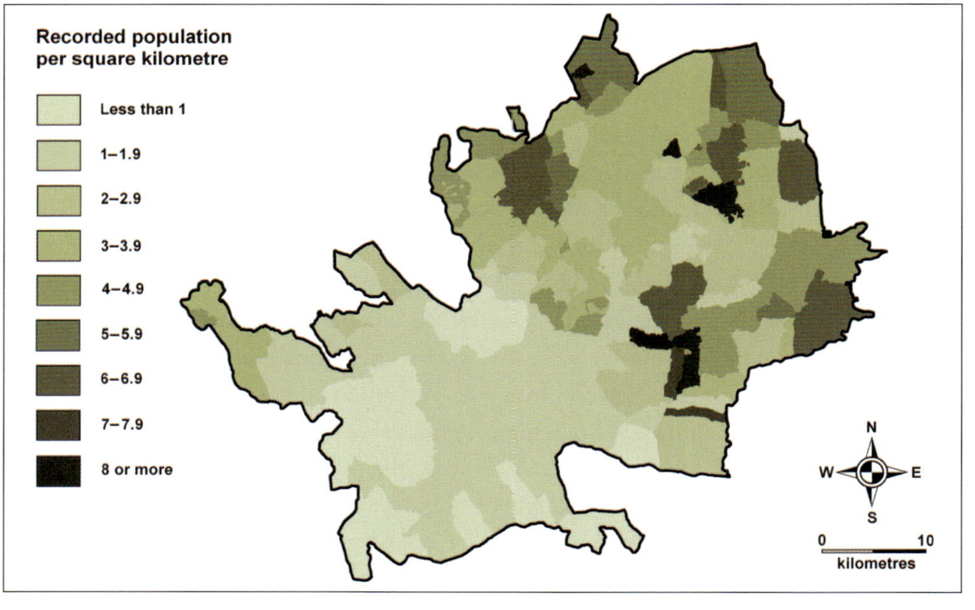

Figure 1.8. Hertfordshire: the distribution of recorded population in Domesday. This map is an approximation only, compiled by equating Domesday vills with later parishes and adding the inhabitants of 'lost' townships to those of the parish within which these places probably lay. It does, however, clearly highlight the contrast between the densely settled north and east of the county and the sparsely inhabited south and west.

Berkhamsted, Bishop's Stortford, Wheathampstead, Hadham, Ashwell and Cheshunt.[51] Of course, in some cases we may be dealing not with continuity but rather with the reoccupation of favourable sites following a period of abandonment. Moreover, as with the Iron Age–Roman transition, continuity was accompanied by change. The main centre of power in the north of the county in middle Saxon times was Hitchin – a great Mercian royal estate complete with a minster which is referred to in the will of the noblewoman Æthelgifu of *c.*980 and in Domesday Book.[52] This developed from the territory of the *Hicce*, but has no obvious, immediate Roman predecessor. The overall pattern is, however, clear enough: major Roman settlements became the foci for early Saxon tribes and evolved through middle Saxon times into major royal and ecclesiastical centres. This does not, however, necessarily mean that a similar degree of continuity is evident at the level of minor local settlements. Indeed, as we shall see, there appears to have been a retreat from the drift-covered uplands in this period, a retrenchment onto the lighter soils of valley and escarpment, and even here settlements tended to be relatively few in number and mobile in character, few remaining in precisely the same place for more than a century.

Into the Middle Ages

The retreat from the more difficult soils, and the mobility of settlement in areas which continued to be occupied, suggest that there was both a major demographic decline and a significant technological recession in the immediate post-Roman period. The heavy plough employed on many Romano-British and, some believe, late Iron Age farms, capable of tilling the difficult upland soils, appears to have gone out of use, just as wheel-thrown pottery ceased to be made and coinage became redundant. But in middle and, especially, in later Saxon times the population appears to have expanded once more, most rapidly perhaps from the ninth or tenth centuries. More sophisticated ploughing equipment came back into use and the heavy clay soils began to be cultivated again.[53] There were also wider political and social changes. England became a unified country for the first time, as the only royal dynasty to survive the Viking incursions, that of the West Saxons, reconquered the east of the country from the Danes, and in the process Hertfordshire itself came into existence in the early tenth century as a territory created to provision and garrison the fortress and defended market place at Hertford. A market economy gradually developed once again, together with new systems of taxation and administration, and the new shire, like others, was divided into units called hundreds, many of which were based, at least in part, on earlier territories (Figure 1.7). All this was accompanied by significant changes in the character and status of the peasant population. Old patterns of social organisation, in which a population mainly comprising free *ceorls* rendered tribute to royal estate centres such as Hitchin or St Albans, were gradually replaced by a system based on manors – estates owned by local lords or great magnates and containing populations of mainly dependent cultivators who held their land in return for labour services rendered on the lord's 'demesne', or home farm. Domesday Book, compiled in 1086, twenty years after the

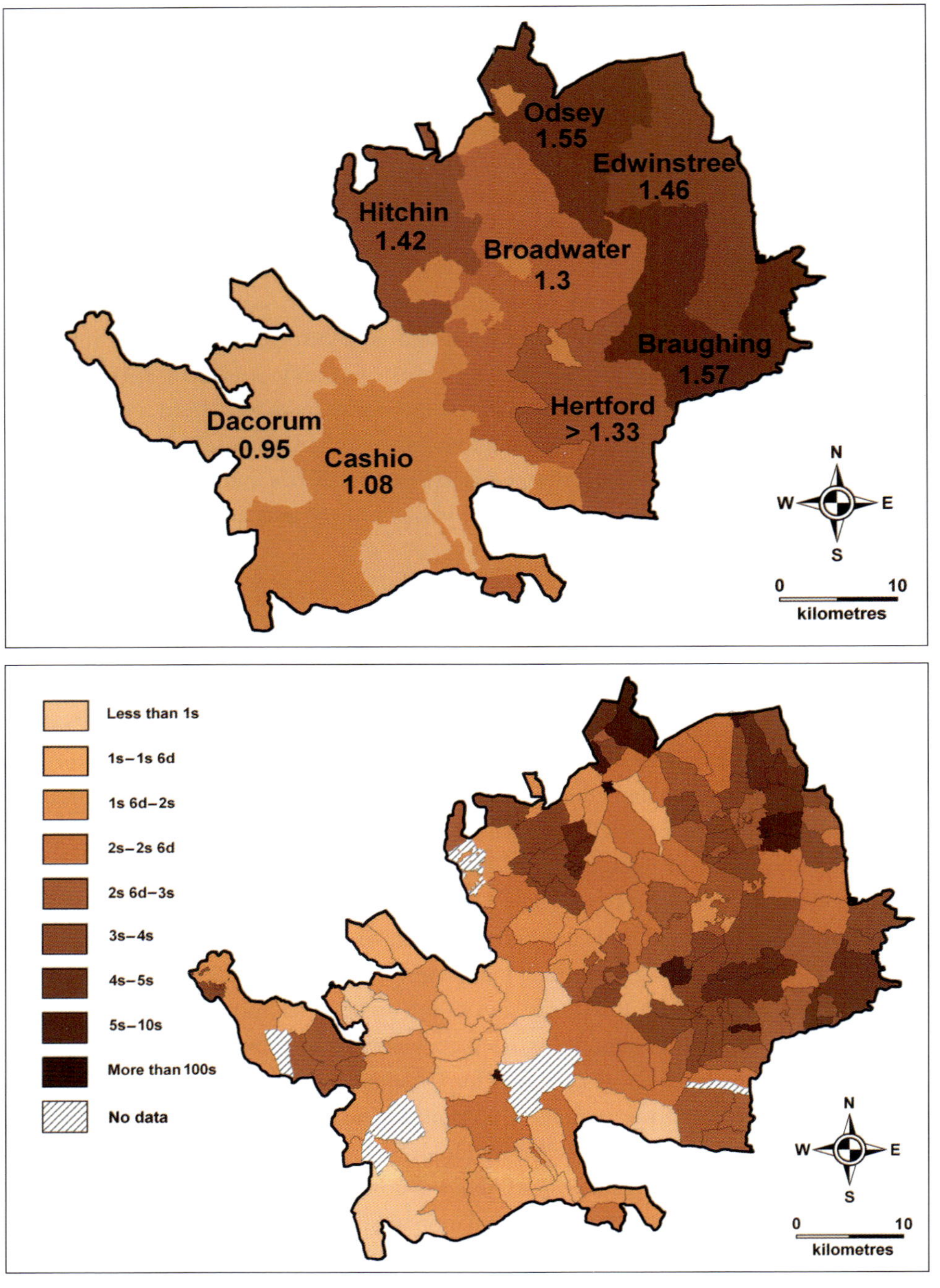

Figure 1.9. Hertfordshire in 1307, from the lay subsidy returns. (a) Density of taxpayers per 100 acres, by hundred; (b) tax payable per 100 acres, by vill (equated with parish). As with Figure 1.8, these maps represent approximations but clearly indicate the continuing economic and demographic dominance of the north and east over the south and west of the county, although the gap between the two appears to have narrowed to some extent.

Norman Conquest, is organised within the framework of such estates, although these are also described in terms of the townships or vills – taxable units, usually bearing the name of later villages and towns – in which they lay. Local landowners erected and endowed churches on their estates, leading to the break-up of the old minster territories and the emergence of a system of parishes.

Domesday provides our first overall view of the county's economy and landscape and, although much of its record is vague or open to different interpretations, there is no doubting the contrast it reveals between the south and west of the county and the north and east (Figure 1.8). On the Chiltern dipslope and the Southern Uplands there were extensive tracts of woodland, population densities were well below the national average and relatively few named townships are recorded. Indeed, many places which by the later Middle Ages had developed into sizeable villages are passed over in silence. Large tracts of land were, moreover, in the hands of great ecclesiastical houses, principally St Albans Abbey but also the abbeys of Ely and Westminster – land donated by kings and magnates during the Saxon period because of its undeveloped character.[54] The north and east of the county, in contrast, had population levels which were generally above the national average and more land under cultivation. There were more named townships and larger numbers of manors, relatively few of which were controlled by monasteries. There was much less recorded woodland, especially in the far north, on and below the Chalk escarpment, where many townships had none. This contrast in the density of settlement and in patterns of ownership reflects the fact that the soils formed in the Chalk and in the chalky boulder clay were inherently more fertile than those of the Plateau Drift which covered much of the west of the county, or the poorly draining, acid clays of the Southern Uplands.

Already, by the time of Domesday, many of the key elements of the county's countryside had come into existence, especially in the more populous north and east. Some places were already being described as towns, and large numbers of churches had evidently been built by local landowners, although most were still of wood, rather than stone, and have thus disappeared without trace. But much more of the detail of the landscape crystallised out, as we shall see in the course of the following chapters, over the following two centuries or so, as the county's population continued to rise, probably more than doubling from around 30,000 in 1086 to nearly 70,000 in 1307.[55] The extensive wooded 'wastes' on the upland interfluves steadily dwindled, eroded by the expansion of farmland: fragments were enclosed as coppiced woodland or as deer parks by the feudal elite, with the remainder surviving as greens and commons, often surrounded by girdles of farms and cottages. Indeed, it was in this period that the vast majority of the scattered farms and hamlets so characteristic of the county first came into existence, as the uplands between the major valleys were gradually brought into cultivation. But as population rose the regional differences in the extent of clearance and cultivation apparent in Domesday, while they did not disappear, grew less marked, something which becomes clear when the evidence of the latter source is compared with that provided by the lay subsidy returns of 1307 (Figure 1.9). Although the inhabitants of the boulder clays of north-east Hertfordshire continued to be more prosperous than those on the Chiltern dipslope and the Southern Uplands, the differences 'narrowed during the twelfth and thirteenth centuries, when rising population pressure and the growing demand for fuel and food in London resulted in some new settlement, the reduction of wood-pasture, and a shift towards arable farming in the latter area'.[56]

It was in the course of the twelfth and thirteenth centuries that most of the county's parish churches were built: medieval monuments so ubiquitous that we sometimes forget how impressive and sophisticated they are, given their antiquity. Larger monuments raised by a powerful feudal elite, and paid for by the surpluses extracted from the peasant population, have also come down to us from this period of expansion. These include the great soaring mass of St Albans Abbey, the royal castle at Hertford and the various castles built by great magnates, lay and ecclesiastical, at Berkhamsted, Pirton, Benington and Bishop's Stortford, together with shorter-lived examples at Anstey and Therfield. Minor examples elsewhere were often, like Walkern, in remote upland locations and perhaps intended to protect manorial flocks and herds grazed in wood-pastures during times of unrest, such as the 'Anarchy' of 1135–53. The majority of the county's 160 or so moats – artificial islands defined by water-filled ditches – also date from this period. Most were the sites of manor houses but some, especially in the north-east of the county (where they are most numerous) were occupied by the homes of wealthy peasants. Although often described as 'defensive' they were primarily intended to improve drainage, as well as functioning as fish ponds and as fashionable statements of status, although the larger ones would doubtless have served to deter casual intruders. We will return to some of these features, especially parish churches, at a later stage. But our main concern, especially in the four chapters that follow, is with the main fabric of the landscape: with the distribution of villages and other settlements, the pattern of fields and roads and the character and location of woods and commons.

Less tangible features also became fixed in this period, such as parish boundaries, although their layout in part reflects the configuration of the secular estates and territories from which they developed. In broad terms, variations in size reflect differences in early medieval population density, with parishes in the south and west being larger, in general, than those in the east and north. Minor details of their boundaries reflect early medieval economic interests rather than spiritual concerns, with distortions and projections indicating the need to embrace particular resources, such as the westward extension of the parish of Sacombe which incorporates a short stretch of the river Beane, presumably to provide a site for a mill. Tracts of woodland, formerly shared between a number of communities and lords, were now partitioned between them, leading to the curious pattern of isolated 'enclaves' found on the clay uplands to the north-west of Tewin, where the parishes of Welwyn, Digswell and Datchworth were intermingled in complex ways and a detached portion of Knebworth formed an island within Datchworth, embracing an isolated farm whose name – Swangley, 'the clearing of Swan's people' – still recalls its woodland origins. Such patterns would repay further study: they can tell us much about the allocation and division of resources at a time when the countryside was filling up as never before.

Rural expansion was accompanied by the growth of the county's towns, most of which came into existence and gained their basic form and structure in the period between 1100 and 1300. Most were placed on the main roads leading to London, already a dynamo of economic growth. Some of these routes were of Roman date – Ermine Street and Watling Street continued to be important – while others, such as the old A411 through Berkhamsted, Watford and Bushey, perhaps developed only in the early Middle Ages. The Lea was the only river navigable by large vessels and, because the highest navigable point and the lowest crossing point on any river were of great economic importance, there were major disputes between Hertford and Ware. The bridge carrying Ermine Street across the river at the latter place had disappeared in Saxon times, and traffic diverted to the former, where the Lea could be more easily forded. By 1191, however, a bridge had clearly been re-established at Ware, as it was reported to have been 'broken down by the men of Hertford'. In 1211–14 Sayer de Quincey, Justiciar of England and lord of the manor of Ware, removed the blocks and chains which the bailiff of Hertford had erected to obstruct the bridge, while in 1274 the bailiffs at Ware were said to have deliberately neglected the weirs there, so that boats could no longer pass upriver.[57]

The demographic expansion which created the medieval landscape ground to a halt in the early fourteenth century when, from 1314, a series of bad harvests caused by particularly heavy rainfall were followed by periods of famine and outbreaks of disease among both people and livestock.[58] In 1348–9 the population declined dramatically as a result of the Black Death. Many books on landscape history concentrate on the desertion of settlements, which resulted not so much from the direct impact of the disease but from its longer-term consequences. As population declined, villages located in more agriculturally marginal terrain often shrank, sometimes to extinction, as their inhabitants drifted away to take up more productive holdings elsewhere. Such places were sometimes finally destroyed, as were other small places, by a shift from arable to pasture farming which occurred in the course of the fifteenth century. This is usually explained in terms of changes in the relative prices of arable and livestock produce, and in the costs of their production, for wheat prices were now low and wage costs high, while wool, leather and meat continued to command high prices yet required less manpower to produce. It is, however, arguable that this emergence of specialised livestock enterprises was part of a wider development in the late medieval economy of agriculturally specialised regions: either way, the 'deserted medieval villages' so often encountered in Midland counties are rare in Hertfordshire. Some villages in the 'champion' north of the county declined, but few disappeared completely. Contraction was, moreover, usually extended over several centuries and the places in question had seldom ever been anything more than small hamlets. Caldecote, for example, to the west of Ashwell, shrank considerably in the later fourteenth century, but there were still six families living there in the seventeenth century and the place continued to exist as a parish even when only the main farm and associated cottages were left in the nineteenth century.[59] Other shrunken or deserted sites are clustered in the north-west around Tring – most notably Tiscot – but lacked parish churches and had always been small settlements, eking out a precarious living in challenging terrain. These places were much like the classic deserted and shrunken Midland settlements, contracting because their soils could now be more economically farmed as grazing ranches than by communities of arable farmers. But in the 'woodland' parts of the county the decline in population usually led, at most, to the shrinkage rather than the desertion of villages, or the abandonment of small hamlets and isolated farms. Much of this settlement contraction, moreover, was the consequence of more complex social and economic changes occurring through the post-medieval period, as we shall see. In the north-east of the county Broadfield, Berkesden and Wakeley all appear, at first sight, to be classic 'deserted medieval villages', complete with abandoned churches. All, however, seem to have been little more than hamlets and in the case of Berkesden and Wakeley certainly, and Broadfield probably, the church was in fact never more than a chapel-of-ease, rather than a parish church *sensu stricto* (below, p. 78). The more complex tenurial structure in this part of the county, where most townships had no one controlling lord, meant that few could be laid entirely to pasture. As Munby recognised long ago, in a county generally characterised by soils of reasonable fertility there were usually many tenants keen to take on holdings vacated by the death of individuals or families: he quoted Levett's study of the St Albans manors in west Hertfordshire and the case of Abbot's Langley, where, in spite of severe losses of population, 'new tenants were found for almost all the empty holdings'.[60]

The history of the late Middle Ages should not, however, be written primarily in terms of disease and decline. In a landscape less clogged with people wealth was spread less thinly, standards of living generally improved and in many ways the economy expanded. The numerous additions and improvements made to Hertfordshire's parish churches, and the wealth of farmhouses surviving from the fifteenth and early sixteenth centuries, are clear testimony to this.

Post-medieval developments

The radical social changes of the early modern period, with the growth of a centralised state and the dissolution of the monasteries, have left an indelible mark on the landscape of England, not least in the decline of great residences with pretensions to defensibility and the rise of true 'country houses', accompanied by extensive parks and gardens. Of equal importance, but extending into the eighteenth and nineteenth centuries, was the development of a more sophisticated economy which was both more urban and more industrial in character, and in which agriculture became more specialised and carried out by larger farms, developments which in Hertfordshire's case were closely associated with the continued growth of London. These agricultural changes, so important in moulding the character of the rural landscape, are particularly complex and are discussed in more detail in the chapters that follow, but, in essence, as time went on local farming became more and more adjusted to the particular circumstance of soils and topography, a process which sometimes generated continuities of land use but sometimes radical change. The generally fertile and calcareous soils formed in Chalk and chalky boulder clay in the north and east of the county continued, for the most part, to be farmed as arable, albeit with some pasture land on the heavier clays of the level plateaux. In agrarian terms the main development here was an increasing dependence on the cultivation of barley, much of it processed into malt either locally or in London; and, in landscape terms, it was the disappearance of the open fields. These were particularly extensive in the main river valleys and on or below the escarpment, where the main exposures of Chalk were to be found. They often contracted gradually through informal, piecemeal enclosure, but were mainly swept away by a series of enclosure acts in the late eighteenth and early nineteenth centuries. This was the period of the 'Agricultural Revolution', when farmers raised cereal yields by employing a range of new techniques, especially crop rotations featuring turnips and clover. On the narrower strip of heavy land associated with the Gault Clay, to the north – that is, in the two 'salients' which extended the county boundary out onto the Midland plain around Tring and Ashwell – 'champion' landscapes likewise tended to disappear through large-scale planned enclosure, although here, as we shall see, often at a rather earlier date, and largely to create grazing farms – a continuation of the process of enclosure and adjustment on this difficult land which had begun in late medieval times.

In the west of the county, on the Chiltern dipslope, the enclosure of open fields tended to occur even earlier, and again through informal methods, many townships losing their open arable completely before the end of the seventeenth century, although the vast commons on the crest of the Chilterns generally survived until enclosed by parliamentary acts – and some still remain today. In this district mixed arable farming tended to predominate throughout the post-medieval period, albeit with the widespread adoption of new crops and rotations in the course of the eighteenth and nineteenth centuries, and some expansion of pasture on the poorer land. The most striking developments in farming were on the Southern Uplands. Most of the London Clay and Reading Formation soils had been under cultivation in the Middle Ages, but the extent of pasture on the former increased steadily through the seventeenth and eighteenth centuries and, following enclosure in the decades after 1800, the great heaths on the Pebble Gravels, occupying the highest ground, were also laid to grass. By the time the tithe award surveys were made in the years around 1840 some parishes in the south of the county were almost entirely occupied by pasture (Figure 1.10). This was partly to provide meat and milk for the capital, but partly to supply the crops of hay necessary to sustain the many draft horses kept there.

This was one of many ways in which London's influence on Hertfordshire grew steadily in the course of the post-medieval period, a development epitomised in the landscape by the 'New River', an artificial watercourse constructed between 1608 and 1613 to bring water

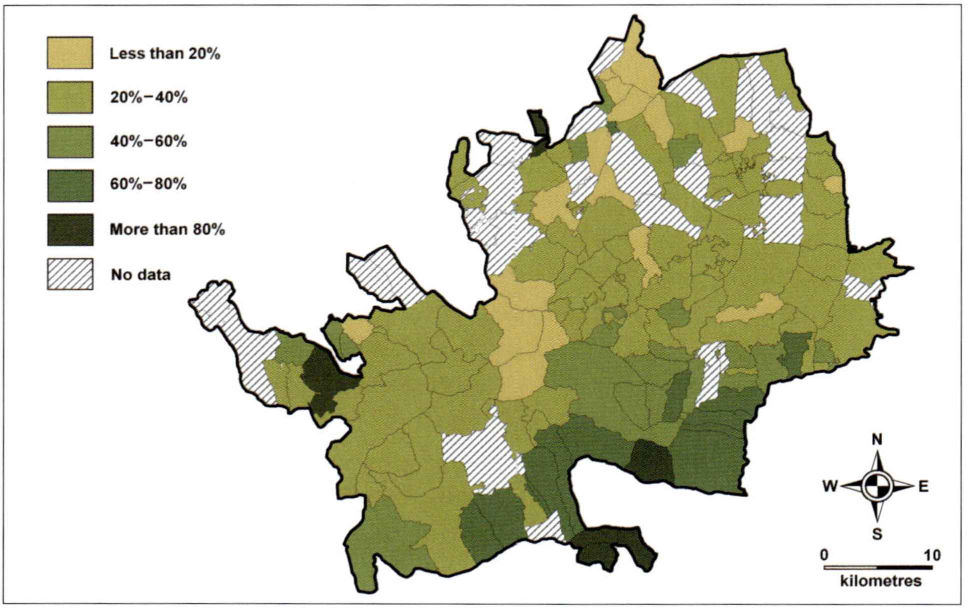

Figure 1.10. Percentage of land area occupied by pasture in *c.*1840, based on the tithe maps. The majority of parishes containing relatively little arable were to be found on the Southern Uplands.

from Chadwell Springs near Ware to the capital.[61] In the sixteenth century the fertile north and east remained the wealthiest and most densely settled area of the county (Figure 1.11), but by the late seventeenth century a clear distinction was beginning to appear between towns in the south of the county and those in the north and east, the former expanding to a greater extent than the latter. This trend was exacerbated, and the influence of London further increased, by improvements in transport which occurred in the course of the eighteenth century, with the establishment of turnpike trusts and the construction of the Grand Junction canal through the west of the county. One key aspect of this, explored in some detail in Chapter 8, was the growing tendency for wealthy Londoners – politicians, administrators, lawyers, merchants and eventually industrialists – to purchase property in the county in order to build country seats within easy reach of the capital. Although a number of large estates did emerge in Hertfordshire in the post-medieval period, such as Hatfield, Gorhambury, Ashridge and Cashiobury – principally in the south and west, where land was poorer and cheaper – this influx of London money limited their development (Figure 1.12). Arthur Young, writing in 1804, described how 'Property in Hertfordshire is much divided: the vicinity of the capital; the goodness of the air and roads; and the beauty of the country' had led 'great numbers of wealthy persons to purchase land for villas'.[62] The impact of large landed estates on Hertfordshire's landscape was also limited, when compared with some other parts of the country, by the simple fact that the county was, by the seventeenth century, already largely divided into fields and woods and filled

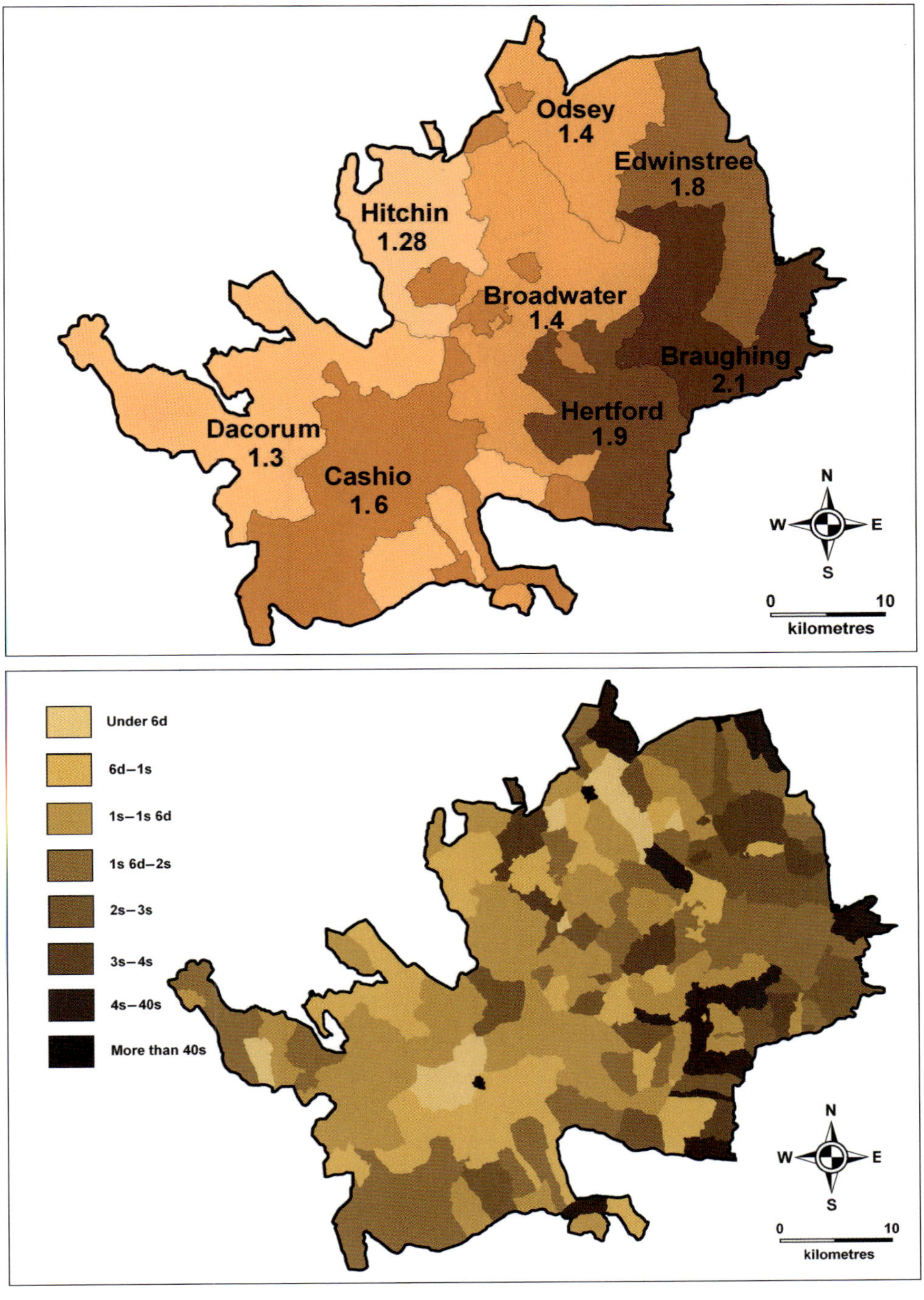

Figure 1.11. Hertfordshire in 1543, from the lay subsidy returns. (a) Density of taxpayers per 100 acres; (b) Tax payable per 100 acres, by vill (equated with parish). By the sixteenth century there had been a further narrowing of the gap in wealth and population density between the north-east and south-west of the county.

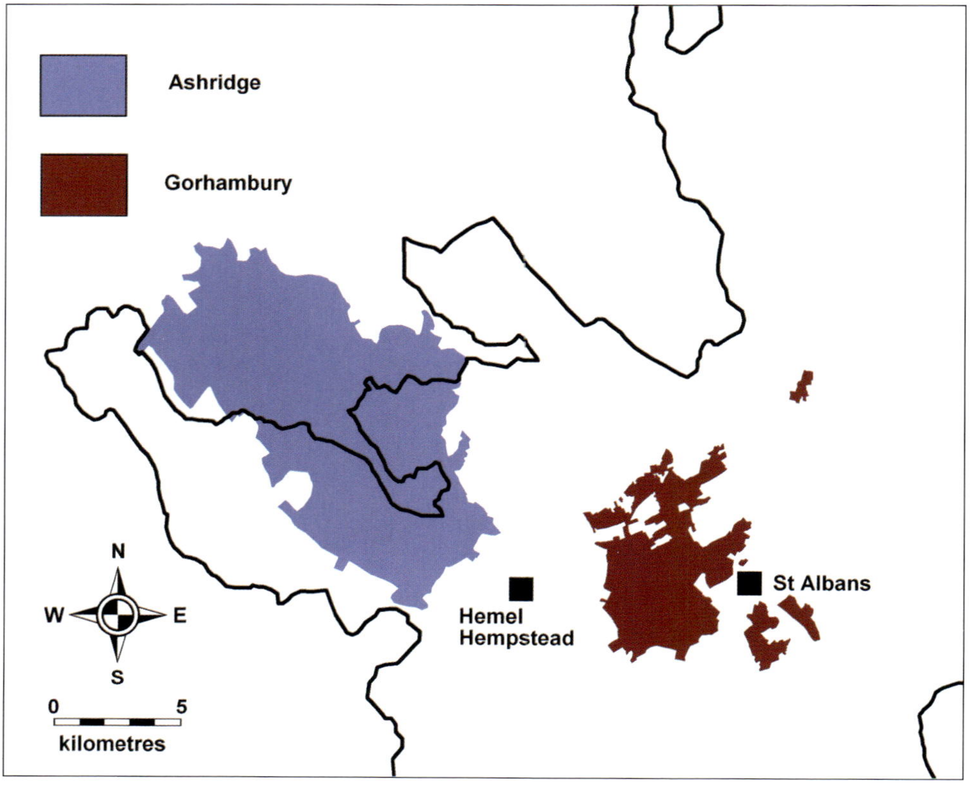

Figure 1.12. Although Hertfordshire was not for the most part a county of very large estates, some extensive landed properties did emerge in the course of the post-medieval period. Two of the largest were based on Ashridge and Gorhambury, both in the west of the county where, because of relatively infertile soils and distance from London, land was comparatively cheap in the seventeenth and eighteenth centuries. The bounds of Ashridge are as depicted on an estate map of 1853, those of Gorhambury on a map of 1768. Sources: National Trust and HALS DE/V/P2.

with numerous well-built houses. This was very different to the situation in the Yorkshire Wolds, for example, where the enclosure of vast tracts of open-field land and heathland ensured that the Sykes family had effectively a blank slate on which to create their estate, involving extensive schemes of building and tree-planting around West Heslerton. On most Hertfordshire estates, in contrast, extensive areas of woodland had existed from medieval times, and farms and cottages tended to be provided with picturesque details, or were rebuilt only gradually, in the course of the eighteenth and nineteenth centuries.

Throughout the later eighteenth century, and for most of the nineteenth, agriculture flourished and landowners benefitted from healthy rental incomes. But from the late 1870s imports of cheap grain from America, and subsequently of meat from a range of countries, led to a severe agricultural depression which affected most parts of the country. At this point in time Hertfordshire was, in Munby's memorable phrase, 'on the eve of transformation':[63]

for the later nineteenth and twentieth centuries saw the county irrevocably changed. The industrial revolution and the transition to a fossil fuel economy did not affect it directly to any marked extent although, as we shall see, particular industries – malting, brewing and paper manufacture – expanded in certain areas. But the arrival of the railways and a more general growth of the nation's population saw the steady urbanisation and suburbanisation of the south of the county and the growth of towns more widely, developments which intensified in the course of the last century and which have had far-reaching effects on landscape and environment.

Notes

1. L. Munby, *The Hertfordshire landscape* (London, 1977), p. 25.
2. Domesday shows that a number of other places now in Bedfordshire were then included in the county: Meppershall, Polehanger and Westoning. These were detached at an early date and will not be discussed here.
3. H.W. Gardiner, *A survey of the agriculture of Hertfordshire*, Royal Agricultural Society of England County Agricultural Surveys 5 (London, 1967), pp. 1–4.
4. We shall limit ourselves to the 'historic' boundaries as these existed before the changes of 1965 – including this southern 'salient' but excluding South Mimms, which was transferred *from* Middlesex at the same time.
5. J.E.B. Gover, A. Mawer and F.M. Stenton, *The place-names of Hertfordshire* (Cambridge, 1938), p. xiii.
6. R. Niblett and I. Thompson, *Alban's buried towns: an assessment of St Albans' archaeology to AD 1600* (Oxford, 2005); J. Crick (ed.), *Charters of St Albans* (Oxford, 2007); D. Short (ed.), *An historical atlas of Hertfordshire* (Hatfield, 2011); J. Catt (ed.), *Hertfordshire geology and landscape* (Welwyn Garden City, 2010). Other important publications include T. James, *Flora of Hertfordshire* (Welwyn Garden City, 2009), T. Slater and N. Goose (eds), *A county of small towns* (Hatfield, 2008) and H. Prince, *Parks in Hertfordshire since 1500* (Hatfield, 2008).
7. J. Catt, 'Introduction', in Catt (ed.), *Hertfordshire geology and landscape*, p. 11; H. Bailey and C. Wood, 'The Upper Cretaceous Chalk', in Catt (ed.), *Hertfordshire geology and landscape*, p. 37.
8. J. Catt and J. Doyle, 'The Palaeogene period', in Catt (ed.), *Hertfordshire geology and landscape*, pp. 62–3; D. Curry, 'Tertiary', in D. Duff and A. Smith (eds), *The geology of England and Wales* (London, 1992); D. Millward, R.A. Ellison, R.D. Lake and B.S. Loorlock, *The geology of the country around Epping: memoir for the 1:50,000 Sheet 240 of the British Geological Survey* (London, 1987), pp. 10–12.
9. Catt and Doyle, 'The Palaeogene period', p. 64.
10. *Ibid.*, pp. 69–70.
11. *Ibid.*, p. 75.
12. J. Catt, 'Neogene deposits and early landscape development', in Catt (ed.), *Hertfordshire geology and landscape*, pp. 87, 90–91, 107. This is shown as 'Clay-with-flints' on British Geological Survey maps.
13. Catt and Doyle, 'The Palaeogene period', pp. 63, 67.
14. J. Catt and A. Cheshire, 'Quaternary deposits and later landscape development', in Catt (ed.), *Hertfordshire geology and landscape*, pp. 140–41.
15. *Ibid.*, p. 140.
16. The lake also extended southwards down the valley of a major tributary of the proto-Thames (the proto-Mole/Wey), which later became the Lea valley.
17. *Ibid.*, pp. 149–50.
18. *Ibid.*, pp. 152–5.
19. O. Rackham, *The history of the countryside* (London, 1986), pp. 1–5; T. Williamson, *Shaping medieval landscapes: settlement, society, environment* (Macclesfield, 2003), pp. 1–12.
20. A. Rowe, 'Rabbit warrens', in Short (ed.), *Historical atlas*, pp. 112–13; J. Wallace, 'Sand and gravel quarrying', in Short (ed.), *Historical atlas*, pp. 104–5.
21. R. Holgate, 'Early prehistoric settlement of the Chilterns', in Holgate (ed.), *Chiltern archaeology*, pp. 3–16; D. Roe, *The Lower and Middle Palaeolithic periods in Britain* (London, 1981), pp. 184–200.
22. Holgate, 'Early prehistoric settlement', p. 12; C. Partridge, *Foxholes Farm: a multi-period gravel site* (Hertford, 1989); C. Le Quesne *et al.*, 'Post-excavation assessment of The Grove Estate, Watford', unpublished report, Hertfordshire County Council Environmental Department (2001).
23. Holgate, 'Early prehistoric settlement', p. 10; HHER 40; HHER 7312.
24. HHER 4261 and 4290.
25. S. Bryant, 'The late Bronze Age to the middle Iron Age of the north Chilterns', in Holgate (ed.), *Chiltern archaeology*, pp. 17–27; p. 24.
26. Le Quesne *et al.*, 'Grove Estate'.
27. S. Bryant, 'Iron Age settlements', in Short (ed.), *Historical atlas*, pp. 70–71; Bryant, 'Late Bronze Age to middle Iron Age', pp. 17–27.
28. B. Cushion, 'Braughing: woodland archaeological earthwork surveys', unpublished report for Hertfordshire County Council Environmental Department (2008).
29. Crick, *Charters of St Albans*, pp. 158–61.
30. HHER 25.
31. J.D. Hill, *Ritual and rubbish in the Iron Age of Wessex*, BAR British Series 242 (Oxford, 1995).

32. S. Bryant and G. Burleigh, 'Late prehistoric dykes of the eastern Chilterns', in Holgate (ed.), *Chiltern archaeology*, pp. 92–5.
33. M. Morris and A. Wainwright, 'Iron Age and Romano-British settlement and economy in the Upper Bulbourne Valley, Hertfordshire', in Holgate (ed.), *Chiltern archaeology*, pp. 68–75.
34. Bryant, 'Late Bronze Age to middle Iron Age', pp. 26–7; S. Bryant and R. Niblett, 'The late Iron Age in Hertfordshire and the north Chilterns', in A. Gwilt and C. Haselgrove (eds), *Reconstructing Iron Age societies* (Oxford, 1997), pp. 271–81; Niblett and Thompson, *Alban's buried towns*, pp. 23–5.
35. C. Saunders, 'Some thoughts on the *oppida* at Wheathampstead and Verulamium', *Hertfordshire Archaeology* 8 (1982), pp. 31–9; Bryant and Niblett, 'Late Iron Age in Hertfordshire', pp. 273–5; J. Hunn, *Reconstruction and measurement of landscape change*, BAR British Series 236 (Oxford, 1994), pp. 26–41; Niblett and Thompson, *Alban's buried towns*, pp. 23–40.
36. B. Cunliffe, *Iron Age Britain* (London, 1995), pp. 79–87.
37. Bryant and Niblett, 'Late Iron Age in Hertfordshire', pp. 271–3.
38. R. Niblett, 'A Catuvellaunian chieftain's burial from St Albans', *Antiquity*, 66 (1992), pp. 917–29; R. Niblett, 'A new site at Verulamium (St Albans)', in Holgate (ed.), *Chiltern archaeology*, pp. 96–102; R. Niblett, *The excavation of a ceremonial site at Folly Lane, Verulamium, Britannia* Monograph 14 (London, 1999).
39. Niblett and Thompson, *Alban's buried towns*, p. 148.
40. R. Niblett, *Roman Hertfordshire* (Wimbourne, 1995).
41. R. Niblett, 'Roman settlements', in Short (ed.), *Historical atlas*, pp. 72–3; Niblett, *Roman Hertfordshire*, pp. 64–76; K. Branigan and R. Niblett, *The Roman Chilterns* (Great Missenden, 2005); T. Williamson, 'The development of settlement in north-west Essex: the results of a recent field survey', *Essex Archaeology and History*, 17 (1986), pp. 120–32; J.R. Hunn, *Settlement patterns in Hertfordshire: a review of the typology and function of enclosures in the Iron Age and Roman landscape*, BAR British Series 249 (Oxford, 1996).
42. Munby, *Hertfordshire Landscape*, pp. 70–71.
43. *Ibid.*, pp. 106–16.
44. D. Hinton, *Archaeology, economy and society: England from the fifth to the fifteenth century* (London, 1990), pp. 7–12; D. Hall, *The open fields of Northamptonshire* (Northampton, 1995), pp. 129–30; T. Brown and G. Foard, 'The Saxon landscape: a regional perspective', in P. Everson and T. Williamson (eds), *The archaeology of landscape* (Manchester, 1998), pp. 67–94.
45. S.J. Bassett, 'In search of the origins of Anglo-Saxon kingdoms', in S.J. Bassett (ed.), *The origins of Anglo-Saxon kingdoms* (Leicester, 1989), pp. 3–27.
46. Friel, 'The Hicce – an Anglo-Saxon tribe of the Hitchin area', *Hertfordshire's Past*, 13 (1982), pp. 2–18.
47. T. Williamson, *The origins of Hertfordshire* (Hatfield, 2010), pp. 90–102.
48. M. Biddle and B. Kjolbye-Biddle, 'The origins of St Albans Abbey: Romano-British cemetery and Anglo-Saxon monastery', in M. Henig and P. Lindley (eds), *Alban and St Albans: Roman and medieval architecture, art and archaeology* (Leeds, 2001), pp. 45–77.
49. Niblett and Thompson, *Alban's buried towns*, pp. 166–95.
50. D. Short, 'Braughing: a possible Saxon estate?', *Hertfordshire's Past*, 23 (1988), pp. 8–15.
51. Williamson, *Origins of Hertfordshire*, pp. 125–63.
52. Crick, *Charters of St Albans*, pp. 144–54; Friel, 'The Hicce', pp. 2–18.
53. P. Fowler, *Farming in the first millennium* (Cambridge, 2002), pp. 182–204. See also D. Hill, '*Sulh* – the Anglo-Saxon plough *c.*1000 AD', *Landscape History*, 22 (2000), pp. 7–19; D. Banham, 'In the sweat of thy brow thou shalt eat bread: cereals and cereal production in the Anglo-Saxon landscape', in N.J. Higham and M.J. Ryan (eds), *The landscape archaeology of Anglo-Saxon England* (Woodbridge, 2010), pp. 175–92.
54. N. Rushton, 'Religious houses', in Short (ed.), *Historical atlas*, pp. 130–32.
55. M. Bailey, 'Introduction', in J. Brooker and S. Flood (eds), *Hertfordshire lay subsidy rolls 1307 and 1334* (Hertford, 1998), p. xxii.
56. *Ibid.*, p. xxi.
57. R.J. Kiln and C.R. Partridge, *Ware and Hertford: the story of two towns from birth to middle age* (Welwyn Garden City, 1995), pp. 108–14.
58. W. Page (ed.), *VCH Hertfordshire*, vol. 4 (London, 1914), p. 187.
59. G. Beresford, *Caldecote: the development and desertion of a Hertfordshire village*, Society for Medieval Archaeology Monograph 28 (London, 2009).
60. Munby, *Hertfordshire landscape*, p. 128.
61. W. Branch Johnson, *The industrial archaeology of Hertfordshire* (Newton Abbot, 1970), p. 99.
62. A. Young, *General view of the agriculture of Hertfordshire* (London, 1804), p. 18.
63. Munby, *Hertfordshire Landscape*, p. 190

– 2 –

Hertfordshire's 'champion' landscapes

Introduction

Although most of Hertfordshire is characterised by 'ancient' or 'woodland' countryside, some parts in the north have landscapes more akin to those of the Midlands, with nucleated villages which, in the Middle Ages, farmed extensive open fields. These areas lie on the Chalk escarpment, on the mixed soils found at its base, and extend onto the strip of poorly draining Gault Clay which just clips the northern fringes of the county. Henry Chauncy described the hundred of Odsey as being 'for the greatest part champion', and the same could have been said for the areas immediately to the east, the northern section of Edwinstree Hundred, and to the west, around Hitchin. Further to the south-west the county boundary swings southwards, excluding the escarpment and the level lands below it, but in the extreme west of the county Puttenham and Tring – the latter a particularly large parish containing a number of separate medieval townships – again extend across the Chalk and out onto the Gault Clay. Most of this northern part of the county – at least those parts lying astride the Chalk escarpment – appears to have been densely settled in prehistoric and Roman times. Numerous early settlements are known from fieldwalking surveys and aerial photography, and there are a number of prominent hillforts crowning the escarpment and a major late Iron Age *oppidum* and Roman urban centre at Baldock. But while some elements of the modern landscape may have their origins in these remote periods, most of what we see today appears to have been created in the medieval and, in particular, the post-medieval periods.

As we explained in Chapter 1, this narrow strip of open-field country forms part of a much larger landscape zone extending far into the Midlands. In medieval times settlement was almost entirely concentrated in villages or large hamlets and agriculture was carried out on highly communal lines. The holdings of farmers were intermixed in a myriad of unhedged strips, scattered evenly and sometimes very regularly through a number of great 'fields', one or more of which lay fallow (or uncultivated) each year and was grazed by the village livestock. Beyond the fields areas of common grazing land existed, and there were usually extensive areas of hay meadow which were also, like the arable fields, usually divided into strips and managed communally. Such landscapes, with their complex intermixture of properties and communal and customary regimes and routines, were difficult to unscramble and reorganise, especially where proprietors were numerous. They usually survived unenclosed, or only partially enclosed, well into the post-medieval period and, as a result, now have relatively recent field patterns (and to some extent road patterns), in many cases the result of comprehensive replanning by eighteenth- and nineteenth-century surveyors – hence the term 'planned countryside' often used to described their appearance today (Figure 2.1). Hertfordshire is not the best county in which to study landscapes like these, given that they characterise only a small portion of its area. But, while superficially less interesting than the kinds of 'ancient countrysides' found elsewhere in the county, they nevertheless have many features worth examining in some detail.

When we begin to discuss these districts, however, we immediately encounter problems of definition. Landscapes seldom have neat boundaries, and it is hard to draw a firm line between Hertfordshire's 'champion' and the various 'woodland' areas lying to the south. In part this is because the definitions of landscape types often employed by historians are oversimplifications, and well into the eighteenth century large areas of open fields also existed to the south of the Chiltern escarpment, and especially in the larger valleys of the north-east of the county. But it was also because many of the parishes which were located

Figure 2.1. Typical 'planned countryside' in north Hertfordshire: a straight road bounded by a flimsy hawthorn hedge created when an earlier track across the open fields was 'tidied up' at the parliamentary enclosure of Ashwell in 1863. In the distance, the parish of Bygrave has never been formally enclosed.

astride the escarpment, as well as possessing extensive and continuous areas of open fields – true 'champion' country – on this light land, also included areas of scattered settlement, small open fields and enclosures on the heavier land of the clay plateau above it. Such places lay, that is, across a major landscape divide, with the 'champion' lands extending north, far into the Midlands, and the 'woodland' stretching southwards to the English Channel.

Perhaps 16 parishes fall into this category – places such as Barley, Barkway, Therfield, Reed and Great Offley – slightly more than the 14 or so lying below the escarpment which were entirely 'champion' in character, without early outlying settlements, and with arable wholly cultivated, at least in medieval times, in open fields. In addition to this, the two large parishes of Tring and Hitchin included a number of distinct townships, ecclesiastically dependent but farming their own agricultural territories, which were effectively separate 'champion' villages with their own open fields: large and compact hamlets such as Walsworth in Hitchin and Long Marston and Wilstone in Tring, rather different from the kind of small 'green' hamlets or loose scatters of farms around more extensive areas of commonland which are found in the centre and south of the county. Indeed, here again we encounter problems of definition, for – aside from the fact that it lacked its own church – there was little difference between a 'hamlet' such as Walsworth, surrounded by open fields and with a territory covering over 350 hectares, and many of the smaller 'villages' in the area, such as Radwell (302 hectares), Newnham (388 hectares) or Caldecote (134 hectares).

The character of the open fields

Although broadly 'champion' in their characteristics these scarp and scarp-foot settlements displayed a number of distinct features, in terms of agrarian organisation, which distinguished them both from each other and, in some cases, from the 'normal' forms of open-field agriculture found in the Midlands. In the latter districts the arable was usually divided into two or three great 'fields', one of which lay fallow each year, but by the time that they were enclosed many Hertfordshire townships contained four or more named fields. The most complex arrangements were generally found in parishes lying astride the escarpment. Seventeenth-century Weston, for example, had extensive areas of open arable along the escarpment itself – Baldock Field, Gibbet Field, Weston or Gages Field, Atterhill Field and Lannock Field – but also smaller, more discontinuous areas on the sides of valleys and embayments or recesses in the escarpment, and in particular in valleys beyond the watershed, draining south into the river Beane (Town, Millhill, Marlands, Cheesell, Berryfield, Fontley, Lince and Slowgrove Fields).[1] Where the clay plateau gave way directly to Chalk scarp – where the topography was uncomplicated by valleys cutting through the Chalk and where the area of the township did not include such south-draining valleys – arrangements were generally simpler, although still rather different from the Midland 'norm'. The parish of Barley was surveyed between 1593 and 1603 by John Norden.[2] On the clay plateau in the south of the parish the land lay in enclosed fields but in the centre and north holdings were splintered into 1,798 strips with an average area of just over an

Figure 2.2. Part of the open fields of Ickleford, as shown on a map of 1771. Note the serpentine shapes of the individual strips and the way that these are grouped into numerous furlongs or 'shotts'. By this stage many of the strips had been consolidated into larger – but still unhedged – units. Reproduced by kind permission of HALS (DE/Ha/P1).

acre. These were grouped into six 'fields' of very unequal size, ranging from Little West Field of 113 acres to Waterden Field of 678.[3] It is possible that this was a modification of an earlier three-field arrangement, given that the largest field comprised 33.8 per cent of the entire area of open arable, while the combined area of Church Field and Picking Field made up a further 32.5 per cent, the final rough third comprising Great and Little West Field and Cobdell Field. Certainly, the distribution of holdings through the fields appears to have been fairly even, although not regular, except that while the manors of Hores and Greenbury (identifiable with two of the Domesday manors) had demesne land in all the fields, Abbotsbury and Mincinbury (likewise identifiable with Domesday holdings) had theirs more clustered – in Church Field, Picking Field and Waterden, and in Church, Picking, Waterden and Cobdell, respectively.

Villages lying on the chalk loams at the foot of the escarpment, or on the clays beyond, generally had field systems similar to that at Barley. In typical 'Midland' fashion, holdings were spread relatively evenly across the open fields; yet there were, at the same time, often more than three fields. Ashwell, for example, had four when finally enclosed in 1863 – Quarry Field, Claybush Field, Redland Field and North Field – all roughly the same size.[4] Ickleford similarly had four named fields when mapped in 1771 (West Field, Home Field, Mildens and North Field), as had Great Wymondley in 1811 (Roxley, Middle, Little Middle and Purwell), while Pirton had no less than seven when enclosed in 1818 (Dane, Highdown, White, Hambridge, Windmill, North Mead and Green Ley Hill).[5] How far such arrangements were the consequence of post-medieval reorganisation of the cropping boundaries is uncertain. The boundaries and the numbers of fields could certainly be changed over time. Hitchin had nine or ten open fields listed in a survey of 1676 (the precise number is uncertain owing to ambiguities in nomenclature): by 1819, however, there were six, which were grouped into three 'seasons' for the purposes of cropping (the customs of the manor, drawn up in that year, noted that they had 'immemorially been and ought to be kept and cultivated in three successive seasons of tilthgrain, etchgrain and fallow').[6] Similar changes occurred elsewhere. In Offley, what remained of the open arable after a period of steady attrition was reorganised into three cropping 'seasons' less than fifty years before the parish was finally enclosed in the nineteenth century.[7] It may well be that some places which, by the eighteenth century, had four or more fields, possessed two- or three-field systems in the Middle Ages; and some settlements were certainly associated with such 'normal' arrangements. Norton, for example, had three fields in the seventeenth century but four at the time of enclosure in 1796.[8] Caldecote appears to have had a two-field system of normal Midland form in the fourteenth century, while Walsworth still farmed three fields of roughly equal size when it was enclosed in 1767.[9]

Whatever their precise number, the open fields were divided into furlongs, which were usually termed 'shots' or 'shotts' in Hertfordshire (Figure 2.2). These formed the basic unit for rotation: that is, all the strips in any one furlong grew the same crop in any year. Furlongs, like the individual strips, were unhedged. They sometimes butted up directly onto an adjoining furlong, but often had field lanes as boundaries. If they joined the next furlong at right angles, the first strip within it was used as a 'headland'. As the *Victoria County History* described in 1912 at Bygrave, where the open fields still survived at this date, 'Where there is no road the plough is turned on transverse strips known as "headlands". In consequence the owner of the "headland" must wait until the adjacent land is ploughed before ploughing his own.'[10] The strips or 'lands' were usually slightly sinuous in plan, having been distorted over the years as the plough moved across steep slopes, or by the tendency of the ploughman to begin his 'turning circle' some way before each end of the strip was reached (thus producing an elongated 'reversed S' shape). In some places early maps show two, three or more furlongs arranged end-to-end, suggesting that they had developed through the subdivision of what had originally been bundles of much longer strips, a phenomenon which has been noted in a number of places in the Midlands. In most cases this may be an illusion, caused by the tendency for strips to be aligned at right angles to the contours (except on the steepest slopes, where they were ploughed parallel with them) but in Barkway and Ickleford some of the strip boundaries appear to be in precise alignment over three furlongs.[11]

On the Chalk the furlongs, and often the individual strips, were defined by unploughed grassy banks or 'balks', locally described as 'walls'; where they were absent the strips were separated by a single furrow. There were discontinuous 'hedges' growing on some of the balks, and early twentieth-century accounts describe how 'here and there, especially in the neighbourhood of the village, bushes growing on the balks have formed small hedges'.[12] Where there were steep slopes, and the strips were laid out along the contours, they were sometimes cut into the hillside as terraces which archaeologists refer to as 'strip lynchets', and which early maps and documents call *lynches* or *linces*. On the more level clay ground below the Chalk there were no lynchets or balks but the strips were often ploughed, as was normal in the Midlands, in ridges (as 'broad rig' or 'ridge and furrow') to improve the drainage of this heavy land. Light, thin soils were seldom ploughed in ridges. Not only was this unnecessary but, in addition, ploughing in this manner, because it moved soil gradually towards the centre of each strip, would eventually have cut into the infertile subsoil on its margins.

The origins of villages and fields

The origins of champion landscapes – of nucleated villages and extensive common fields – have been debated by archaeologists, historians and geographers for over a century. It is now clear that neither were introduced, fully formed, by Anglo-Saxon settlers in the fifth and sixth centuries, as used to be believed. Most scholars agree that they developed in the course of the middle or later Saxon periods, as a dispersed pattern of settlement was abandoned, voluntarily or otherwise. Scattered farms were suddenly reorganised into compact villages or large hamlets and communal field systems were laid out around them. This reorganisation might have been associated with the growth of local lordship, with demographic pressure and the need to farm more efficiently, or with both. According to one popular view, as the population grew holdings became progressively subdivided and intermingled, partly through the action of partible inheritance (each son receiving a portion of his father's farm) and partly because, as the area of arable expanded at the expense of grazing land, the new intakes were divided fairly between all those who had formerly pastured their stock there. In both cases, the parcels took the form of strips because this was the shape most conveniently cultivated, especially with the heavier ploughs which were coming into widespread use in the later Saxon period. Eventually, as resources of pasture dwindled, farmers needed to make more efficient use of the grazing afforded by the fallows and by the weeds growing on the stubbles after harvest. But this was difficult where lands lay intermingled in small, unhedged plots, for one cultivator might wish to graze his land before his neighbours' crops had been harvested.

> A peaceful option for a long-term resolution of their difficulties involved the inhabitants reorganising their numerous farms and hamlets into common fields where the problems of competition would be minimised. The animals of the whole community were pastured together on the land which lay fallow or awaited spring cultivation.[13]

In fact, there are many difficulties with this popular model. In national terms 'champion' landscapes are not clearly or closely associated with areas which, to judge from the evidence of Domesday, were the most densely populated. In Hertfordshire, it is true, the landscapes of nucleated villages and open fields do correspond in general terms with areas in which the Domesday population was relatively dense: but equally dense populations could also be found in many places in the 'woodland' districts in the north-east of the county. Such a model is, moreover, hard to square with the presence of villages which lay astride the woodland/champion divide, which could boast extensive open fields *and* large areas of dispersed settlement. More importantly, the idea that early Saxon settlement was highly dispersed, and that in each area this scattered pattern was replaced at a stroke by one of nucleated villages, is based on limited fieldwork from a handful of parishes in the Midlands. Even here there are usually no more than one or two outlying abandoned Saxon sites away from modern villages, and these were not necessarily all occupied (or abandoned) at the same time. The early Saxon settlement pattern was fairly mobile and such sites were often occupied for less than a century before being abandoned, the inhabitants setting up a farm elsewhere. What happened in the course of the Saxon period was not so much a process of settlement 'nucleation' but rather of settlement 'stabilisation'. Sites stopped moving around the landscape and began to grow *in situ*, and medieval villages were the end result.

In north Hertfordshire relatively few of these abandoned Saxon sites, away from villages, are known. Some of those which have been discovered evidently date from early in the Saxon period – like those near Baldock at Clothall Common and Blackhorse Farm.[14]

Two others, in Ashwell and Letchworth, are rather *later* in date than is usual for such sites in the Midland counties, where outlying sites had generally been abandoned before the middle of the ninth century. At Ashwell, the site near Shire Balk, discovered in 1991, featured two sunken-feature buildings and appears to have been occupied in the tenth and eleventh centuries, while that at Green Lane in Letchworth (strictly speaking, in Norton parish), again with sunken-feature buildings, appears to have been in use between the ninth and the eleventh century. This suggests that the medieval pattern of fields and villages may, in this area of Hertfordshire, have finally developed at quite a late date. This said, there is good evidence that settlements of middle–late Saxon date also underlie most existing villages. At Bygrave, for example, examination within the parish church during building work revealed that its twelfth-century walls were cut through at least two earlier floors. The later of these produced tenth- and eleventh-century pottery, the earlier produced none, but residual eighth- and ninth-century pottery was found both inside and outside the building.[15] In most cases these settlements were, like those now lying well outside villages, probably small, occupying a fraction of the modern settlement area – often close to the sites later occupied by the manor house ('Bury') and church, which generally stand beside each other. Villages grew gradually, either from a single middle–late Saxon nucleus or – in the case of some of the larger villages with complex plan forms – from the expansion and fusion of more than one. At Ashwell, for example, there appears to have been at least two areas of middle–late Saxon occupation: at Westbury, at the western end of the village; and in the area around Nine Wells.[16] In some cases it is possible that neighbouring farms never fully fused, perhaps explaining places such as Norton, where the 'Bury' or manor house, unusually for this part of Hertfordshire, lies over 600 metres from the village and the parish church: 'Bury' and church, that is, may mark the sites of separate middle–late Saxon farmsteads.

So far as the evidence goes, villages reached their maximum extent during the thirteenth century, many then experiencing contraction due to the demographic and economic changes in the wake of the Black Death. But perhaps 'village' is not quite the right word for, as we have already noted, in technical terms some of these nucleated settlements, with their associated field systems, remained as hamlets within a larger parish and never gained a church of their own. It is noteworthy that such places are all found in just two parishes, Tring and Hitchin, both of which in middle and later Saxon times seem to have been major royal estates, as they gave their names to the hundreds within which they lie. Walsworth, Langley and Preston in Hitchin and Aldwick, Pendley, Dunsley, Wilstone and Long Marston in Tring were all like this, their development in the Middle Ages as separate parishes prevented by the fact that they continued to form part of a large secular estate.

The precise manner in which any of these nucleated settlements, whether villages or hamlets, developed over time is difficult to ascertain because the areas of most are still occupied by houses and gardens. Even where they have become shrunken or deserted – the fate of several in north Hertfordshire – and can be excavated on some scale, the areas examined are not necessarily representative of the whole history of the settlement. In the case of the shrunken village of Caldecote, for example, excavations suggested that there had been no occupation on the site before the middle of the eleventh century but this is hard to square with the fact that Domesday Book describes nine villein farms and four cottagers there, as well as a priest, implying that the church was already in existence.[17] Some indications of the way villages came into being can, however, be derived from an analysis of their plans, although such an approach is often problematic because many suffered radical shrinkage and other changes in the course of the late medieval and post-medieval periods and few have maps dating from before the eighteenth century. Expansion from one or more middle–late Saxon nuclei, as in other 'champion' areas, seems sometimes to have involved the spread of settlement across adjacent areas of arable land, in which case the village enclosures or 'tofts' exhibit a strippy, parallel form, as in the southern end of Lilley. Occasionally, as at the western end of Norton, they preserve the sinuous or 'reversed S' layout of the former strips. Settlement also clearly expanded along the sides of pre-existing roads, as at Ashwell, where the plan of the village – originally a major estate centre and early borough – is structured by the ancient Ashwell Street and the various tracks running roughly parallel to it, as well as by the arable strips laid out at right angles to these.

Often, however, village plots are more irregular in character, settlement presumably having expanded into some adjacent area of as yet unploughed ground. In some cases there are signs that farms were initially scattered around the margins of the latter, creating a large central common or green which was then steadily fragmented as further properties were cut out of it. Hertfordshire village plans, both in the 'champion' area and elsewhere, often feature two or more greens, sometimes so small that they can have served little practical use as pasture for livestock. It is often possible to see how these result from the erosion or fragmentation of a once much larger area of common land. Sometimes, for example, two greens are separated by an area of housing, but share a continuous boundary line which extends behind the area of enclosed land, sometimes as a footpath or road, marking the former common edge. Such intakes are usually occupied by cottages and shops, with the principal farms, by post-medieval times at least, lying towards the periphery of the settlement on the fringes of the former common. At Ickleford, for example, the church and rectory occupy an oval of land clearly taken out of Upper Green, and presumably before the twelfth century, to judge from the church fabric. The properties immediately to the west appear to occupy two further encroachments (the former common edge visible as a continuous boundary line further to the west) which are separated by a small surviving fragment of open land. Some 500 metres to the north of Upper Green lies the rather larger Lower Green, the two now separated by a line of farms and a number of enclosed fields. In 1771, however, there were patches of common land scattered among these irregular enclosures, clearly indicating that the two greens had once formed a single large tract of grazing beside the river Hiz, from which *inter alia* Pound Farm and the associated moated site had been enclosed (Figure 2.3).

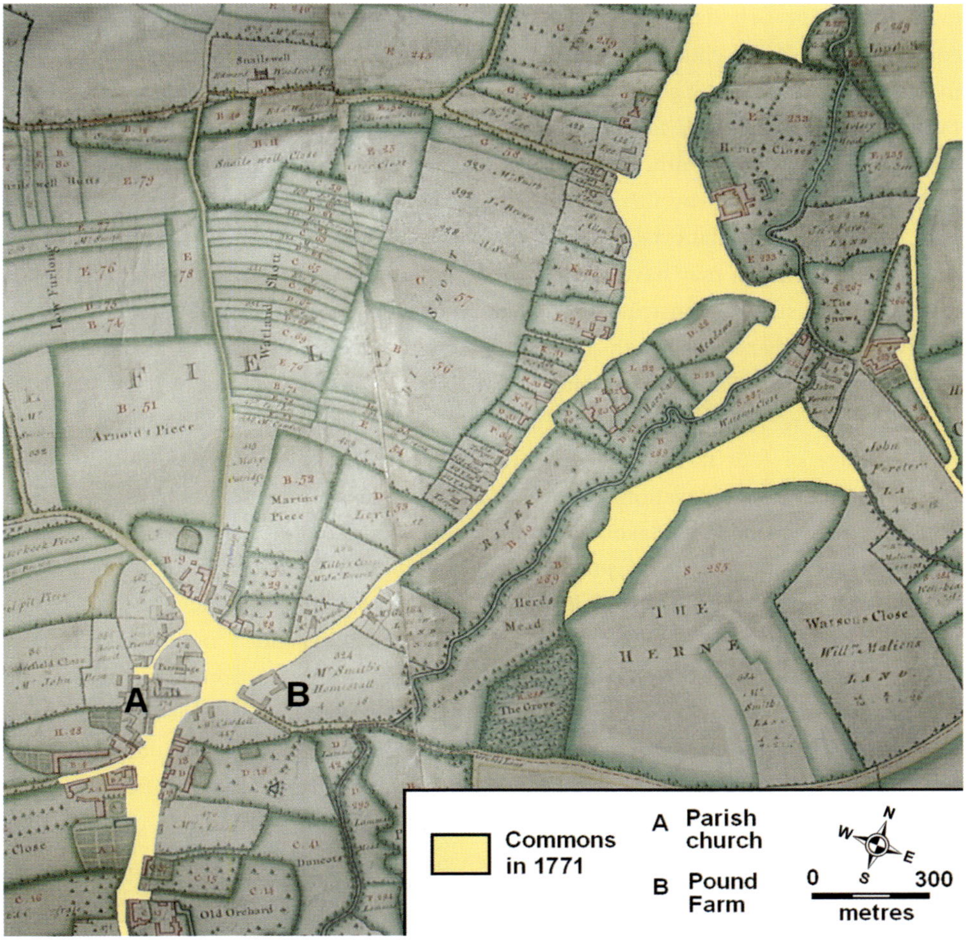

Figure 2.3. The village of Ickleford, first mapped in 1771, is clustered around a number of greens. These had evidently been created by the fragmentation of a single, much larger tract of common grazing lying beside the river Hiz, through the encroachment of fields, farms and cottages. One intake is occupied by the parish church of St Katherine and the rectory ('A'). The damp, low-lying location probably saved the remaining areas of common from further encroachment: in the case of other settlements the process continued until only small 'pocket-handkerchief' greens remained within a large village with an irregular plan.

Usually fragmentation went further, leaving only a number of small, pocket-handkerchief greens connected by winding roads, as appears to have happened at Willian, for example (although the extent of later shrinkage makes it hard to be sure), or at the hamlet of Wilstone in Tring, where the probable filled green is mainly occupied by enclosed pastures, rather than house tofts. Great Wymondley is an intriguing case. The enclosure map shows a number of small greens here, some with encroachments taken out of them; fields whose shapes suggest that they were once greens, and had been taken into private ownership at an early date; and a number of curving boundaries which look like former common edges.[18] The church and the adjacent fields, as well as part of the village, appear to have been taken out of a single large common, into which the roads from Hitchin and Little Wymondley all once 'funnelled', in the characteristic way that roads do as they enter areas of common land, both being part of the 'waste' of the manor. A castle – probably constructed in the early twelfth century and which may have developed, as was often the case, from an earlier manorial complex – was clearly placed on the southern edge of the common. The church may have been erected on an intake from the common around the time that the castle was constructed. Here, as at Ickleford, no priest is listed in Domesday.

Pirton, another village with a castle, is another interesting case. A programme of 'test pitting' – the excavation of numerous small trenches – carried out over several years revealed scant remains of early or middle Saxon settlement, but by the tenth or eleventh century three or four farms existed beneath what is now the eastern side of this large village.[19] In addition, excavations in 1993 and 1995 on land to the north of The Fox public house revealed a small late Saxon cemetery associated with a probable wooden church.[20] These various settlements were probably, as apparently at Ickleford, scattered round the margins of common pastures, which were here subsequently fragmented and infilled not only with closes, farms and cottages as the settlement expanded and coalesced in the course of the twelfth and thirteenth centuries, but also by the erection of the large motte and bailey castle, with a new church within it, some time after the Norman Conquest. The surviving greens at either end of the castle enclosure – which, to judge from the configuration of building lines and property boundaries, were once significantly larger – are clearly fragments of a once much more extensive common: the lord of the manor would, of necessity, have chosen a piece of open ground as the location for the castle, rather than taking up valuable land already cultivated or occupied by houses (Figure 2.4).

The tendency for settlement to expand *in situ* around and into commons or across adjacent fields, with minimal dispersal, as farms proliferated and increasing areas of land were brought into cultivation, stands in marked contrast to developments in most parts of the county, where settlement grew increasingly dispersed in the course of the later Saxon and early post-Conquest periods. The reasons for the difference may have been largely environmental. Across most of Hertfordshire drinking water was widely available from shallow wells, but in these northern districts, as across much of the scarp-and-vale countryside of the Midlands, reliable sources were much more limited. Water was available only at the spring lines found at the junction of the Middle and Lower Chalk, and the Lower Chalk and the Gault Clay, or from wells sunk in the boulder clay on the plateau above the escarpment, or in the intermittent patches of such clay at its foot. Settlement was thus to some extent obliged to cluster, as farmland spread out across the effectively waterless land around, so that holdings became steadily more intermixed in the manner already described, drawing farmers ever further into mutual cooperation and eventually encouraging the communal organisation of farming. In addition, where townships

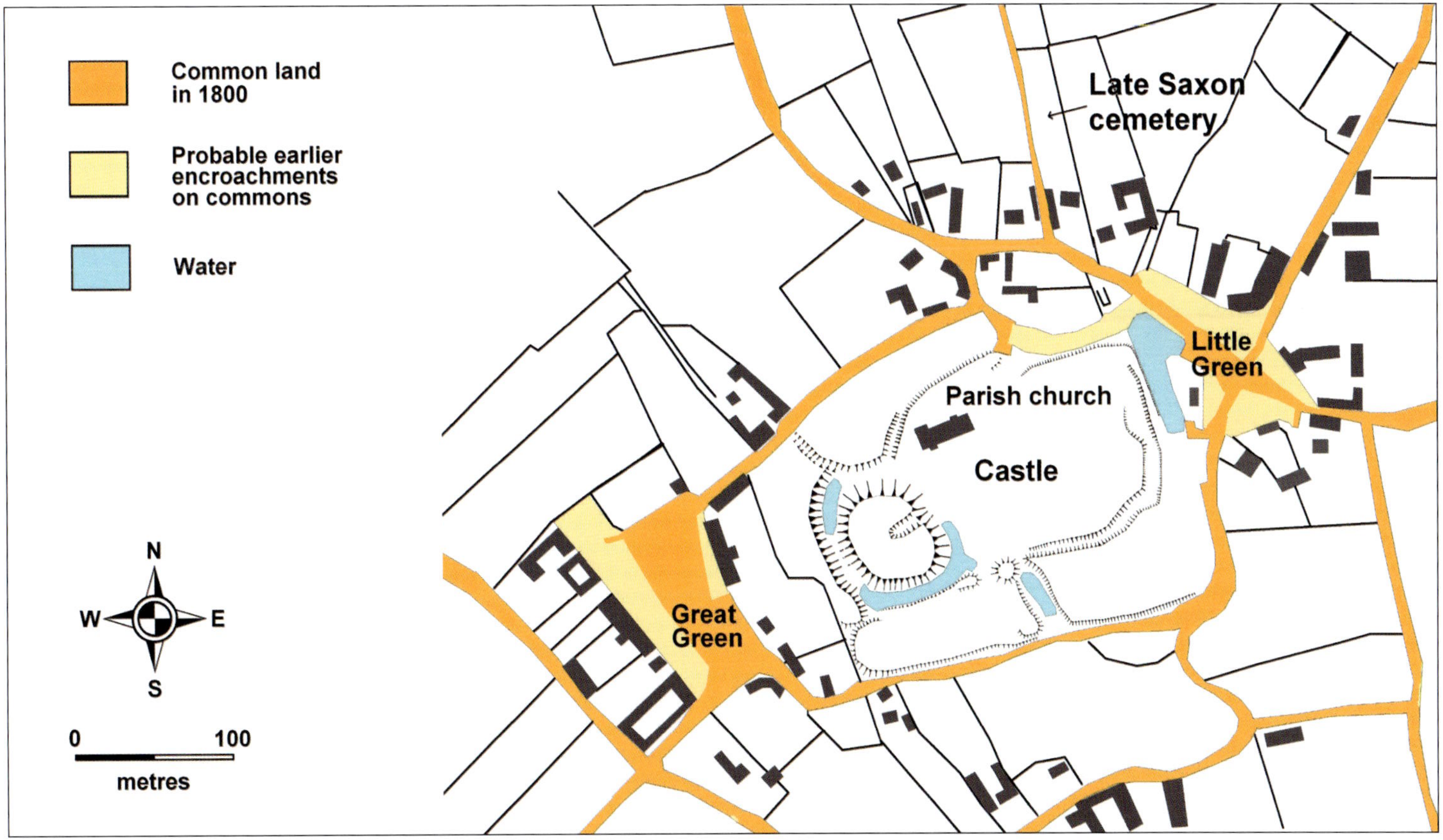

Figure 2.4. The village of Pirton, based on a map of 1800 in Hitchin Museum. The castle was apparently placed on a large tract of common pasture, the remaining fragments of which became two small greens flanking the eastern and western ends of the bailey: these in turn were subsequently reduced in size by encroachment. The present parish church stands within the castle bailey: what was probably a pre-Conquest church and associated cemetery lay to the north, on the edge of the former common.

included soils formed in Gault Clay, ploughs and teams needed to be mobilised rapidly to take advantage of the short windows of opportunity available for ploughing in the spring, when these easily impacted soils dried out sufficiently to be cultivated in safety. But early medieval tenants shared ploughs, contributing oxen to the common team, and it therefore made sense to live in reasonable proximity rather than widely scattered across the landscape. Clustered settlement allowed rapid mobilisation of shared ploughs and teams, while intermingled holdings, scattered across land of varying aspect and drainage potential, allowed each cultivator a reasonable chance of getting their holding ploughed in time for seeding. Some of the light soils formed over the Chalk, especially those of the Upton Association, are silty and likewise prone to impaction, so that cultivation has to be carefully timed.[21] More importantly, the chalk soils leach nutrients rapidly, and fertility could be maintained only by folding sheep on the land – that is, penning them up at night so that they trod dung and urine into the soil – when it lay fallow or the crops were young. Again, patterns of splintered holdings which had developed organically may have been reorganised, perhaps through piecemeal exchanges, to create a more even distribution, in order to allow all farmers to receive equal benefits from the fold as it was moved gradually through the land of the township. Unfortunately, it is difficult to test these suggestions, just as it is hard to test other explanations for the origins of 'champion' landscapes. The development of villages and open fields had been completed before local documents begin to be abundant from the end of the twelfth century.

Enclosure

In England as a whole the enclosure of open fields and common grazing began in late medieval times and was largely completed by the 1820s. It was achieved in a number of ways, and often in more than one way in any one township or parish. Historians sometimes make a broad distinction between 'piecemeal' and 'general' enclosure. The former was a gradual process. Owners bought and sold land until they had a number of contiguous strips, which they could then fence or hedge and remove from the routines of communal cultivation. The field boundaries, established on the edges of the former 'lands', perpetuated their slightly sinuous outline. In Hitchin, at least by the seventeenth century, the manorial court often granted tenants permission to enclose portions of the fields in this way. In October 1676 Lawrence Tristram agreed to pay 1s 11d per annum for leave to enclose with hedges and ditches four acres of arable land at Robin Hood's Hill, the court having ruled that this would cause no harm to the lady of the manor or her tenants. The latter would be denied a portion of the grazing on the fallows, but Tristram would have to reduce the number of animals which he could graze on the common lands.[22] The customs of the manor of Hitchin, as set out in 1819 but almost certainly recording ancient practice, noted that:

> Every owner and every occupier of land in any of the common fields of this township may at his will and pleasure enclose and fence any of his land lying

> in the common field of this township (other than and except land in Bury Field and Welshman's Croft), and may, so long as the same shall remain so enclosed and fenced, hold such land ... exempt from any right or power of any other owner or occupier of the land in the said township to common or depasture his sheep on the land so enclosed and fenced (no right of common on other land being claimed in respect of land so enclosed and fenced).[23]

The areas of land enclosed in this manner varied greatly in size. In Weston, for example, a terrier (or survey of property) drawn up in 1633 describes how Francis Humberston held 'a peece closed in w[i]th a hedge' covering a mere acre and a half in Stonley Field; but in Neather Down Field Mr Kympton held just under 24 acres 'being lately taken in & severalled'.[24] Over large areas of north Hertfordshire, as on the dipslope of the Chilterns and the East Anglian Heights (in 'woodland' country), numerous field boundaries display the familiar sinuous shape characteristic of piecemeal enclosure. In some cases, however, enclosure of this kind was associated not with the actions of small proprietors but with large landowners gradually 'engrossing' their holdings, buying out neighbours and enclosing as they did so, often ending up as the sole owner of a parish, as, for example, at Letchworth.

Some sinuous boundaries like this were also established by 'general' enclosures, but for the most part these created field patterns less closely related to the layout of strips and furlongs. In a general enclosure the proprietors of a parish or township, or a single proprietor, decided to enclose all or most of the commons and open fields at a stroke. The holdings and rights of each owner were surveyed and the land then reallotted in the form of enclosed parcels. There were several forms of general enclosure. The earliest were the depopulating enclosures of the fifteenth and sixteenth centuries, where a manorial lord, holding a manor in which the tenure of his copyhold tenants lacked any degree of long-term security, enforced enclosure, usually so that the land could be laid to pasture, often as large fields with slightly serpentine boundaries. More common were the enclosures 'by agreement' of the seventeenth and early eighteenth centuries, where proprietors all agreed to enclose and often secured confirmation of the new arrangements by bringing a fictitious case in the court of Chancery. Most important were enclosures by parliamentary act. Landowners petitioned for an act, which was followed by a survey – to ascertain both the extent of possession and the nature of the rights owned by the inhabitants – and then an award, which laid down the new pattern of properties and often rearranged the pattern of roads and footpaths. With this form of enclosure the termination of common rights and the consolidation of intermixed holdings in the fields did not require the agreement of the majority of landowners, but only that of those holding a majority of the land – generally three-quarters by value – in a parish or township.[25] In the Hammonds' famous words, 'the suffrage was weighed, not counted'.[26] Parliamentary enclosure began in the early seventeenth century but was mainly a feature of the post-1750 period, peaking in two great waves – in the 1770s and during the Napoleonic Wars.[27] The first of these peaks characteristically affected open-field parishes on the heavy clay soils of the Midlands and was often – like most earlier forms of enclosure in this district – associated with the laying of land to pasture. The second, which coincided with the high prices of the Napoleonic War years, was principally concerned with wastes and commons, especially in the highland areas of England, but also affected much open arable, mainly on lighter soils in the east of the country. The main motive for enclosing was now to allow farmers freedom to adopt the new husbandry techniques of the 'agricultural revolution' and to bring common grazing land into more intensive (and usually arable) use.

Although most champion parishes in Hertfordshire were enclosed by parliamentary act, a significant minority were enclosed by other means, in some cases before the eighteenth century. Even where parishes were enclosed by act their fields had often undergone a significant degree of prior attrition through piecemeal enclosure. Moreover, although parliamentary enclosure was concentrated in the Napoleonic War years, as we might expect given the nature of the local soils, some also took place in the 'first wave', mainly but not exclusively on the heavier ground at the base of the escarpment. Conversely, there were some very late enclosures by parliamentary act – Sandon in 1842 and Ashwell in 1863. In Therfield an act of 1849 enclosed the open fields but another, as late as 1893, removed the greens and commons. This was part of a wider pattern. In several parishes enclosure came later still, with some open fields surviving into the twentieth century. In Hitchin, Bygrave, and Clothall in particular large tracts of open arable were still in existence when the maps of land ownership were prepared in accordance with the terms of the 1910 Finance Act.[28]

Early enclosures

Like many complex patterns, Hertfordshire's varied enclosure history is the consequence of a number of factors. Perhaps the most important were soils and geology. Places which appear to have been enclosed before the eighteenth century tend to be located not on the Chalk escarpment but at its foot, on the deposits of boulder clays overlying the Lower Chalk and on the Gault Clay beyond it (Figure 2.5). Most, moreover, were rather small places: diminutive townships within larger parishes, such as Pendley in Tring or Preston in Hitchin; or parishes which were themselves small: Newnham (388 hectares), Letchworth (350), Radwell (302) and Caldecote (134). Both characteristics fit in well with wider national patterns. Enclosure tended to come early to 'champion' townships in areas where soils were heavy because it was often motivated by a desire to lay land to grass – something which applies to Puttenham, to Long Marston and other Tring townships and perhaps to some of the parishes in the north, such as Letchworth or Radwell. It also tended to come early where townships were small, because in such circumstances they were more likely to become dominated by a single owner or by a small number of owners, who could then enclose or agree to enclose. Tenurial composition was an important determinant of chronology because enclosure was not generally in the interests of the smaller proprietors

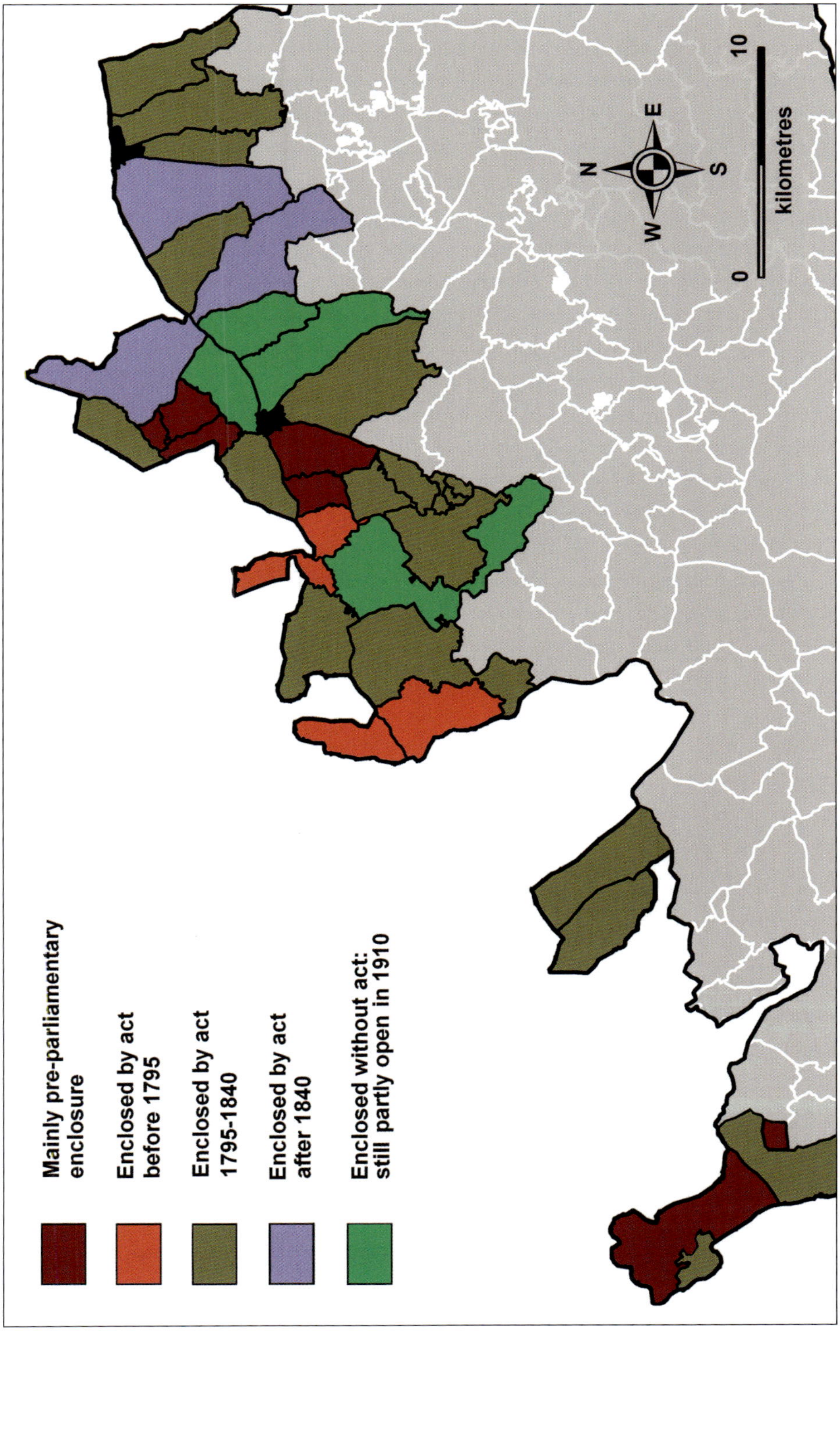

Figure 2.5. The chronology of enclosure in the 'champion' parishes of north Hertfordshire.

and many of the larger parishes lying immediately below the Chiltern escarpment, on what were generally good soils, had substantial populations of freeholders. What small landowners gained, in terms of being able to farm as they wished free from communal controls, they generally lost through the actual costs of enclosure and the loss of access to common and fallow grazing. In Pirton in the seventeenth century the parishioners drew up a document agreeing to defend each other if they were prosecuted for asserting their grazing rights on land which had been enclosed from the fields by Thomas Docwra.[29] Certainly by the nineteenth century early enclosed places generally had a single dominant owner. At Newnham in 1847, for example, Samuel Mills owned all the land in the parish other than the glebe; at Caldecote in 1841 William Hall was in a similar position; while at Letchworth in 1837 John Allington owned 829 acres and Thomas Dimsdale just over 6 acres, with the glebe constituting a further 43.[30]

Most of these early enclosed places, if parishes, have rather small churches, those at Radwell and Letchworth with bell turrets rather than towers. Many, moreover, display clear signs of settlement shrinkage: that is, they have earthworks indicating that they were once rather larger. Particularly striking are the holloway and house platforms to the west and north of the village pond at Willian and the toft boundaries and abandoned roadway to the north-east of Norton church. At Caldecote the village has been reduced to a handful of cottages, a large farm and the church: until relatively recently extensive earthworks, now destroyed, indicated its former, never very great, extent. Whether population decline came before or after enclosure, or both, often remains unclear, although the 1545 tax returns suggest that many of the smaller villages in the district, including Letchworth and Willian, had become relatively impoverished. Where parishes came to be dominated by large landowners they were not only vulnerable to early enclosure but also tended to decline demographically. They became 'closed' parishes in post-medieval times, in which the owner attempted to limit the poor rates by reducing the pool of potential claimants, demolishing cottages and importing any additional labour required during the busy times of the farming year from neighbouring 'open' settlements. Under the terms of his lease of 1771, for example, the farmer of Newhouse Farm in Newnham was obliged to 'make as few parishioners or Inhabitants in the said parish of Newnham by hiring Servants for a year as he or they can avoid' and 'shall not out of Malice or Ill will make any one a parishioner of the said parish who is or shall be a married Man with Children'.[31] At Radwell in the early nineteenth century there was a major legal case concerning the rights of the lord of the manor to grant permission for poor people to erect cottages on portions of the roadside waste. The principal landowner, Samuel Mills, was (unusually) not the lord of the manor and opposed the construction of one cottage, and the manorial lord's rights in this regard, on the grounds that 'cottages to an unlimited extent may be erected on the vacant or waste ground adjacent to his estate', the poor rates would rise and, in consequence, 'his tenants will not be able to pay so great Rent as they otherwise would if the poor rates were less'.[32] Enclosure might itself, however, also lead directly to a reduction in the number of farms and cottages, if it was associated with a shift to pasture. Enclosure chronology, tenurial structure, parish or township size and the demographic behaviour of particular places were thus linked in a number of complex ways.

We do not know precisely when most of these places were enclosed. Pendley, originally a separate township in Tring, was depopulated in the fifteenth century in order to create a park. The Enclosure Commission were informed in 1506 that some eighty years earlier the place had been:

> A great town, whereof part lay in the parish of Tring and part in the parish of Aldbury. The part in the parish of Tring was held of the archbishop of Canterbury as of his manor of Tring and the part in the parish of Aldbury of the manor of Aldbury. At that time there was no great mansion house there, but there were in the town above thirteen plows besides divers handicraft men, as tailors, shoemakers and cardmakers with divers others. The town was afterwards cast down and laid to pasture by Sir Robert Whittingham, who built the said place at the west end there as the town sometimes stood, for the town was in the east and south part of the same place.[33]

There is little doubt that the deponents exaggerated: tax returns suggest that the place had long been a relatively small settlement. Tiscot, Aldwick, Long Marston and Wilstone, other small townships within the large Tring parish, were also probably enclosed in the fifteenth or early sixteenth century and laid to grass as part of an expansion of pasture farming on heavy clays at a time when – with population levels low in the wake of the Black Death – little money was to be made from grain production, or from the rents of grain farmers. Most places, however, appear to have been enclosed gradually, and piecemeal, probably as ownership became increasingly consolidated in fewer and fewer hands. Enclosure was thus extended over decades or centuries, and is poorly documented, but we ocasionally have passing references in the documents. In the case of Radwell, a terrier of the glebelands dating to 1664 refers to their enclosure by exchange thirty years earlier.[34]

Parliamentary enclosures

Interesting though these early enclosures are, most 'champion' land in Hertfordshire was enclosed by parliamentary act and – with the exceptions of Hexton (enclosed in 1766), Walsworth (1766), Lilley (with part of Offley) (1768) and Ickleford (1776) – during or after the Napoleonic Wars. That is, north Hertfordshire's parliamentary enclosures are mainly concentrated within the second 'wave' identified by Michael Turner, as we might expect given the light, freely draining nature of most of the soils located on and below the escarpment. These parishes were on average larger than those affected by pre-parliamentary enclosure, and those enclosed in the second wave of parliamentary enclosure were larger on average than those enclosed in the first.

The amount of land enclosed by act in each of the parishes affected varied. In general, those which included large areas of boulder clay above the escarpment had less of their area affected than those lying below it, for the simple reason that the former had more mixed landscapes, with much of their land enclosed from an early date, and with open arable largely restricted to the lighter soils. Around 60 per cent of Reed and Barkway, for example, were enclosed by act; 49 per cent of Sandon; and 27 per cent of Weston. All of these parishes lay mostly above the escarpment. In contrast, most villages at the base of the escarpment had more than three-quarters of their area enclosed, although in Ickleford (with around 60 per cent) and Ashwell (*c*.65 per cent) rather more prior enclosure had taken place.

Ashwell, enclosed in 1863, is a large parish, of over 4,000 acres, and a large settlement which had developed (as we shall see: below, pp. 241–3) as a small urban centre rather than as a village. The enclosure map shows the new pattern of boundaries but also the numerous 'old enclosures', areas of land taken in piecemeal from the open fields during the previous centuries.[35] What is striking is the very large number of individuals who

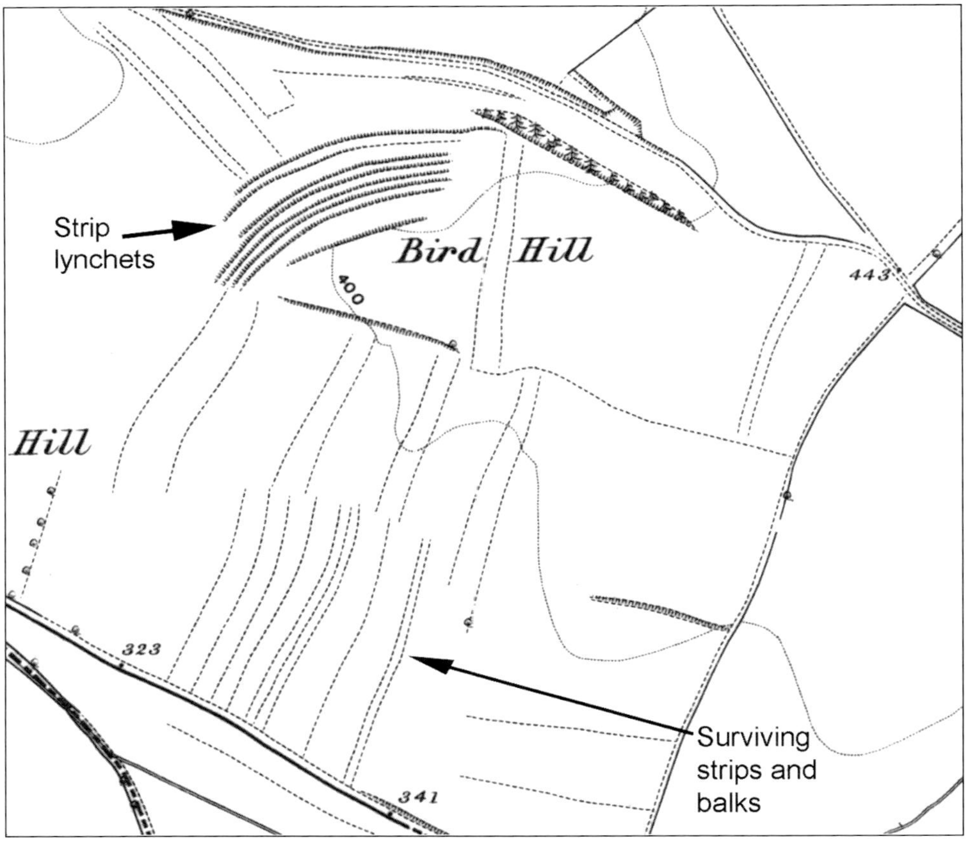

Figure 2.6. Part of the open fields of Clothall, which still remained largely unenclosed when the 6-inch Ordnance Survey map was made in 1883. The map shows a series of strip lynchets on the northern slope of Bird Hill and surviving balks between strips. See also Figure 2.8.

received allotments of land: no fewer than 114. In addition to the larger landowners, such as John Chapman (who was allotted 250 acres), Margaret Snow (489 acres) and Edward Rogers (916 acres), there were numerous individuals receiving very small parcels in lieu of their common rights. Nearly a third of allottees received an acre or less. These tiny plots took the form of narrow rectangles laid off the public roads, ensuring ease of access. They were mainly located to the south-east of the village (on Ashwell Street Way and the road to Royston) or to the south-west (rather fewer, on Claybush Road and Newnham Way). It is noteworthy that the first edition Ordnance Survey 6-inch map, published only 20 years later, shows few of these small plots. The allottees evidently sold out quickly to the larger landowners: indeed, it is doubtful whether the complex mesh of properties shown on the enclosure map was ever fully laid out on the ground. Although some of the earlier piecemeal enclosures from the open fields remained, the majority did not. They were thrown into the melting pot, as it were, and reallotted together with the unenclosed land, thus ceasing to exist.

The late survival of open fields

Ashwell was a very late enclosure, in national terms, but in a local context not unusual. Both Sandon and Therfield were enclosed by parliamentary acts after 1840; but more striking were Hitchin, Bygrave, Wallington and Clothall, where open fields remained into the twentieth century, and were widely noted by historians and antiquarians as intriguing survivals from the medieval past. Slater discussed Clothall and Bygrave in his *The English peasantry and the enclosure of common fields* of 1907 (Figure 2.6), while Hitchin formed a major part of Frederick Seebohm's discussion of open-field farming in *The English village community* of 1883.[36] The *Victoria County History* also described the appearance of the fields at Bygrave and Clothall at some length.

> Between Baldock and Clothall Church lies Clothall Field, containing about 600 acres, a 'common-field' of open arable land famous for its barley, and divided into irregular strips by 'balks', or narrow banks of grass, sometimes grown with bushes. The strips are still divided among the three chief landowners: the rector, the Marquess of Salisbury and Miss Cotton Browne. On the hill-side the scarped terraces, or 'lynches', form a distinctive feature of the parish. The high ridges between these terraces have the appearance of artificial defences, but are in reality due to the custom of turning the sod down-hill in ploughing.[37]

The same volume recorded how in Bygrave:

> Beyond the pasture land are the great common fields which occupy not far short of 1,000 acres undivided by hedge or fence. They stretch in every

> direction upon the slopes which surround the village, and consist of open arable land broken into irregular strips of uneven size.[38]

Both the *Victoria County History* and Slater were understandably struck by the fact that these rare survivals were to be found next door to the place where that great symbol of modernity, Letchworth Garden City, was at that time being laid out.

It is not entirely clear why so many parishes in north Hertfordshire were finally enclosed at such a late date: in other counties open fields had normally been entirely removed by this time. In part it may have been due to particular circumstances of ownership. Although not all proprietors had to agree to a parliamentary enclosure, a majority of three-quarters by land value still had to be obtained, and in some places this was difficult. In Ashwell, apart from the fact that there were large numbers of small proprietors, two sizeable estates were held by Cambridge colleges, which were renowned for their antiquated land management practices and general conservatism. Conversely, in some places enclosure was held up because there were only a few proprietors – ownership was evenly balanced and one objector was enough to block the process. Slater in 1907 described how in 1885 an attempt was made by the Marquis and the vicar to enclose Clothall Field, but the other owner dissented. He suggested that similar reasons obtained in Bygrave, 'the Marquis of Salisbury being here again the lord of the manor, and the other Clothall proprietor the next largest landowner'.[39]

One important reason why enclosure came late may have been that there was often little incentive to enclose. This was arable country, so no fences or hedges were required to constrain livestock, and, by the start of the twentieth century, as Slater noted, 'the absence of hedges of course facilitates the use of certain kinds of machinery'.[40] And, as he also observed, while the open fields might have survived in physical form the institutions of communal farming had long since gone, through the steady growth in the size of – and concomitant fall in the number of – farms. In Clothall the entire area of the open field, 600 acres in all, was let to a single farmer, and the cottagers exercised no common rights there. The separate strips were still divided by balks or 'walls', but they were no longer farmed in a communal fashion. At Bygrave there were more farms, and rights of common grazing were still exercised, but the manorial rights of sheepwalk were no longer in operation because the farmers bought an annual exemption. Lastly, we should note that those parishes which still had surviving open fields in 1910 did not have their whole area occupied by them. Piecemeal enclosure had removed most of the open arable, leaving only a relatively small rump; or, at the very least, consolidation of holdings had removed many of the intermixed parcels, leaving larger blocks of notionally unenclosed land. In Clothall the open-field land constituted less than a fifth of the total area of the parish. The open fields in Hitchin covered over half the parish in 1816, but had been whittled away to around a fifth by the end of the century, and much consolidation of strips had taken place, although extensive areas of common grazing also remained. In the case of Bygrave, while over half the land lay technically open most comprised consolidated blocks owned by the main proprietors, while the unenclosed land at Wallington included virtually no intermixed parcels and was cultivated as two huge farms: 'the only surviving feature of the externals of the common field system is the wide, breezy stretch of open land, under wheat, roots and grass; and of the spirit of the "village community" there is nothing'.[41]

Open fields in the modern landscape

Although the last open fields effectively disappeared from Hertfordshire in the early twentieth century, their traces remain. In some places, as to the west of Hitchin, many of the field boundaries exhibit the sinuous and often 'reversed S' shapes created by piecemeal enclosure, and thus preserve, in simplified form, the essential layout of the medieval furlong pattern. More striking, perhaps, although fairly rare, are the earthworks of 'ridge and furrow' or 'broad rig': low ridges, normally around seven or eight metres in width and half a metre or so in height, each corresponding to a single strip or 'land' in the open fields. Each strip was ploughed individually, as an elongated spiral, and the mouldboard turned the soil to the right, towards the centre. Over the years a low ridge would build up, unless the land was 'ploughed out' at intervals – that is, ploughed from the ridge outwards, redistributing the soil back towards the edges. Ridging was thus at least partly intentional and the classic high-backed ridges of the Midlands certainly so. Ploughing in this way improved drainage and thus allowed the soil to warm up more quickly in the spring.

It is, unfortunately, very hard to map the distribution of ridge and furrow in Hertfordshire for two main reasons. One, to which we shall return in a subsequent chapter, is that not all low, parallel ridges in pasture fields represent 'true' ridge and furrow of the kind just described, associated with open-field agriculture. In enclosed land much quite recent ploughing created rather low and narrow (*c*.3.5 metre) ridges – too narrow to represent the cultivation of separate open-field strips, which were almost always seven metres or more across. Much of the 'medieval ridge and furrow' recorded in the county is, in fact, of this type. Secondly, there are many records of 'cropmarks of ridge and furrow', especially in the north-east of the county. These usually take the form of parallel dark lines in the soil, or in growing crops, which display the classic sinuous form of open-field strips. But in the majority of cases, while clearly related to open-field cultivation, such cropmarks and soilmarks cannot represent ridge and furrow *sensu stricto*, because most come from chalk soils where documentary evidence makes it clear that individual lands were not ridged, but instead ploughed flat and bordered by narrow balks. It is in fact the latter which, when finally levelled, show up from the air as the dark lines. If we consider *only* ridge and furrow preserved as upstanding earthworks, and remove from consideration examples which appear to represent various forms of ridging in enclosed fields, then it is clear that Hertfordshire examples are strongly clustered on the heavy Gault Clays to the north of Tring, with only scattered examples in the north-east of the county, generally on boulder clay. Of course, this pattern is also in part a function of later land use, for the plough ridges would only survive after enclosure where the land in question was laid to pasture and

Figure 2.7. Well-defined 'ridge and furrow' surviving on Norton Common in the centre of Letchworth Garden City.

remained as such into the modern period. The character of the soils, that is, encouraged *both* the ridging of arable strips *and* the conversion of land to pasture following enclosure. One of the most remarkable examples of ridge and furrow in the county lies on Norton Common, in the middle of Letchworth Garden City, on one of the thin deposits of boulder clay found at the base of the Chalk escarpment (Figure 2.7). At first sight a common – an area of grazing land used by the village community – seems an odd place to find evidence of arable cultivation, but this was not in fact the traditional common grazing of the parish, but an area which was (rather unusually) put aside for the use of the minor landowners when Norton was enclosed in 1796.[42] The ridges of what had formerly been part of the open fields were thus preserved under grass and, more recently, undergrowth. Elsewhere in the parish they were levelled by post-enclosure ploughing or – if they survived under grass – were later destroyed by urbanisation.

Although open-field strips were not usually ridged in areas of lighter land, other evidence of former open-field cultivation can be found in these locations. The most striking – although there are relatively few surviving examples – are the 'strip lynchets' formed where lands on steep slopes were ploughed parallel with the contours, creating a series of steps or terraces varying in height from a few centimetres, as at Bury Barn in Sandon, to six metres or more, as, for example, on Bird Hill in Clothall (Figure 2.8).

Figure 2.8. This great bank of scrub is growing on the sheer face of a massive 'strip lynchet' on Bird Hill in Clothall (shown as a strip of trees on the map in Figure 2.6).

Figure 2.9. A low ridge (arrowed) running through this arable field in Clothall marks a former headland at the junction of two open-field furlongs.

Unless – as here – they are particularly massive, lynchets are vulnerable to destruction by modern ploughs, no longer confined within the bounds of particular strips: those to the south-west of Arbury Banks in Ashwell, for example, are currently in the process of being levelled. Many of the examples shown as extant on the first edition Ordnance Survey 6-inch maps of the late nineteenth century now survive only as cropmarks, especially in Clothall. Cropmarks elsewhere indicate where examples were removed at an earlier date, following enclosure in the late eighteenth or nineteenth century. The best examples survive within secondary woods and plantations, or in the few patches of permanent pasture on the Chalk: as at Slipes Wood, and in and around Fordham's Wood, in Therfield; in Lodge Side Plantation in Graveley; or in Howletts Wood in Aldbury. Much less noticeable, although seemingly less prone to destruction than balks and lynchets, are the low, diffuse and sinuous banks running through the fields – most visible when the land is under young grain crops – marking the junctions between furlongs and caused by the way in which, over the centuries, soil was gradually piled up at the end of the strips by the movement of the plough, and by the way that the ploughman cleaned the share and coulter as he turned (Figure 2.9). But perhaps the most striking way in which Hertfordshire's open fields have left their mark on the modern landscape is the one which we notice least. The countryside to either side of the A505, approaching Baldock from the east, is remarkably open and prairie-like. It is easy to assume that this is the consequence of large-scale hedge removal in the second half of the twentieth century. But it is not: in Clothall Field and in the open fields in Bygrave ownership was gradually consolidated and intermixed parcels removed, but no hedges were ever planted.

Notes

1. Hertfordshire Archives and Local Studies (hereafter HALS) Acc 4283 map of the township and parish of Weston by Henry Liley, 1633; British Library (hereafter BL) Add. MS 33,581 Terrier of the manors of Weston, Argentein, Newbury and Lannock, undated but contemporary with the 1633 map at HALS.
2. BL Add. MS 42,508 Surveys and Perambulations, with Plans, by John Norden, of the manors of Mincinbury, Abbotsbury and Hoares, in Barley, co. Hertf., made for Sir John Spencer, Lord of the Manors, 1603.
3. J.C. Wilkerson (ed.), *John Norden's survey of Barley, Hertfordshire 1593–1603* (Cambridge, 1974).
4. HALS QS/E9 Ashwell enclosure map, 1863.
5. HALS DE/Ha/P1 Plan of the Manor and Parish of Ickleford and other Lands, 1771; QS/E80–QS/E81 Great Wymondley enclosure map and award, 1811; PC528 Pirton enclosure map, 1818.
6. B. Howlett (ed.), *Survey of the royal manor of Hitchin, 1676*, Hertfordshire Record Publications 16 (Hitchin, 2000), pp. xv–xvi.
7. HALS 57594–5 list of lands of John Ireland, 1705 and Terrier, 1705–53.
8. HALS DP/75/3/1 copy of a 1637 terrier of Norton church lands; HALS QS/E54 Norton enclosure map, 1796.
9. Page (ed.), *VCH Hertfordshire*, vol. 3 (London, 1912), p. 217; HALS 58882 'A Plan of the Estates of the Revd Thomas Whitehurst and others; in the Hamlet of Walsworth in the Parish of Hitchin, Inclosed by Act of Parliament, 1767'.
10. Page (ed.), *VCH Hertfordshire*, vol. 3, p. 212.
11. HALS DE/Ry/P3 plan of manor and lands of Newsells in Barkway and Barley, 1788; HALS DE/Ha/P1.
12. Page (ed.), *VCH Hertfordshire*, vol. 3, p. 212.
13. C. Lewis *et al.*, *Village, hamlet and field: changing medieval settlements in central England* (Manchester, 1997), p. 199.
14. HHER 13177 and 13178; 6826.
15. HHER 4299 St Margaret's church, Bygrave.
16. HHER 6979.
17. G. Beresford, 'Excavations at the deserted medieval village of Caldecote, Hertfordshire: an interim report', *Hertfordshire's Past*, 4 (1978), pp. 3–14.
18. HALS QS/E81 Great Wymondley enclosure map, 1811.
19. C. Lewis, 'Test pit excavations within currently occupied rural settlements: results of the HEFA CORS project in 2009', *Medieval Settlement Research*, 24 (2009), pp. 43–58.
20. HHER 9677.
21. Throughout the text we use the system of classification of soils employed and developed by the Soil Survey of England. This divides soils into *series* – a soil exhibiting particular characteristics of structure, drainage, etc – and *associations*, which are groups of series frequently occurring together – on level and sloping ground within the same geological formation, for example. See C.A.H. Hodge *et al.*, *Soils and their use in eastern England* (Harpenden, 1984).
22. Howlett, *Survey of Hitchin, 1676*, p. xvii.
23. F. Seebohm, *The English village community* (London, 1883), pp. 443–53.
24. BL Add. MS 33,581, ff. 67 and 65.
25. G.E. Mingay, *Parliamentary enclosure in England: an introduction to its causes, incidence and impact* (London, 1997), p. 60.
26. B. Hammond and J.L. Hammond, *The village labourer 1760–1832* (London, 1911), p. 104.
27. M. Turner, *English parliamentary enclosure* (Folkestone, 1980).
28. HALS IR1/90, 91, 105, 106, 1910 Finance Act maps, Hitchin; IR1/36–7, 1910 Finance Act maps, Bygrave; IR1/64 and 79, 1910 Finance Act maps, Clothall.
29. HALS 72036 agreement by parishioners of Pirton, 1692.
30. HALS DSA4/72/1–2 Newnham tithe award and map, 1847; DSA4/29/1–2 Caldecote tithe award and map, 1838–41; DSA4/66/1–2 Letchworth tithe award and map, 1837.
31. HALS DE/X58/E1 Lease and tenancy agreement of Newhouse Farm, 1771–93.
32. HALS DE/X58/E9 correspondence to Towse (solicitor), case papers and related papers in a plea of trespass and ejectment, Mills v. William, 1810–11.
33. TNA:PRO SC2/176/120 Aldbury court roll, 1506.
34. HALS DP/82/1/1 terrier of glebelands of 1664 inside back cover of register of baptisms, marriages and burials, 1590–1718.

35. HALS QS/E9.
36. G. Slater, *The English peasantry and the enclosure of common fields* (London, 1907); Seebohm, *The English village community*.
37. Page (ed.), *VCH Hertfordshire*, vol. 3, p. 220.
38. *Ibid*., p. 212.
39. Slater, *English peasantry*, p. 43.
40. *Ibid*., p. 44.
41. *Ibid*., p. 45.
42. HALS QS/E54.

– 3 –

The landscape of east Hertfordshire

Introduction

The boulder clay landscape of eastern Hertfordshire occupies an area of roughly 440 square kilometres and constitutes one of several versions of the 'ancient' countryside which characterise the majority of the county. Unlike the districts considered in the previous chapter, this extensive area boasts a highly dispersed settlement pattern and, while nucleated villages can be found, there are in addition numerous isolated farms and hamlets, many of which are clustered around areas of common land or small 'greens' (Figures 3.1 and 3.2). There are many semi-natural woods, formerly managed as coppice-with-standards; the fields are mainly bounded by ancient, species-rich hedges; and miles of bridleways and footpaths form a dense, intricate web. Narrow lanes, sometimes sunk below the surface of the adjacent fields, wind through the countryside in a seemingly aimless manner, connecting the various settlements and commons. This is a complex landscape, difficult yet rewarding to read. Although the ancient fabric of the countryside was damaged to some extent by agricultural intensification in the second half of the twentieth century, this district is still remarkably attractive, and the most rural part of the county. It is often hard to believe that its southern boundary lies a mere 20 miles (35 kilometres) from central London (Figure 3.3).

We noted in the opening chapter how landscapes are no respecters of county boundaries, and that of east Hertfordshire is no exception, flowing seamlessly into western Essex. Indeed, in 1941 the Land Utilisation Survey described the district as forming 'the western fringe of the great East Anglian arable area' and today, if anything, local land use is even more intensively arable in character.[1] But most visitors will not be reminded of the landscapes of East Anglia, largely because the countryside here is for the most part undulating, for the boulder clay plateau which extends in a great arc from Norfolk and through Suffolk and north Essex, becomes increasingly dissected as it progresses south-westwards. In Hertfordshire it is so interrupted by valleys – those of the rivers Rib, Ash, Stort, Quin and their tributaries – that it scarcely reads as a 'plateau' at all. But it is this

Figure 3.1. A typical east Hertfordshire greenside settlement: Bury Green, Little Hadham.

Figure 3.2. Levens Green, Great Munden, from the air, looking north. The green is the small area of pasture in the shape of three conjoined triangles in the centre of the picture. It is ringed on all sides except the north by farms and cottages. Courtesy Eng ish Heritage.

Figure 3.3. Typical east Hertfordshire countryside between Throcking and Sandon, with wide arable fields, scattered woodland and ancient hedges.

essential contrast, between valleys and the narrow plateaux between them, which is the key to unravelling the history of this complex landscape.

The principal valleys have cut through the boulder clay to reveal the older deposits beneath: mainly Chalk, but in a few places (as to the south of Standon) the mottled clays and sands of the Lambeth Group and the London Clay. Although often covered by thin layers of glacial till, the Chalk gives rise to light, freely draining and fertile soils of the Melford and Elmton 1 Associations. The valleys of tributary streams and the higher reaches of the river valleys themselves are not in general cut deeply enough to reveal the Chalk but their soils are nevertheless often fairly light and freely draining, the sloping valley sides ensuring that water, the great enemy of the clayland farmer, can be removed with relative ease, conducted along furrows and via ditches into the natural watercourses. The soils here, moderately well-drained and calcareous, are mainly those categorised as the Stretham Series by the Soil Survey.[2] The more level areas on the higher ground are different. Here it is harder to remove surface water and the layers of till are thicker, giving rise to the poorly draining soils of the Hanslope and Ragdale series, which were particularly challenging to early farmers.

Landscapes with scattered settlements, abundant woodland and numerous winding lanes have traditionally been interpreted as the consequence of the 'assarting', or clearance, of extensive tracts of woodland in the course of the twelfth and thirteenth centuries. No less an authority than W.G. Hoskins chose the area around Bishop's Stortford as a typical example of such a late-settled landscape in his classic *The Making of the English Landscape*, describing how the:

> Tangle of lanes and paths winding from one isolated farmstead to another, the score or more of farms named after medieval owners or occupiers … and the still-frequent remnants of woodland, are all deeply characteristic of country enclosed direct from former forest.[3]

Writing with L. Dudley Stamp, he described the small green-edge settlements so characteristic of this kind of countryside in similar terms, as having 'pretty certainly originated as clearings in densely wooded country, either as natural glades in an old forest or as deliberately cleared ground'.[4] Some of the isolated farms found in the area are surrounded by moats – water-filled ditches defining rectangular, sub-rectangular or ovoid islands – and many further examples can be found in fields or woods which no longer surround a dwelling. These features, too, have often been interpreted as evidence of 'late' clearance and colonisation. There is a measure of truth in these suggestions. But to interpret all the distinctive features of this landscape as the simple consequence of the twelfth- and thirteenth-century clearance of 'dense forest' is to simplify a more complex history, and a much longer one.

Early settlement

Chance finds and fieldwork have, over the years, revealed that later prehistoric and Roman settlement was dense in this fertile region. In part this is because of the presence, close to its centre, of one of the most important archaeological sites in Hertfordshire, at Braughing: a major late Iron Age *oppidum* which subsequently developed into a substantial Roman 'small town'. *Oppida* is the term used, as we noted in Chapter 1, by archaeologists to describe a certain kind of late Iron Age site which comprised extensive but usually discontinuous areas of occupation. Earlier generations saw these places as 'proto-towns' but, while they display some apparently 'urban' characteristics (especially metalworking and the use and production of coinage), *oppida* are best thought of as concentrations of tribal settlement around the residences of major leaders – petty kings. The most striking example in Hertfordshire is, as we saw earlier, at St Albans,[5] but that at Braughing was earlier, and initially richer. It comprised a large and complex area of settlement around Braughing and Puckeridge, in the Rib valley. Exotic Mediterranean goods, including amphorae containing wine and fine pottery from Gaul and Italy, were being imported here in significant quantities by *c.*25 BC, and by the time of the Roman Conquest settlement covered an area of *c.*200 hectares, although by this time the scale of foreign imports appears to have diminished.[6] The site was characterised – typically – by dense concentrations of settlement, like that around Skeleton Green, interspersed with more open areas, and it featured a discrete zone of ritual activity recovered at Station Road.[7] Large quantities of coin moulds have been recovered from the eroding bank of the river Rib.[8] Following the Conquest the status of the settlement declined markedly in comparison with Verulamium, and it received only a few public buildings and no walls. But it remained a vibrant 'small town', benefiting from its position as a route centre where Ermine Street, a new Roman road heading north from London, met Stane Street, leading to Camulodunum (modern Colchester), while other roads led north-west to the Roman settlement at Baldock and north-east to Great Chesterford. A grid of streets was laid out on Wickham Hill and occupation extended from the river Rib westwards as far as Ermine Street, and although the buildings were mainly of wood with thatched roofs there was a masonry bath house beside the Rib and a substantial L-shaped stone building on Wickham Hill, which may have been a temple. Aerial photographs also suggest the presence of others.[9] Braughing presumably had both economic functions and social and ritual ones, as the focus for a section of the Catuvellauni tribe.

The large number of Iron Age and Roman sites known in the locality, including a very extensive area of kilns in Much Hadham which produced a distinctive and widely traded form of pottery, in part reflects the social and economic prominence of Braughing. But the presence of this settlement itself indicates that the fertile and easily cultivated land in the principal valleys was capable of supporting a significant population, and the area had probably been densely settled in previous centuries. In this intensively arable countryside earthwork traces of early settlements and field systems seldom survive, except under woodland, but there is at least one possible Iron Age hillfort in the area, at Caley Wood in Little Hadham, adjoining the great area of pottery production in Much Hadham. The present wood falls into two sections. The southern comprises ancient, semi-natural woodland, with a narrow arm projecting to the north-east. On the northern edge of this area, and continuing along the projecting arm, is a huge rampart. This has been mutilated in places by gravel extraction for the wartime construction of Stansted airfield, over which the northern section of the wood, to the north of the rampart, grew up in the twentieth century. This area, which represents the interior of the Iron Age enclosure, is shown as an arable field on nineteenth-century maps.[10] Beyond, to the north, the northern edge of the 'fort' can still be seen as a marked bank running along the edge of the road from Pig's Green to Westfield Farm. The western side of the earthwork has been destroyed, probably centuries ago. The area is described as 'the Castell alias Cales' in a Chancery deposition of 1575, and as 'Castle Field and Bank' on a plan of 1767, but the site is very unlikely to have been a medieval castle and is almost certainly a hillfort of middle or later Iron Age date.[11] Typically, it enjoys magnificent views to the south-east, down the valley of the river Ash. Another possible hillfort site buried within ancient woodland, overlooking and predating the Braughing *oppidum*, has been suggested. Although the name of the wood, Gatesbury – 'the defended enclosure of Gaeta' – would seem to support the suggestion that the earthworks within it represent an ancient fortification, this is in fact derived from the nearby settlement of Gatesbury, and 'bury' is here used in its common Hertfordshire sense of 'manor'. The character of the earthworks themselves is uncertain. The twin banks and ditches which enclose the wood may be wood-banks of medieval date, although their complex arrangement at the eastern end of the wood is reminiscent of the entrances of some hillforts. The site remains a puzzle.

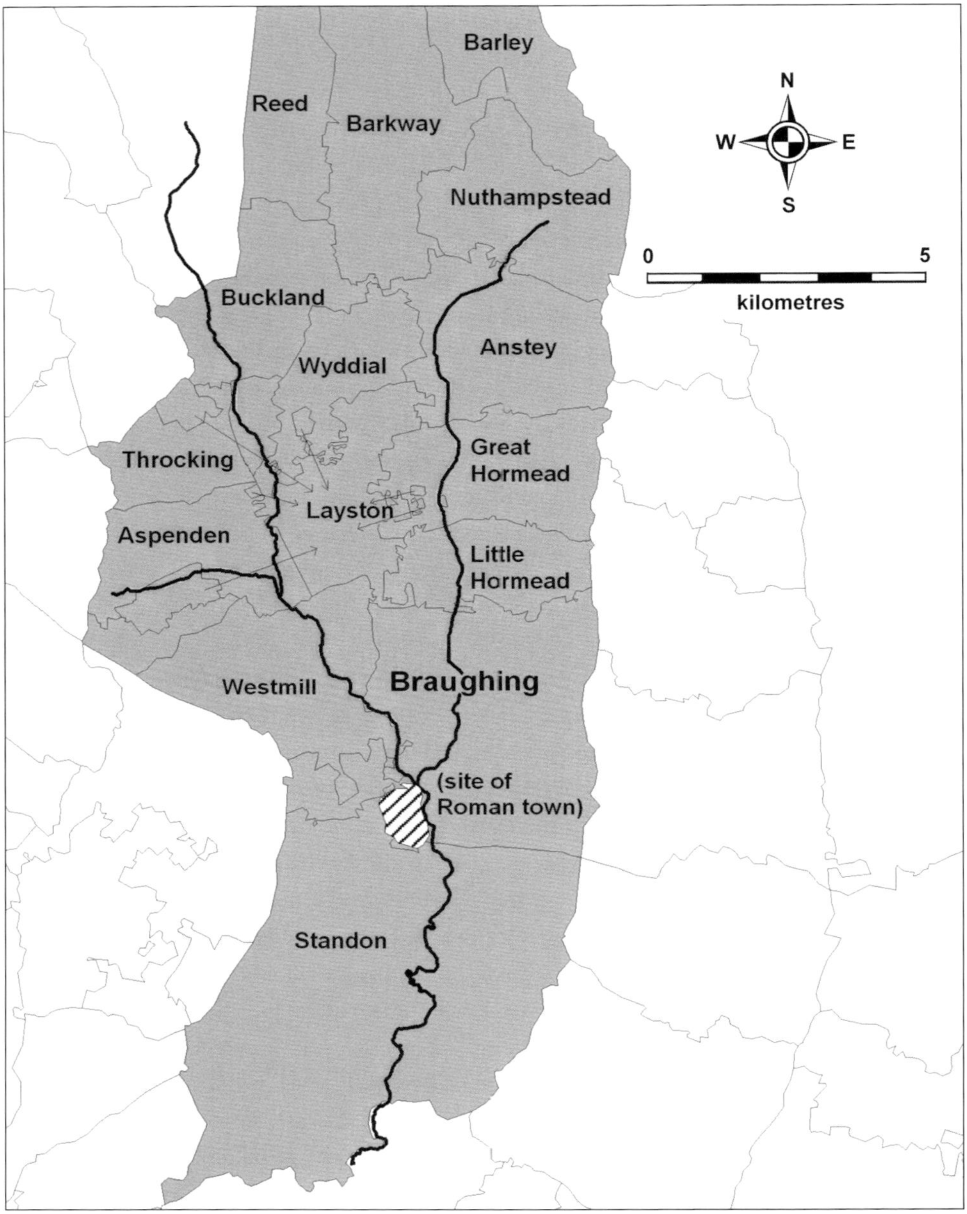

Figure 3.4. The ancient territory focused on Braughing, as reconstructed by David Short.

The area was thus densely settled in late prehistoric and Roman times, and this in itself undermines any assertion that its principal features developed as a consequence of late 'assarting' of 'dense forest'. Moreover, by the time of Domesday this was again among the most densely settled areas of the county, with a population little different from that of the 'champion' lands to the north, most vills having between three and eight recorded individuals per square mile. The population certainly declined in the area in the immediate post-Roman period, recovering again during middle and later Saxon times – as it did elsewhere in the county. But there is no evidence that the district was completely abandoned in the fifth and sixth centuries, so that Saxon colonists had to begin the work of clearance and cultivation from scratch. Settlement probably retreated from the heavier soils, but it appears to have continued within the major valleys.

Braughing itself seems to have maintained the status of some kind of 'central place'. Its name, 'the people of Brahha', probably derives from a tribal group which occupied the area in the early Anglo-Saxon period, and institutional continuity of some kind is suggested by the name of Wickham Hill, which coincides precisely with the main built-up area of the Roman town, for this is an example of a relatively rare kind of place-name which combines Latin and Old English elements, in this case *vicus* – the Roman term for a small settlement – and *ham* – the English for an estate (Wickham Spring, close to Caley Wood and the Much Hadham pottery kilns, is another example). The *Brahingas* were probably absorbed at an early date into the kingdom of the East Saxons, and their territory may have comprised the majority of the area of the modern county which was incorporated within that kingdom (the larger, western, portion becoming, as we have seen, part of Mercia).[12] The boundary between the medieval diocese of London (originally the bishopric for the East Saxon kingdom) and the diocese of Leicester, later Lichfield and then Dorchester, and ultimately Lincoln (serving Mercian territory) runs along the watershed to the west of Braughing, between the catchments of the rivers Rib and Beane. As late as AD 825, when ten hides of land at Braughing were leased by Bishop Ceolberht of London and the community of St Paul's to Sigric, 'minister' of Wiglaf, king of Mercia, one of the witnesses was '*Sigric rex Orientalum Saxonum*' – probably Sigered, last king of the East Saxons.[13] Braughing maintained its importance into late Saxon times, for it gave its name to, and presumably provided the meeting place for, the hundred of Braughing. Indeed, finds of coins covering 'the entire medieval period (apart from the period after the Black Death)', suggest that Wickham Hill itself continued as a significant meeting place and was perhaps the location for the hundred courts.[14] David Short has shown how the configuration of parish boundaries suggests that the Rib and Quin valleys, from Standon northwards as far as Barkway, originally formed a single ecclesiastical territory, presumably that of a minster based on Braughing (Figure 3.4) (above, p. 17). The will of Æthelgifu, drawn up in *c.*980, made bequests to a minster here, and the present church contains at least three fragments of Anglo-Saxon stonework, reused in later walls, suggesting the existence of a major church.[15] Like many minsters, Braughing eventually became the head of a rural deanery.[16]

Braughing was not the only important 'central place' in the district in Saxon times. Place-names, the configuration of parish boundaries and the evidence of early charters suggest that a tribal area, later a territory focused on a *villa regis* within the East Saxon kingdom and a minster *parochia* – was based on Much Hadham and extended north along the valley of the river Ash, embracing Albury, the Pelhams and Meesden. Another, based on Bishop's Stortford, seems to have included Thorley and Sawbridgeworth; while a fourth – absorbed, unlike these, into Mercia rather than Essex – may have been centred on Benington, 'the settlement of the people of the river Beane' (Bengeo, just to the north of Hertford, was the 'spur of land of the people of the river Beane'). With the exception of the last, all of these early territories had a focus located at the point where a Roman road crossed a river: most of the Roman roads in the district seem to have remained in existence.

There were presumably other, less important settlements scattered along the principal river valleys in the early and middle Saxon period but their locations remain uncertain: with the exception of recent discoveries of early–middle Saxon pottery at Station Road, Watton,[17] no certain finds of early or middle Saxon material have been made and even for the late Saxon period, when pottery becomes more abundant and more durable, only a single archaeological site, at Manyons Farm in Barkway, is known, in spite of the fact that Domesday Book provides clear evidence for the existence of numerous settlements in the area by the eleventh century.[18] The churches at Reed, Westmill and Walkern (the latter probably in origin a minster serving the territory based on Benington) contain pre-Conquest fabric, as may that at Little Munden. Again, this scant archaeological evidence stands in sharp contrast to that of our documents, for Domesday lists numerous priests, indicating the presence of churches, in the district. Anglo-Saxon churches were usually built of wood, and were thus completely eradicated when rebuilt in stone at some point after the Norman Conquest. Even for the twelfth century the documentary evidence for the development of landscape and settlement is often sparse. Fortunately, these rather meagre sources can be augmented by treating the landscape itself as a major piece of evidence, for much can be learnt by examining recurrent patterns in the location and morphology of fields, woods and settlements, examining their relationship with the soils and topography and analysing the meaning of place-names. Such approaches, although they rely to a large extent on comparatively recent sources – especially early maps – can do much to tease out the limited character of more conventional forms of evidence.

Unpicking the Saxon landscape

By later Saxon times there were unquestionably large numbers of settlements scattered along the principal valleys of the area. Domesday mentions all the main villages, together with three which have ceased to exist (all of which became separate parishes with their own churches): Broadfield, Wakeley and Berkesden. In addition, it lists a number of small places which never became parishes in their own right. Some, such as Luffenhall, are still hamlets to this day, but most are former manor houses – halls or 'burys' as they are often known, in this part of Hertfordshire especially.[19] A few are now simply described as 'farm', such as Chaldean Farm in Much Hadham. Later evidence, in the form of manorial descents, allows us to say with reasonable confidence that a number of other isolated manorial sites were also in existence by this date. Patient End in Furneux Pelham means *Payne's tun*, and Payne was the name of the holder of one hide of land 'in *Peleham*' listed in Domesday. Upwick in Albury (the 'higher dairy farm'), Abbotsbury and Mincinbury in Barley, Gannocks, Chamberlains and Queenbury in Reed, the manor of Lannock in Weston and Nuthampstead Bury can similarly be identified with some confidence with particular Domesday holdings by tracing back the course of manorial descents. Mincinbury in Barley, for example, is the manor of the *myncen*, or nuns: the Benedictine nuns of Chatteris already held one of the manors here in 1066.

Other probable pre-Conquest settlements may be indicated by the manorial halls and, less frequently, hamlets which bear names of Old English form. Extreme care needs to be taken here. An Old English name is not necessarily a pre-Conquest one, for that language was only gradually replaced by Middle English in the course of the late eleventh and early twelfth centuries. Moreover, many of the minor Old English place-names in the locality contain elements of 'topographic' type: that is, they refer to some feature of the natural topography, such as a hill, ford or valley – and these may have been coined centuries before the settlements themselves were established close to the features in question and named after them. Nevertheless, the majority of places with names containing Old English 'habitative' elements – terms for a farm or settlement – must refer to settlements established by or very soon after the Conquest. Domesday's Alswick (in Layston), for example, can be paralleled by Great and Little Barwick in Standon and Thundridge respectively, by Haultwick in Little Munden and by Upwick in Albury, all having as a second element *wic*, 'dairy farm'. But some place-names of 'topographic' type also mark early settlements. Mentley (formerly Milkley) Farm, 'the milk clearing', is referred to as a settlement in a document of the mid-twelfth century[20] and it is possible that some other manorial sites which first appear in our sparse documentary record within a century or so of the Norman Conquest are, in fact, of pre-Conquest origin. Examples include Quinbury in Braughing, the 'queen's manor', which is identifiable with the manor granted by Queen Maud to Hubert de Anstey, her chamberlain, some time between 1135 and 1153; Whitebarns in Furneux Pelham, almost certainly the manor granted to St Paul's Cathedral before 1152; Monks' in Wallington, granted to St Albans some time before 1175; Standon Friars, which is probably the site of the Preceptory of the Knights of St John of Jerusalem, established following a grant made in the early twelfth century; and Puttocks End, which probably took its name from the family of John Puttock, referred to in a pipe roll of 1175. Some of these places might have been in existence for a century, perhaps several centuries, before they first appear in our documentary sources.

All these probable, or possible, Saxon settlements – Domesday manors and diminutive vills, places bearing names with Old English habitative terms and the rest – occupy the same general topographic positions as the parish churches which, we may reasonably assume, mark the site of the principal centres of late and perhaps middle Saxon occupation: that is, they lie either on the floors of major valleys, generally in their lower reaches; or, more usually, immediately above them, on the edge of the boulder clay plateau (Figure 3.5). Such a location provided access to the well-drained soils of the valley sides as well as a reliable supply of water. The lower reaches of the main river valleys were incised deeply enough to reach the water table in the Chalk, and water could here be obtained from springs or from saturated gravels in the valley floor. Some villages, such as Aspenden or Standon, are partly strung out along the edges of 'bournes' or 'bones' – the seasonal watercourses running down from the clay plateau, which are flanked by spreads of gravel that likewise serve as aquifers – at the point where these meet river flood plains. A more reliable supply of water, however, was available from lenses of sand and gravel within the glacial till above the valleys, and most settlements – even the main villages, other than Braughing, Much Hadham, Westmill, Aspenden and Standon – were consequently located here, on the plateau edge.

It would thus appear that by later Saxon times, while the boulder clay plateau itself remained largely uncultivated, within the valleys themselves settlements were densely

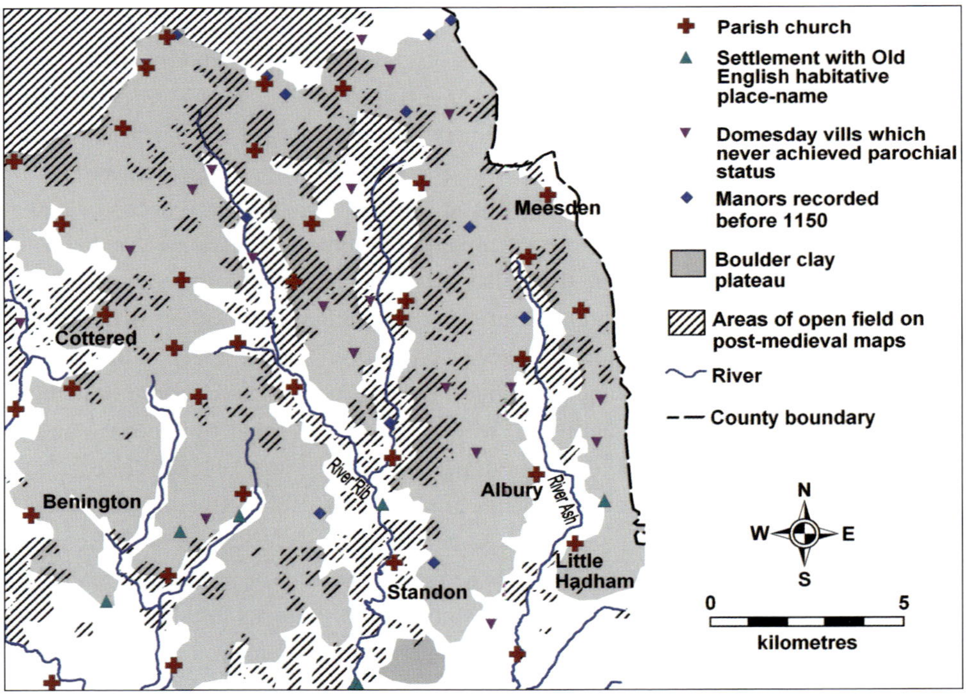

Figure 3.5. Probable late-Saxon settlements, and areas of open field shown on post-medieval maps in east Hertfordshire. Note how both tend to occur in or around the principal valleys, where soils are lighter and more freely draining. By the time the earliest maps were made much open-field arable had been enclosed. In medieval times it would probably have been almost continuous on the lighter soils.

packed, often lying a kilometre or less apart. The implication is that most, although probably not all, of the land within the main valleys was under cultivation. We have no idea how old this busy landscape may have been because, as we have seen, early and middle Saxon sites in this region leave few surface traces, and the subsequent stability of the settlement pattern has ensured that the archaeological evidence for the origins of modern farms and hamlets is generally hidden beneath buildings, gardens and paddocks: local researchers should make every effort to monitor any form of ground disturbance in the vicinity of such places and record any pottery or other material uncovered. It is, however, noteworthy that one of the diminutive Domesday vills, now incorporated within the parish of Bishop's Stortford, was called Wickham, a third example of a *wic–ham* hybrid name in the district. This said, the Roman remains found in the area – much more archaeologically 'visible' than those of the Saxon period – are not for the most part very close to probable late Saxon settlements and, even when they are, they usually lie 300 metres or more away, as in the case of the finds made near the churches at Watton or Little Hadham, or at Luffenhall or Standon Friars. This may suggest that extensive tracts of land, especially on the higher ground, were abandoned in the post-Roman period, settlement retrenching to a few valley-floor sites. More probably, the settlement pattern had continued to develop and change over the five centuries or so between these points of archaeological visibility, in the Roman period, and documentary visibility, in the later Saxon.

Areas of 'ancient countryside' like this, characterised by a fairly scattered settlement pattern, are often said to have had little open-field land in the Middle Ages, farmers mainly cultivating their land in hedged closes. This, however, is a considerable oversimplification and north-east Hertfordshire was by no means unique in boasting extensive tracts of open arable, much of which survived well into post-medieval times. It is possible to map the distribution of open fields from the early maps of the area – mainly dating from the eighteenth and early nineteenth centuries, but with some seventeenth-century examples.[21] Not surprisingly, most open arable was found on the light, freely draining soils of the valley sides, with only small patches occurring on the clay plateau, where the landscape was indeed dominated by enclosed fields, woods and commons. The maps also suggest that areas of open-field land had originally been continuous in the main valleys, but had generally contracted through early piecemeal enclosure, for surviving patches are often shown to be interrupted by fields whose sinuous outlines replicated those of the surrounding open-field strips.

As noted in the previous chapter, these open fields were not like those found in the areas to the north of the escarpment. They were 'irregular' in character, in the sense that there were usually large numbers of named fields in each parish: Aston, for example, had at least 12 in 1540, Benington had 10 in 1628 and Weston had 16 in 1633.[22] In addition, the holdings of individual farmers were often clustered, rather than being widely distributed, possibly reflecting the fact that many of the small Saxon settlements found in and around the valleys had originally farmed their own small systems of fields. The names given to the various portions of the arable certainly suggest that many open fields had originated by the eleventh century, for many contain Old English terms. In Benington, for example, Dane Field is from OE *denu*, 'steep-sided valley'; Lether Field is from *Lyðre*, 'worthless'; while Pop Hill is probably a corruption of *Popel*, 'pebbly land'. In Weston we have Cheesell Field (OE *ceosol*, 'gravel'); and Lynce Field (OE *hlinc*, strip lynchet). Nevertheless, there is no doubt that open fields continued to be created well into the post-Conquest period, mainly, perhaps, as newly reclaimed land was shared between colonists and commoners. Small areas of open-field land, as noted, can be found even in the central parts of the various clay plateaux, while in the upper reaches of the Ash, where the river valley fades gently into the clay plateau in countryside that was still well-wooded at the time of Domesday, one of the fields in Stocking Pelham ('the Pelham with the tree-stumps') was called 'Assett Field', almost certainly from the Old French word *assart*, meaning woodland clearance.

By the time of the Conquest most of the light valley soils were probably under cultivation but when settlement first stabilised – probably in the seventh or eighth centuries – farms and clusters of farms would have been separated almost everywhere by wide ribbons of grazing, on light soils as much as heavy. As cultivation gradually expanded, settlements located in the main valleys, where the most extensive areas of light, well-drained soils were to be found, had more opportunity to grow into sizeable villages. In contrast, those located at the upper end of tributaries running back into the plateau, such as the Domesday vills of Patmore (Albury), or Hixham (Furneux Pelham), had more limited areas of easily ploughable land available to them and thus remained small places, never attaining parochial status. This said, many settlements located beside extensive tracts of valley soil failed to develop into anything other than hamlets, possibly because they originated as grazing farms attached to larger settlements in the valley below: the element 'wic', in the sense of 'outlying farm', which features in the names of many may suggest this (although this term also occurs as a field name, often in the form 'Aldwick', 'the old farm', perhaps indicating where earlier farms, part of the shifting pattern of Saxon times, had once stood). The distinction between what are now nucleated villages and other, smaller settlements is for the most part already apparent in Domesday, although there are some places which have changed their position in the pecking order. The village and parish of Aspenden had, at the time of Domesday, a smaller population than nearby Alswick, which is now a single 'hall'.

As in the areas to the north of the escarpment, many of the villages located in and above the valleys appear to have originated as loose clusters of farms around an area of common grazing. Some of these central greens have remained largely intact, as at Cottered or Westmill: in both cases, the parish church appears to have been placed on an 'island' cut out of it. More usually, they were largely or entirely infilled with houses in the course of the medieval period, a pattern of development particularly clear at Rushden, where the houses to the south of the residual green beside the church display what is clearly a 'double building line', with some set back from the road along the old common edge and others fronting directly on the road, occupying what was once common land. It is probable that the infilling of greens usually occurred in the Middle Ages but in some cases it may have taken place later. At Weston, for example, the small triangular 'island' of land lying south of Maiden Street (and the village shop) clearly represents an encroachment taken from a once continuous area of common which extended southwards towards Damask Green and north-eastwards to the present-day green on the north side of Maiden Street; but a more extensive triangle, surrounded by roads and lying to the north-east of the green, was also apparently cut from a once even larger area of open pasture. A map of 1633 shows that this was already enclosed ground, although not as yet so densely built over as today, but a survey of 1548 describes properties lying *in* Oldfield Green, perhaps suggesting recent encroachment.[23]

Colonising the uplands

Anglo-Saxon settlement was thus concentrated within and on the edges of the valleys, in the largest of which extensive open fields had probably developed by the time of the Norman Conquest. The surrounding uplands, in contrast, appear to have been occupied by almost continuous tracts of uncultivated woodland and grazing, the largest of which survived well into the twelfth or even thirteenth century, and can be picked out on the earliest surviving maps as concentrations of contiguous or near-contiguous woods, commons and former deer parks.[24] Of particular note is the area of high ground between Cottered and Benington, which includes the former medieval parks of Walkern and Benington, numerous woods and complex patterns of interconnected greens and common, especially to the south-east of Ardeley. Similar is the area extending from Albury to Latchford, which features the former parks at Albury, Little Hadham and Standon, much ancient woodland and a particularly complex collection of commons around Westland Green and Pig's Green in the south-west of Little Hadham. Other examples include the complex of greens, commons and woods around Sandon and the cluster of woods and commons around Meesden, including Scales Park, which extends across the county boundary into Essex.

It is noticeable that in a number of places in the south of the district these late-surviving uncultivated tracts extend down from the plateau into the major valleys (Figure 3.6). This superficially breaks the main 'rule' of the landscape just outlined, for we have posited so far a rather simple dichotomy between valley – with light soils, and early settled – and plateau – with heavy soils, and cleared and cultivated at a relatively late date. But this anomaly in fact serves to confirm the importance of environmental factors in the development of the medieval landscape, for these places correspond precisely to areas where the underlying solid geology exposed in the valleys comprises, not Chalk, but deposits of the Lambeth Group (generally infertile sands and clays) or heavy London Clay. Indeed, it is noteworthy that several of the principal concentrations of woods and pastures just noted correspond to extensive areas of upland where the boulder clay itself overlies these formations, rather than the Chalk, presumably making for particularly difficult and poorly draining ground.

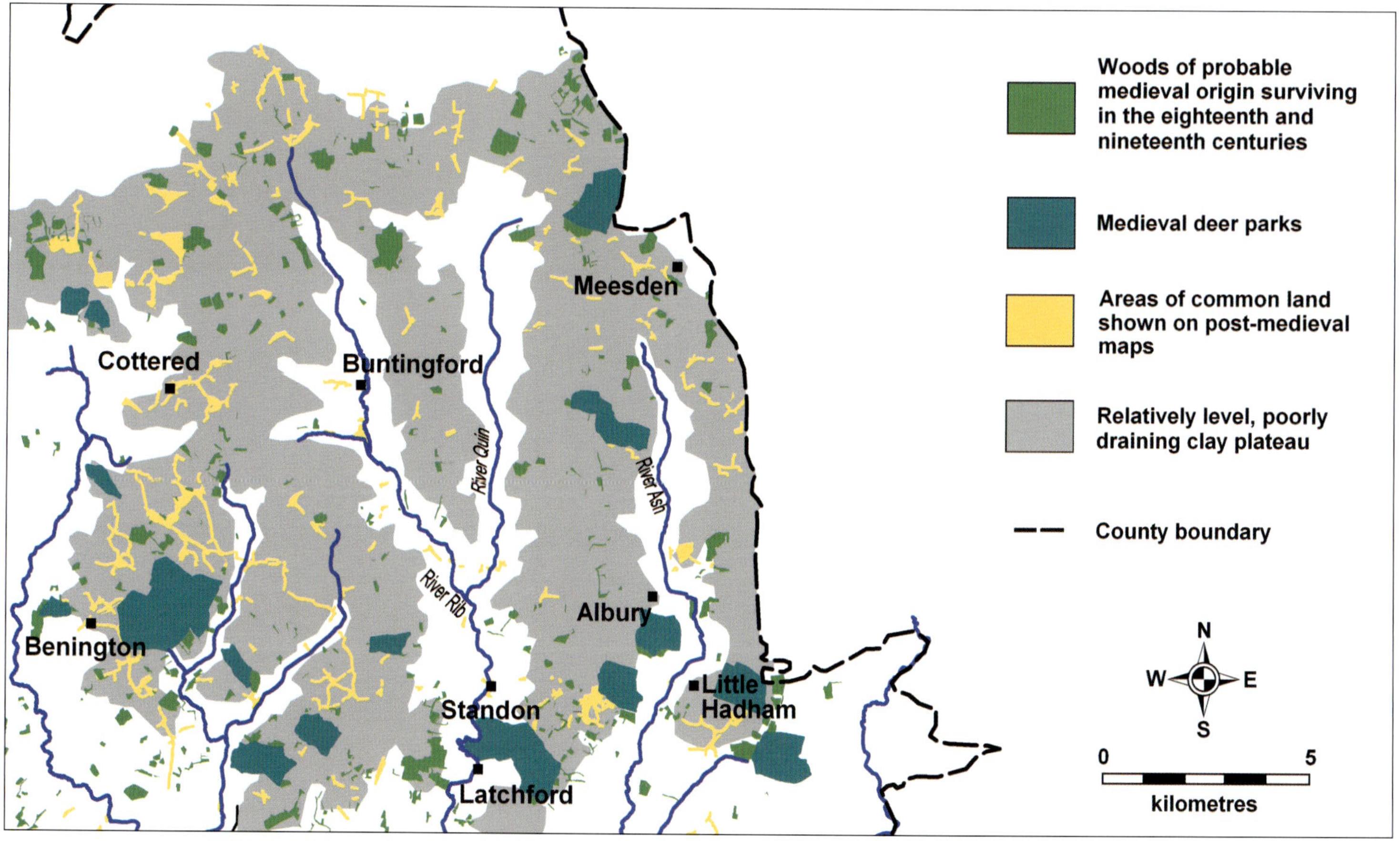

Figure 3.6. The distribution of ancient woods, commons and medieval deer parks in east Hertfordshire. The overwhelming majority lie on the heavy soils of the boulder clay plateau, sometimes forming marked clusters which indicate the main areas of uncleared land in early medieval times. For woods and commons, see sources listed in footnote 21; commons as mapped include obvious encroachments.

The others occur on the higher clay-covered land, generally above *c.*145 metres OD, just above the escarpment on the northern edge of the plateau, where the till was deeper and particularly intractable.

The impression that these various areas originally formed single tracts of woodland and pasture is strengthened when the configuration of neighbouring greens and commons is examined in more detail, and in relation to adjacent field boundaries and, above all, roads. The latter were also part of the common land of the manor, so that there is (for example) no clear distinction between a wide road and a linear common.[25] In the west of Little Hadham Westland Green, the smaller Pig's Green and a number of tiny triangles of common land are all connected, in characteristic fashion, by winding roads (Figure 3.7). Minor encroachments into these areas of common grazing are easily recognised, but closer inspection shows how the various greens represent the progressive fragmentation of a once much larger and continuous tract by the creation of blocks and ovals of enclosed fields.

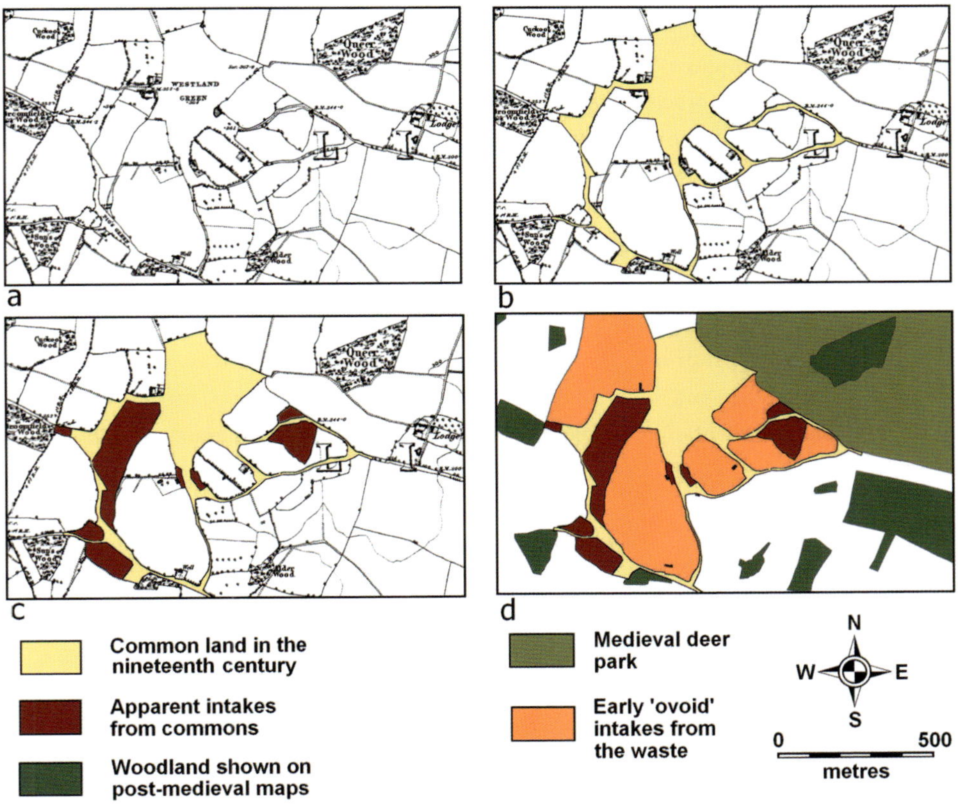

Figure 3.7. The development of the landscape around Westland Green, Little Hadham. Even a cursory inspection shows that Westland Green and Pig's Green were once a single block of common grazing (b and c), and further intakes into a once much larger area of common land can be identified (d). Together with neighbouring areas of ancient woodland and Little Hadham park, this in turn once formed part of a much larger tract of wooded 'waste'.

Evidently, as each new farm was established at the expense of the 'waste' it was placed on the edge of the residue of the former common (now sometimes no more than a narrow ribbon of pasture), with its land extending to the rear; the outline of the intake was generally curvilinear in character, presumably to cut down on fencing costs; and – most importantly of all – where an intake was of any serious size a narrow strip of common was left to its *rear*, running along and fossilising the old common edge, in order to provide access to remaining areas of common grazing and to farms already established along the common margins, farms which thus came to stand beside a winding lane. In short, commons and greens of all sizes are the remnants left over from once more extensive and continuous tracts of common grazing, as are many of the ancient, winding lanes connecting them.

These represent the areas within which, as settlement expanded in the post-Conquest period, peasants were allowed or encouraged by local lords to carve out new holdings. In contrast, they were excluded from areas which lords enclosed as private wood-pastures – deer parks, which functioned as hunting grounds and venison farms, and which also provided a supply of timber as well as grazing for other kinds of livestock. The fragmented complex of greens just discussed was bounded to the north-east by the extensive, continuous block of the bishop of Ely's park at Little Hadham. The various ancient semi-natural coppiced woods of the uplands, which similarly represent blocks of 'waste' enclosed and managed by manorial lords, likewise often form clusters, as in the area to the south-west of Pig's Green (with Standon's medieval park beyond forming another tract of continuous privatised wood-pasture). Commons, woods and parks, in short, are all remnants of the same areas of upland 'waste'. This said, we should not envisage the various clay plateaux in the late eleventh century – before this final phase of expansion and appropriation – as comprising tracts of continuously wooded ground. While still largely devoid of settlement and cultivation, they may already have been fairly open environments. Wood-pastures tend, over time, to degenerate to open ground owing to grazing pressure, as when trees are felled, fall down or die they are unlikely to regenerate naturally and are difficult to replace through deliberate replanting, especially where the thorny underwood which might offer saplings some protection from livestock is being removed on a large scale for fuel.

In some places pressure on resources may already have led to the enclosure and more intensive management of portions of still wooded ground. It is noteworthy that, in most vills, Domesday records woodland in terms of the number of pigs it could support, indicating that it was still unenclosed wood-pasture in which pigs were fattened on the autumn pannage – nuts and acorns. But in several of the smaller vills Domesday describes woodland that was merely 'sufficient for the fences', possibly implying that it was already being coppiced. Woodland, moreover, already formed discrete blocks: Domesday thus describes how on one holding in Munden there was woodland for 150 swine but also 'another woodland where 250 swine might be fed'. Only in the Pelhams, the Mundens, Weston, Meesden and Standon – on the higher boulder clays above *c.*150 metres, or

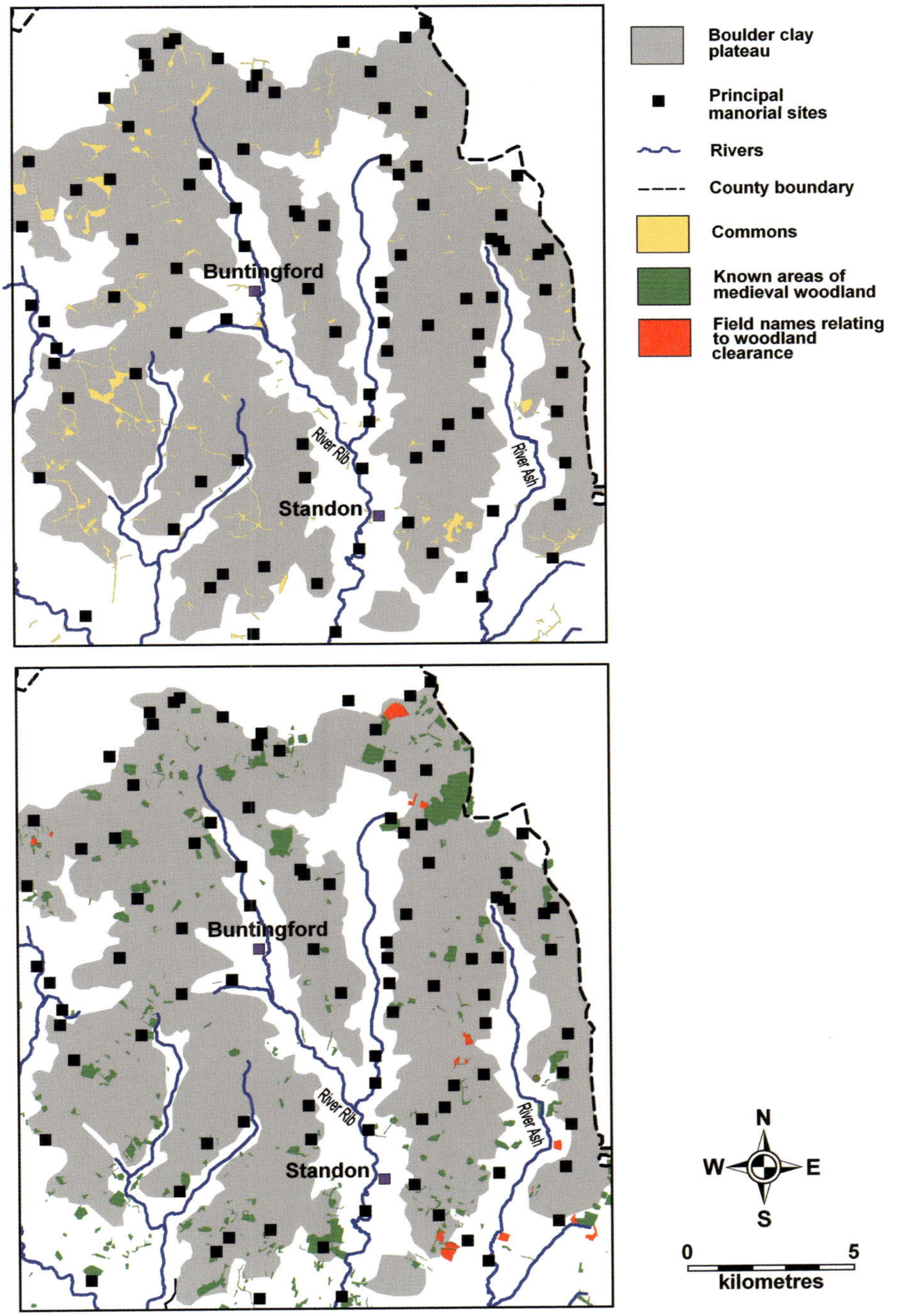

Figure 3.8. Woods, commons, 'clearance' field names and the principal medieval manorial sites in north-east Hertfordshire. Commons (which as mapped include obvious encroachments) tend to occur towards the centres of the boulder clay plateaux and away from manorial sites. Woods and field names bearing names indicative of woodland clearance ('Reading', 'Stocking' and the like) are more likely to be found towards the plateaux margins and closer to manorial sites.

where the till overlies the London Clay or other Tertiary formations – was the sum total of woodland recorded sufficient for more than 300 swine. Walkern had enough to feed 200, Libury 204 and Ardeley 200, but everywhere else had less. Most of the largest medieval deer parks were to be found in these same places and their size, where this can be reconstructed, probably gives us our clearest indication of the extent of these unbroken tracts of unsettled ground in late Saxon times; that at Standon, for example, covered upwards of two square kilometres. Unfortunately, we do not know precisely when any of the parks were created. That at Benington is listed in Domesday Book, and may even have been created before the Conquest, and while most examples are first mentioned in the late thirteenth century, when systematic record-keeping became more prevalent, they could have been established one or even two centuries earlier.

Although woods, commons and former commons often formed contiguous or near-contiguous tracts, there were subtle differences in the locations of each. Woods tend to be concentrated towards the *edges* of the clay uplands, often extending down onto the sloping sides of the valleys and originally abutting directly on the open fields there. Greens and commons, in contrast, tend in general to be located towards their *centres*. The contrast between wooded peripheries and more open interiors is evidently an ancient one, for field names featuring terms such as 'stocking' or 'reading', indicative of the clearance of woodland in early medieval times, are closely associated with the 'marginal' woods and hardly ever lie beside the 'central' greens and commons (Figure 3.8). This kind of 'doughnut' pattern has been noted elsewhere in England.[26] It has been suggested that as population expanded in the early Middle Ages landowners strove to preserve woodland located towards the edges of clayland areas, where wood and timber could be extracted and transported with relative ease, and therefore deflected clearance and colonisation towards the more distant central parts.[27] This was probably the main factor behind the observed pattern in Hertfordshire but there may have been another.

As we have seen, many late Saxon settlements also clustered on the plateau edge and it is possible that their occupants appropriated nearby portions of the upland woods for their own use, managing and to some extent preserving them. The central areas of the 'uplands', in contrast, lay equidistant from large numbers of such settlements, and were presumably exploited in common by them, the intensity of grazing and felling thus encouraging a more rapid degeneration to open ground. By the time of the Conquest many of the plateau-edge settlements had become manorial sites, and the peripheral woodlands brought more firmly under lordly control. Some portions, usually located at a convenient distance from the manorial hall, were enclosed and managed as coppice-with-standards. Ancient woods were invariably manorial property and often (in Hertfordshire as elsewhere) bear the names of manors or their owners: for example, Astonbury Wood, Clothallbury Wood, Alswick Hall Wood in Layston and Chamberlain's Wood in Brent Pelham. Other sections were cleared (hence the clusters of 'stocking' and 'reading' field names) to create sub-manors or large 'ring-fence' farms set within their own compact block of enclosed fields, both of which were often moated to display their superior status. Many of the new manors or farms were also provided with their own nearby wood, such as Mathams in Sawbridgeworth, Patmore Hall in Albury or Sutes in Standon.

As all this was happening colonisation of the 'interiors' was taking place, with these more open tracts gradually becoming fragmented into the complex networks of greens, commons and winding lanes in the manner already described. This process must have been under way by the second half of the twelfth century, but perhaps intensified thereafter. John *de Mora* ('of the moor') of Ardeley, recorded in 1191, probably lived at Moor Green,[28] but there are few other references, direct or indirect, to green-edge settlements in the interior before 1200. A century later, however, increasingly abundant documentary records provide evidence which hints at the names of a few early settlers: an ancestor of a man called Roys Burre, who paid tax in 1307, probably held – and perhaps established – a tenement at Burrs Green (now Burn's Green) in Benington.[29] There were other contrasts between the edges and centres of the clay uplands. For example, many of the manors lying towards the plateau margins farmed particularly extensive fields by the thirteenth century, demesne closes extending over forty, fifty or more acres, much larger than the closes associated with peasant farms on common edges.[30] Such fields were usually subdivided in the later Middle Ages but they can often be picked out on early maps as groups of contiguous closes bearing the same name, such as the three contiguous 'Broad Fields' shown on a late sixteenth-century map, part of the demesne of Abbotsbury Manor in the south of Barley, which together extended over some 40 acres.[31]

In the foregoing pages we have emphasised the subtle differences in the landscapes of north-east Hertfordshire between the early settled valleys, with their open fields and mixture of farms and hamlets; the edges of the upland clays, with manors and ring-fence farms of early origins, abundant woods and enclosed fields; and the interiors, with their complex green-edge settlements. These distinctions are, however, rather more blurred in practice than our account perhaps suggests, owing to the complexity and subtlety of the local topography: the great variations in the shapes and sizes of the upland plateaux, and the way in which tributary valleys gradually fade to level upland, reducing the sharp distinction between each. The contrast between the 'cores' and 'peripheries' of clay uplands is, moreover, often confused by the way in which greens sometimes extend all the way out to parish churches placed on the plateaux margins, from where long strings of farms and cottages extend far into the clayland 'interiors'. Evidently, where a high proportion of a township's territory lay on the clays the expansion of settlement into the interior was sometimes on such a scale that the development of a village around the original focus, marked by the parish church on the edge of the 'plateau', was suppressed. The churches at Rushden, Ardeley, Weston, Benington and to some extent Cottered thus stand to one side of their loosely nucleated villages, fronting on part of the complex network of greens, commons and lanes – shattered fragments of the common pastures – along which houses have spread far onto the plateaux. In the case of Sacombe the pull of the uplands was so great that the church was left entirely alone on the plateau edge, the village clustering around a common (since enclosed) about half a kilometre to the east.

Nevertheless, the patterns we have described although often blurred in detail, are real enough, and are marked in other ways. The complex landscapes of the clay interiors, for example – exemplified perhaps by the stunning, lonely countryside between Rushden, Sandon and Kelshall – are also characterised, to a much greater degree than the clay margins, by a veritable maze of lanes, footpaths and tracks, mostly left to provide access as successive intakes were made from the great upland grazing grounds. Rights of way are less dense around the plateau margins, characterised by manorial sites and ring-fence farms; and fewest on the valley sides, where open fields were often enclosed by parliamentary acts which empowered Commissioners to reduce the number of roads and footpaths.

Late medieval and post-medieval changes

The basic structure of the landscape of north-east Hertfordshire was firmly in place by *c.*1300 and while there were numerous important changes thereafter they occurred within this established framework. The later Middle Ages was a period of demographic decline and economic depression, but also of agrarian change and specialisation. This probably included some expansion of pasture and an increased emphasis on cattle production on the heaviest clays in the locality, but mixed farming was the norm, with a strong emphasis on cereal farming and, increasingly, on the production of good-quality malting barley. As the population began to recover through the fifteenth and sixteenth centuries the district's wealth and population increased relative to the rest of the county. By 1543, according to the lay subsidy returns, it had overtaken all other areas of Hertfordshire – except parts of the south-east, towards London – with Braughing Hundred having an average of 2.1 taxpayers per 100 acres. High levels of late medieval and early modern wealth are clearly reflected in the large numbers of fifteenth- and sixteenth-century farmhouses which still survive in the district, and which are such a characteristic feature of its villages and hamlets (below, pp. 195–7).

We noted earlier that there are few examples of true 'deserted medieval villages' in Hertfordshire, but three of the classic examples can be found in the clay uplands of the north-east, at Broadfield, Wakeley and Berkesden – all of which had been true villages, with their own parish churches. In all three cases earthworks, preserved by the conversion to pasture in the late Middle Ages, survived into the 1970s, although little remains on the ground today. At Berkesden, typically, the parish church stood near the manor house of Tannis Court, on the edge of light valley soils, but most of the houses clustered around greens and commons, principally Berkesden Green, some distance onto the clay plateau. In numerous other places we have evidence for the shrinkage of settlements, as, for example, on the edge of the village of Kelshall, and also for the disappearance of small hamlets or isolated farms – a general 'thinning out' of the sprawling, disorganised patterns of common-edge settlements and isolated farms created in the period leading up to the

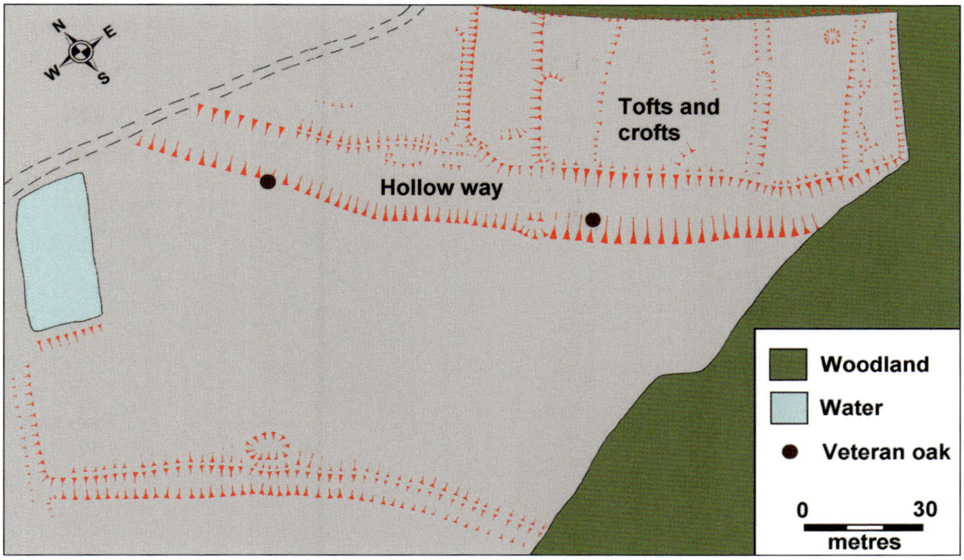

Figure 3.9. The earthworks of a small deserted hamlet at Knights Hill, Westmill.

fourteenth century. The intensity of arable agriculture in the district has, once again, ensured that these have seldom left above-ground traces. But there are some exceptions. For example, in the east of what was formerly the landscape park associated with Knights Hill in Westmill, on clay soils overlooking the Rib valley, are the earthwork remains of a small hamlet fronting on a sunken road (Figure 3.9). This forms part of a wider 'relict landscape' which has been preserved because the land was laid to pasture, probably for the most part long before the park itself was created here in the eighteenth century, and which includes field boundaries, hollow ways and strip lynchets associated with former areas of open-field cultivation. Another extensive area of pasture, containing an even more complex range of earthworks, survives to the north of Warrens Green in Weston (Figure 3.10). In this case, we can compare the archaeological remains with the features shown on a fine map of 1633.[32] This reveals that some of the earthworks represent elements of the landscape which existed at that time (one group, for example, marking where one of the houses fronting on Warrens Green stood) while others are either earlier, or later, in date. In a number of places there are low, parallel ridges which superficially resemble medieval ridge and furrow, but they are not: with a 'wavelength' of only *c.*3.7 metres they are around half the width of a normal open-field strip, and they in fact represent a form of post-medieval ploughing within enclosed fields, known from a number of other places in the county, as, for example, at Burleigh Farm on the western fringes of the district (at TL 221216). Many other odd patches of unploughed ground in the east of the county – and indeed, elsewhere in Hertfordshire – would repay careful inspection by local historians for evidence of early farming and settlement.

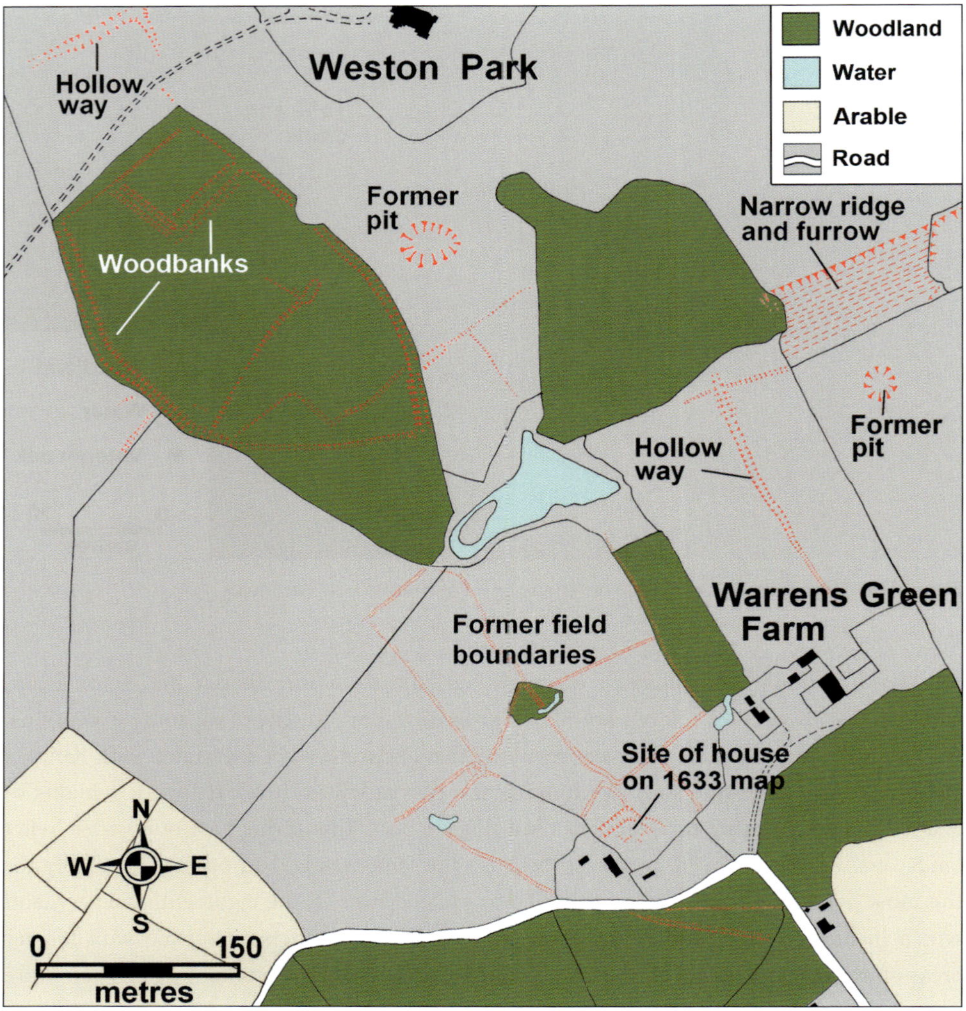

Figure 3.10. Earthworks north of Warrens Green, Weston. A remarkable collection of earthworks is preserved under the woodland and pasture around Weston Park, including former field boundaries, hollow ways (old roads) and narrow ridge and furrow.

The disappearance of one of the houses from Warrens Green in the period after 1633 is a clear indication of the fact that many of the small deserted sites or shrunken hamlets in the area are not simply the result of late medieval population decline but have more complex causes and a longer chronology, related in large measure to the gradual emergence of a market economy and an increase in the average size of (and an associated decrease in the number of) farms. The lost hamlet at Knights Hill in Westmill probably was abandoned in the later Middle Ages but a number of other places within the same parish declined, or disappeared completely, much more recently. The enclosure map of 1819 thus shows two large farms and two cottages clustered around Westmill Green, in the far west of the parish; by the time the Ordnance Survey 6-inch map was surveyed some fifty years later, only the two cottages remained.[33] In the same period the hamlet of Brook End, to the east, halved in size, while in the south-east of the parish a hamlet called Great Hull which straddled the parish boundary with Braughing, and comprising around seven houses, completely vanished, although in this case the disappearance may in part at least be attributed to the construction of the railway line to Buntingford in the 1860s.

Westmill may be an extreme case. Nevertheless, it is clear that farms and cottages continued to disappear in many other parishes into the nineteenth and even twentieth centuries. A farm at Haven End in Braughing was demolished after the Second World War, while Potters Green in Little Munden was a rural hamlet of 11 households in 1861 but had reduced to 8 by 1891 and just 2 by the end of the twentieth century.[34] Yet, at the same time, examination of early maps reveals that many new cottages were being *added* to the settlement pattern, often on small slivers of roadside 'waste'. This in turn was part of a more general post-medieval phenomenon: the gradual removal, through encroachment and appropriation, of many small parcels of common land, a process which continued into the nineteenth century and which further reduced the contrast we have noted between the 'core' and 'peripheral' areas of the clay plateau. At Aston End, in the north of Aston, for example, the tithe map of 1842 shows how farms originally fronted on a network of commons which had been extensively encroached upon, with little strips taken into severalty, leaving only a few places where the road still widened, suddenly, into a small green.[35] Yet the earlier status of many of the intakes was still remembered, the map noting that they were 'waste', although now distinct and severed from the adjacent fragments of 'common' (Figure 3.11). The enclosure map of 1858 dealt with most of the remaining open land shown on the tithe map, but the areas marked as 'waste' it ignored, and all subsequent sources treat them as private property.

While commons and greens might be encroached upon in this manner throughout the post-medieval centuries, and while piecemeal, gradual enclosure eroded the open fields, large areas of unenclosed land still survived in the district up until the time of parliamentary enclosure, which, as in the 'champion' districts to the north, generally occurred at a surprisingly late date. This was in part because of the complexity of land ownership in the locality and in part because there was less incentive to enclose this fertile arable land than there was in districts where land could be more effectively 'improved' by private investment following enclosure, or where conversion of arable to pasture was likely to enhance the fortunes of landowners. The earliest parliamentary enclosures came during the Napoleonic Wars, with the acts for Barkway and Reed in 1808 and Cottered in 1810. But most came some time later: those for Furneux Pelham, Meesden and Sandon were passed in the 1840s, those affecting Walkern, Sacombe, the Mundens, Buckland, Aston and Benington not until the 1850s, and Little Hormead, Layston, Albury and Throcking were enclosed as late as the 1860s. Most of the land affected was open arable. Indeed, in some

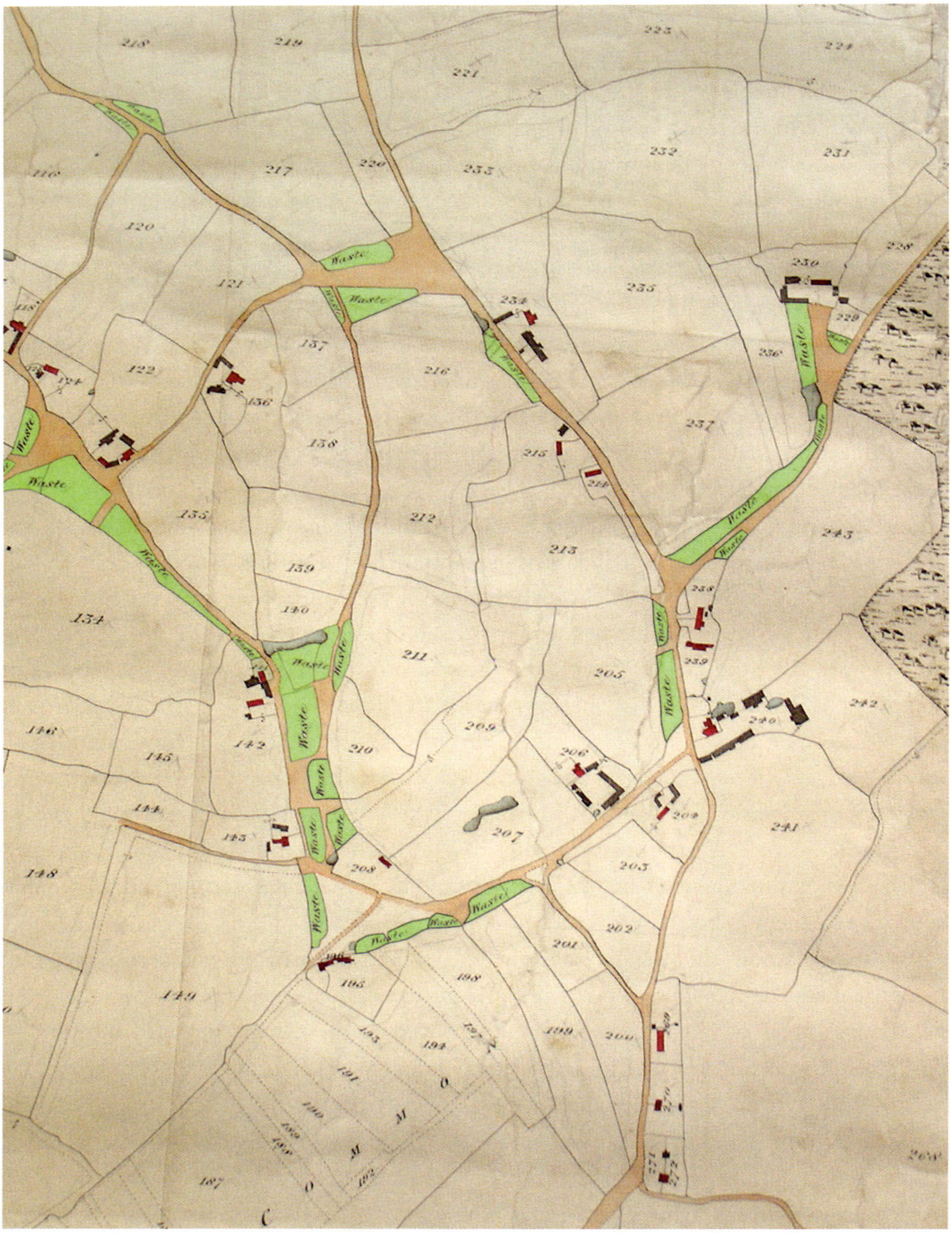

Figure 3.11. Aston End, as shown on the tithe map of 1840 (with green shading added for clarity). Late medieval and post-medieval encroachments often removed the close association of farms and commons in 'woodland' districts. Unusually, this map distinguishes between enclosed land, surviving commons (brown) and areas of 'waste' (green) occupied and improved at some unknown time in the past, making it clear that the farms shown all once fronted directly on a network of linear 'greens'. Reproduced by kind permission of HALS (DAcc 1154).

places, such as Little Hadham and Sandon, the remaining open fields were enclosed but all or most of the commons were left intact. The later enclosures in the sequence often affected quite small acreages. Little Hadham and Albury, for example, had less than 10 per cent of their area enclosed by acts of 1859 and 1869 respectively, but earlier ones often affected a significant area, ranging from around 16 per cent in the case of Braughing or Westmill through 20 per cent at Standon to over 50 per cent at Anstey and Buckland, although these figures also include some small 'greens' and other areas of common grazing. As in the districts to the north, some areas of open-field land still survived even when the first edition Ordnance Survey 6-inch maps were surveyed in 1877/8 in Brent Pelham, Ardeley, the Mundens, Sawbridgeworth, Much Hadham and, especially, Wyddial and Layston. Here, Browns Field, Brownsel Field, Pupsey Hill, Church Field, Great and Little Bartholomew Fields and Downhall Field are all clearly shown on the map and the complex manner in which the lands of the two parishes lay intermingled in these fields may, in part, explain this particularly late survival.

Nineteenth-century agricultural improvement contributed to the field pattern of east Hertfordshire in other ways. In particular, the old, rather irregular pattern of small fields, characteristic in particular of the plateaux between valleys, was inconvenient for ploughing and as more of this land was converted from pasture to arable with the increase in cereal prices there was a tendency to modernise the landscape by amalgamating fields, straightening hedges and removing some of the smaller areas of ancient woodland, all of which was often accompanied by the establishment of networks of sub-surface field drains. Such changes were further encouraged by the adoption, in some places, of steam ploughing in the middle decades of the century, as traction engines and their ploughs could only work effectively when ploughing extensive acreages. John Prout, who farmed at Blount's Farm to the north-west of Sawbridgeworth, was famous for his agricultural experiments in the 1860s, which involved the use of artificial fertilisers on a large scale and significant levels of mechanisation.[36] The field pattern around the farm was transformed, with small and medium-sized irregular enclosures being replaced by fewer, larger fields with straight sides, some of which lacked hedges (curiously, the first edition Ordnance Survey 6-inch shows that this modern landscape was bounded, immediately to the south, by an area of Sawbridgeworth's open fields, which still remained unenclosed). Many landowners made similar alterations, usually on a more piecemeal scale but sometimes more extensively. Typical was Beeches Farm, Brent Pelham, where on the central 255 acres of the estate average field size rose from 8.8 to 12.5 acres between 1769 and 1877. By the later nineteenth century the large freehold farms, often former manors on ancient sites located on the margins of the clay plateau, often boasted larger fields and more 'modern' boundary patterns than their neighbours positioned on the edges of commons or greens or within the principal villages. Some had always been associated with large fields, the large 'demesne closes' of medieval times. Piecemeal enclosures taken from the open fields in the vicinity of such places were also often formed into particularly extensive closes. In 1768 the land to the east of Wyddial Park formed part of North Field, and the holdings of the estate there comprised a few scattered strips. By the 1870s the entire area had been acquired and enclosed and mostly comprised two vast arable fields, one of 64 acres and another of 125.

Recent history

North-east Hertfordshire, with its generally fertile soils and a climate well suited to cereal cultivation, was less drastically affected by the great agricultural depression that began in the late 1870s than were most parts of England. There was some increase in the area under pasture on the more difficult clay soils, associated with an expansion of dairy farming, especially in the west of the district; arable farming continued more strongly in the east, in part because of the expansion of sugar beet cultivation and the proximity of the beet factory at Felsted in Essex.[37] Wheat was now the most important cereal, the cultivation of barley retreating to the immediate vicinity of the malting towns of Bishop's Stortford and Ware. Hedge removal and field amalgamation continued, normally at a lower level but with exceptions. In 1901 Rider Haggard, the writer and rural campaigner, was shown the improvements carried out by Alexander Crossman at Cokenach near Barkway, which involved extensive drainage and the laying out of new roads and fields.[38] But the scale of landscape change increased with the advent of the Second World War. The intensity of cultivation was raised to meet food shortages and, even after peace returned in 1945, the government, and subsequently the European Economic Community, introduced a range of subsidies to increase levels of production in order to feed a hungry population. At the same time the widespread adoption of tractors and combine harvesters, coupled with the low cost of artificial fertilisers, ensured that arable farmers no longer had any incentive to keep animals for manure or traction. Hedges were thus no longer required as stock-proof barriers and, as large machines work most economically in large fields, they were removed wholesale, a task made easier by the availability of mechanical diggers and government subsidies. Not only were hedges removed but also innumerable field ponds, small woods, even green lanes (almost certainly, in most cases, illegally). Only from the 1970s did the rate of destruction ease, and then reverse, as subsidies began to be made available for replanting hedges and woods.

North-east Hertfordshire probably retains just over a third of the hedges that it had in the late nineteenth century, although the precise figure varies greatly from place to place.[39] While in some areas there are still around half the number, in others the reduction has been frightening, as, for example, in parts of Aspenden, where less than 15 per cent of the hedges shown on the first edition Ordnance Survey 6-inch map now survive. Even where hedges remain they are often now gappy and outgrown. Hedges have generally survived better in the vicinity of the larger settlements (as around Cottered or Much Hadham), and where pockets of pasture still exist in this otherwise intensively arable landscape. Conversely, percentage losses have been greatest where the nineteenth-century field pattern was most irregular, and on the more level or gently sloping terrain. But there is no very clear pattern and for the most part the extent of removal has been dependent on the whims of successive farmers and owners.

In spite of the progressive simplification of the landscape over the last few decades, much of historical importance survives. While many surviving boundaries were created only in the eighteenth or nineteenth century a high proportion have medieval origins; significant areas of ancient, semi-natural woodland remain, if often in a neglected condition; and the complex networks of lanes and greens on the clay 'uplands' show where boundaries and rights of way were fixed at some point in the twelfth or thirteenth centuries. This is indeed an ancient landscape, and it deserves to be treated with care and respect.

Notes

1. L.G. Cameron, *The land of Britain, Part 80: Hertfordshire* (London, 1941), p. 308.
2. Hodge *et al.*, *Soils and their use in eastern England*, pp. 140–43, 209–12.
3. W.G. Hoskins, *The making of the English landscape* (London, 1955), p. 92.
4. W.G. Hoskins and L. Dudley Stamp, *The common lands of England & Wales* (London, 1963), p. 32.
5. Saunders, 'The *oppida* at Wheathampstead and Verulamium'; Bryant and Niblett, 'Late Iron Age in Hertfordshire', pp. 273–5; Niblett and Thompson, *Alban's buried towns*, pp. 23–40.
6. Bryant and Niblett, 'Late Iron Age in Hertfordshire', p. 276.
7. C. Partridge, *Skeleton Green: A late Iron Age and Romano-British site*, *Britannia* Monograph 2 (London, 1981); C. Partridge, 'Excavations at Puckeridge and Braughing 1975–79', *Hertfordshire Archaeology*, 7 (1979), pp. 28–97.
8. M. Landon, 'Making a mint: the archaeology of a late Iron Age industry', *Current Archaeology*, 234 (2009), p. 5.
9. B.C. Burnham and J. Wacher, *The 'small towns' of Roman Britain* (London, 1990), p. 109.
10. HALS DSA4/44/1–2 Little Hadham tithe award and map, 1843–4.
11. HHER 797; HALS D/Z44/P1 plan of Westfield Farm, 1767.
12. Williamson, *Origins of Hertfordshire*, pp. 90–102.
13. M. Gelling, *The early charters of the Thames valley* (Leicester, 1979), pp. 81–2.
14. I. Thompson, 'Braughing: Extensive Urban Survey Project assessment report', unpublished, for English Heritage and Hertfordshire County Council (2002), p. 10.
15. Crick, *Charters of St Albans*, no. 7.
16. T. Astle *et al.* (eds), *Taxatio Ecclesiastica Angliae et Walliae Auctoritate P.Nicholai IV circa AD 1291* (London, 1802).
17. P. Boyer, paper delivered as part of 'Archaeology in Hertfordshire: Recent Research. A conference to celebrate Tony Rook's 80th birthday', Welwyn Garden City, 14 July 2012.
18. HHER 6516; There are one or two finds of metalwork, such as the middle or late Saxon arrowhead from Pishiobury in Sawbridgeworth (HHER 6533).
19. Including Alswick Hall, Beauchamps Hall, Boxbury, Broadfield Hall, Cokenach, Cockhamstead, Corney Bury, Hicham Hall, Hodenhoe Manor, Hyde Hall, Libury, Newsells, Patmore Hall and Stonebury.
20. Gover *et al.*, *Place-names of Hertfordshire*, p. 197; Page (ed.), *VCH Hertfordshire*, vol. 3, p. 359.
21. HALS DE/B1622/P1 map of Albury & Furneux Pelham, 1781; DE/X696/P1 map of Anstey, 1730; 57415 map of Anstey Decastrum, mid-eighteenth century; D/Z110/P1–2 map of Ardeley, 1744; DE/H/P3 map of Aspenden Hall estate, 1810; DE/Ry/P1 map of farms in Barkway, 1728; DE/Ry/P3 plan of Newsells estate, Barkway 1788; DE/Ry/P4 map of Nuthampstead Bury Farm, 1789; DE/P/P6 plan of Rushingwells Farm, Barkway, 1732; 63631 map of Barkway, Reed & part of Wallington, 1807; private collection, map of the manor and parish of Benington, 1628; HALS D/Z61/P1 map of Bishop's Stortford, 1823; QS/E82 Bishop's Stortford inclosure map, 1822; DE/B2351/P1 maps of Braughing Bury Farm, Upp Hall Farm, Warren Farm and Gatesbury Mill, Braughing, *c.*1800; 76830 map of Stonebury in Braughing, Little Hormead and Layston, 1712; DE/Nc/1 map of Braughing, *c.*1820; XIII/51 plan of Broadfield, 1775; DE/X526/P1 plan of Broadfield Hall estate, 1810; 26293 map of Bygrave & parts of Clothall, 1808; 74423 map of land in Clothall, Wallington & Cottered, 1729; DE/X251/P3 map of Cottered, *c.*1806; DE/P/P4 plan of Earl Cowper's estate in Hertingfordbury, 1704; DE/P/P7 plan of Hertingfordbury Farm, 1732; DE/H/P6 plan of Little Hormead Bury & Balons Farm, 1835; 63716 Hormead Dane Farm, Gt Hormead, pre-1815; DSA4/56/2 Hunsdon tithe map, 1842; DE/H/P14 plan of Briggens in Hunsdon Stanstead Abbots & Roydon, 1781; 54835 plan of the manor of Corneybury in Layston & Wyddial, 1744; 26872A map of Meesden, 1815; 56457 plan of Brent Pelham Hall & farms, 1778; 56458, 56459 & 56461 plans of estates in Brent Pelham, 1787; DE/Cn/P7 plan of Beeches, Brent Pelham, 1769; DE/Bc/P6 & P7 plan of an estate in Brent, Furneux & Stocking Pelham, *c.*1800; 79699 plan of Furneux Pelham Hall estate, *c.*1800; DE/H/P15 plan of Tinkers Farm, Furneux Pelham, 1808; 74776A map of Rushden & parts of Clothall, Wallington & Cottered, 1802; DE/H/P5 map of land in Standon & Braughing, 1809; A/2837 plan of Colliers End Farm, 1809; DE/B/1768/P1 plan of farm at Colliers End, 1767; DE/B/1768/P2 plan of estate at Colliers End, 1794; A/2834 plan of Fabden Farm, Standon, 1790; A/2832–3 plan of Great Barwick estate, Standon, 1778; 83280–1 plan of Marshalls Farm, Standon, *c.*1730; 83353 plan of Marshalls, 1804; 43754 plan of estate at Puckeridge, 1778; A/2838 plan of Standon Green End, 1774; DE/H/P17 plan of Newgates alias Little Briggens Farm, Stanstead Abbots, 1781; DE/Sa/P1 plan of estate in Stapleford, *c.*1690; DP/107/29/1 map of Therfield

& Buckland, 1724; 71182 plan of Thorley Hall, 1673; DP/114/3/4 map of land in Walkern, 1810; 76828A–B plan of an estate in Westmill *c.*1670; CP/107/26/2 Westmill inclosure map, 1819; 54027 plan of farms in Wyddial & Barkway, 1768.

22. HALS DE/AS/91 extent and rental of Aston, 1540; private collection: map of the manor and parish of Benington, 1628; HALS Acc 4283.
23. HALS Acc 4283; 49464 survey of the manor of Weston, 1548.
24. The outline of the medieval parks has been reconstructed from early documentary references, field patterns and field names: see A. Rowe, *Medieval parks of Hertfordshire* (Hatfield, 2009).
25. Rackham, *History of the countryside*, p. 265.
26. P. Warner, *Greens, commons and clayland colonization* (Leicester, 1987), pp. 5–9; K.P. Witney, 'The woodland economy of Kent, 1066–1348', *Agricultural History Review*, 38 (1998), pp. 20–39.
27. *Ibid.*, p. 20.
28. Gover *et al.*, *Place-names of Hertfordshire*, p. 152.
29. J. Brooker and S. Flood, *Hertfordshire lay subsidy rolls 1307 and 1334*, Hertfordshire Record Publications 14 (Hertfordshire Record Society, 1998), p. 83; Burrys tenement and Burres Grene were recorded in a survey of 1556 (TNA:PRO E315/391 survey of Hertfordshire manors, lands and possessions of King Philip and Queen Mary, 1556); Burrsherne furlong lay close to Burrs Grene in 1628 (private collection, map of the manor and parish of Benington).
30. D. Roden, 'Field systems of the Chiltern Hills and their environs', in A.H.R. Baker and R.A. Butler (eds), *Studies of field systems in the British Isles* (Cambridge, 1973), pp. 342–3.
31. Wilkerson, *Norden's survey of Barley*.
32. HALS Acc 4283.
33. HALS CP/107/26/2.
34. V. Tott, *The Braughing of Victor Tott: the visual and written memories of a country gentleman* (Quebec, 2010), pp. 106–11; A. Rowe, *The protected valley: a history of Little Munden* (Buntingford, 1999), pp. 21–2.
35. HALS DAcc 1154 Aston tithe map, 1840.
36. Cameron, *Land of Britain*, p. 323.
37. *Ibid.*, pp. 336–40.
38. J. Wilkerson, *Two ears of barley* (Royston, 1969), pp. 16, 27, 89–90.
39. Based on a random sample of ten 1 km by 1 km squares, and on a generous definition of what constitutes a hedge. The figures are: 8.8 kilometres reduced to 3.8 (TL3429); 6.4 reduced to 0.7 (3427); 7.8 reduced to 2 (3826); 8.5 reduced to 0.46 (4125); 9.9 reduced to 5.6 (4319); 16.4 reduced to 5.3 (4418); 10.8 to 6.1 (4223); 6.1 to 1.3 (3718); 7.1 to 4.1 (3334); and 9.1 to 4.4 (3232).

– 4 –

The landscape of west Hertfordshire

Introduction

The landscape of west Hertfordshire, the area of the county which corresponds with the eastern section of the Chiltern Hills and in particular with the Chiltern dipslope, is strikingly different from that of the districts discussed in the previous chapters. Although extensive pockets of attractive countryside remain, some of it densely wooded, much of the area is today more industrialised and suburbanised. But even before these modern additions the landscape of west Hertfordshire had its own distinctive character, the consequence in large measure of the fact that, in comparison with the north and east, it was cleared and settled at a relatively late date (Figure 4.1). The pattern of parishes in the area immediately alerts us to the fact that this was, in the early Middle Ages, sparsely settled ground. They are larger – on average around 50 per cent more extensive – than in the north or east. Many, moreover, were formed by the division of larger territories at some unknown date in the past (Great and Little Gaddesden, King's and Abbot's Langley, Ayot St Lawrence and Ayot St Peter); in the later Saxon period (St Michael's, St Peter's and St Stephen's in St Albans); in the course of the Middle Ages (Northchurch created out of Berkhamsted); or in the post-medieval period (Bovingdon and Flaunden, sundered from Hemel Hempstead in 1841; Harpenden formed from Wheathampstead in 1859; and Markyate, created from Flamstead, Caddington and parts of neighbouring Bedfordshire parishes in 1888).[1] In a similar manner, while Domesday names around 70 separate vills in the 440 square kilometres of the boulder clay countryside in the east, in the Chilterns only around 30 are listed within an equivalent area, and even some of the medieval villages – Sarratt, Flaunden, Bovingdon, Markyate and Harpenden – fail to appear.

The Chilterns comprise, as we have already noted, a great tilted mass of Chalk with a long dipslope extending south-eastwards from the dramatic escarpment. The Chalk is exposed on the escarpment itself and in a complex pattern of often steep-sided valleys, some of which – the 'through' valleys – cut right through the escarpment and form passages of low ground running through the hills to the clay plains to the north. Most of

Figure 4.1. The typical landscape of the eastern Chilterns near Breachwood Green: arable fields extend up the sides of a dry valley to wooded hills above.

the valleys are dry or contain streams which flow only in the winter, but the larger ones contain major rivers whose sources likewise often vary in location with the seasons: the Chess, Gade, Ver and the upper reaches of the Lea and Mimram. The soils on the sides of the main valleys, which are mainly derived from the Chalk or a mixture of Chalk and till (the Charity, Hamble, Moulton, Combe and Swaffham Prior Associations) are calcareous, freely draining and moderately fertile.[2] This, together with the fact that water was freely available from the rivers here, ensured that from early times the valleys were well settled: something which explains, of course, the fact that the great Iron Age *oppidum* and Roman town of Verulamium developed at what is now St Albans. Between the major valleys are wide interfluves where the Chalk is covered by a thin layer of 'clay-with-flints' topped with Plateau Drift (p. 6), a light but acidic clay formation containing flint nodules and, in places, sand, giving rise to soils which are acidic and infertile in character. There is no running water here: the Plateau Drift, unlike the boulder clays of the east, is relatively porous. Nevertheless, water is freely available from wells sunk into the Chalk within the deeper dry valleys, or into the Plateau Drift itself, which acts as a weak aquifer. The lower parts of the dipslope – south and east of a line from Sarratt to Sandridge, and then north through Wheathampstead to Kimpton – are, in turn, blanketed with extensive beds of sands and gravels laid down in the Quaternary Period by the proto-Thames, when this flowed along the line of the Vale of St Albans, and by glacial meltwater – principally the Gerrards Cross, Smug Oak and Westmill Lower Gravels.[3]

Much of west Hertfordshire was densely settled in the late prehistoric and Roman periods and there were a number of substantial settlements in the principal valleys, where the Chalk was exposed – most notably the *oppidum*, and later *municipium*, of Verulamium. But the other geological formations in the district give rise to soils which are acidic and comparatively infertile and by late Saxon times this had become a relatively marginal area. Domesday Book shows it as densely wooded, sparsely settled and with relatively few plough teams, forming part of a wider zone of under-developed land which extended westwards into the Chiltern portions of Buckinghamshire and Oxfordshire. Except in the area to the north of the river Lea there were, on average, only 2–3 recorded tenants per square mile, compared with 9–13 in the northern and eastern parts of the county which we have so far discussed. Moreover, whereas the latter areas generally had around three ploughs per square mile, suggesting that about two-thirds of the land was under the plough, in the Chilterns the figure was less than half of this. There were vast tracts of uncleared woodland, with Berkhamsted, for example, having woodland sufficient for 1,000 swine and Hemel Hempstead enough to feed 1,200. We have to be slightly cautious with these figures, given that the district was also characterised by large estates – Hemel, for example, being rated at 10 hides, Berkhamsted at 13. However, even when allowances are made for this, it is clear that the district was perhaps twice as wooded as the north-east of the county: well into the post-medieval period extensive tracts of common land and woodland survived, especially on the higher ground immediately above the Chiltern escarpment. Large areas remain here to this day, and are vital for recreation and wildlife conservation. This distinction, between the wilder countryside on the high ridge, with its woods and commons, and the more tamed and cultivated (and today more suburbanised) landscape of the dipslope, provides a useful conceptual framework for understanding the development of this complex piece of landscape.

The dipslope

The places which Domesday names as separate vills are all located in the principal valleys, beside the major rivers, and while some prehistoric and Roman sites are known from the uplands the majority of major settlements were found in, or immediately adjacent to, the major valleys (above, pp. 12–15). It is thus easy to assume that the Plateau Drift soils and the spreads of river and glacial gravels were an untenanted wilderness in Saxon times, only becoming cleared and settled in the course of the eleventh, twelfth and thirteenth centuries, as the population grew and local towns, especially St Albans, expanded, and as London, lying no more than 30 kilometres to the south-east, became an increasingly important market for agricultural produce.[4] Not surprisingly, Levett suggested in 1938 that 'a large part' of the Liberty of St Albans, the administrative unit which embraces much of the dipslope, 'must have been almost continuous woodland' before the twelfth century.[5] But while this is largely true, it is not the entire story. The uplands were already being exploited, and were to some extent settled, before the eleventh century, and some features of the modern landscape appear to have come into existence long before the Norman Conquest.

The evidence of local place-names is of crucial importance here, for it shows clearly that the places mentioned by Domesday also included numerous other settlements within their boundaries. The villages of Flaunden and Bovingdon, for example, may have been ignored but their names are unquestionably of Old English form and the fact that, until the mid-nineteenth century, they formed part of the large parish of Hemel Hempstead clearly indicates that they were subsumed by Domesday within that manor and vill. Indeed, Bovingdon's dependence upon Hemel Hempstead probably explains its name, *bufan dune*, 'above the down'.[6] It lies high above the steep slopes of the valleys of the Gade and Bulbourne, when viewed from Hemel itself. Sarratt, likewise omitted by Domesday, is also an Old English name, probably meaning 'the dry place'.[7] Its church contains early twelfth-century fabric and the thirteenth-century *Gesta Abbatum* describes, probably reliably, how it was alienated from the monastery of St Albans for a short period during the abbacy of Paul of Caen (1077–93).[8] What is more important, however, is that many minor settlements, even ones located well away from the major river valleys, likewise bear Old English names (Figure 4.2).[9] Once again the caveats expressed in the last chapter should be noted concerning the survival of Old English into the twelfth century, and the way in which names featuring topographic elements might have been coined to describe features of the landscape long before any actual settlements were established at the places in question. This said, there is no doubt that even topographic names may mark settlements which were centuries old by the time they first appear in documentary sources. We might assume that the obscure hamlet of Cockernhoe, in Offley, was a relatively new creation when first mentioned in a document of 1221, and it is only the chance survival of the will of the great noblewoman Æthelgifu, drawn up around 980, which allows us to see it was already over two centuries old, for she instructed that, following her death, a slave called 'Wulfstan of Cockernoe' was to be freed.[10] Hamlets and isolated farmsteads bearing Old English habitative names such as *-tun*, *-wic* and the like are probably more reliable as indicators of early settlement, and a number of these – omitted from Domesday – appear in earlier charters and similar documents, such as Westwick near St Albans, the 'west farm', which is likewise mentioned in Æthelgifu's will and in a charter of 996, or Childwick, *Childewicam*, the 'dairy farm of the young monks', which the *Gesta Abbatum* suggests was in existence by the late eleventh century.[11] Some of these charters – and there are several for the district – have boundary clauses which note, in passing, what appear to be isolated farms located well away from the major river valleys: the 'mercing dwellings' in that for Gaddesden, dating to the late tenth century; or the *heale wic*, 'corner farm' in that (pre-Conquest, but unfortunately undated) for Cashio.[12]

Many of these clauses also refer to individual trees, named valleys, defined areas of woodland, occasional patches of ploughland and clearings, suggesting a varied and often open landscape, not one composed of continuous forest. The fact that the dry valleys, well

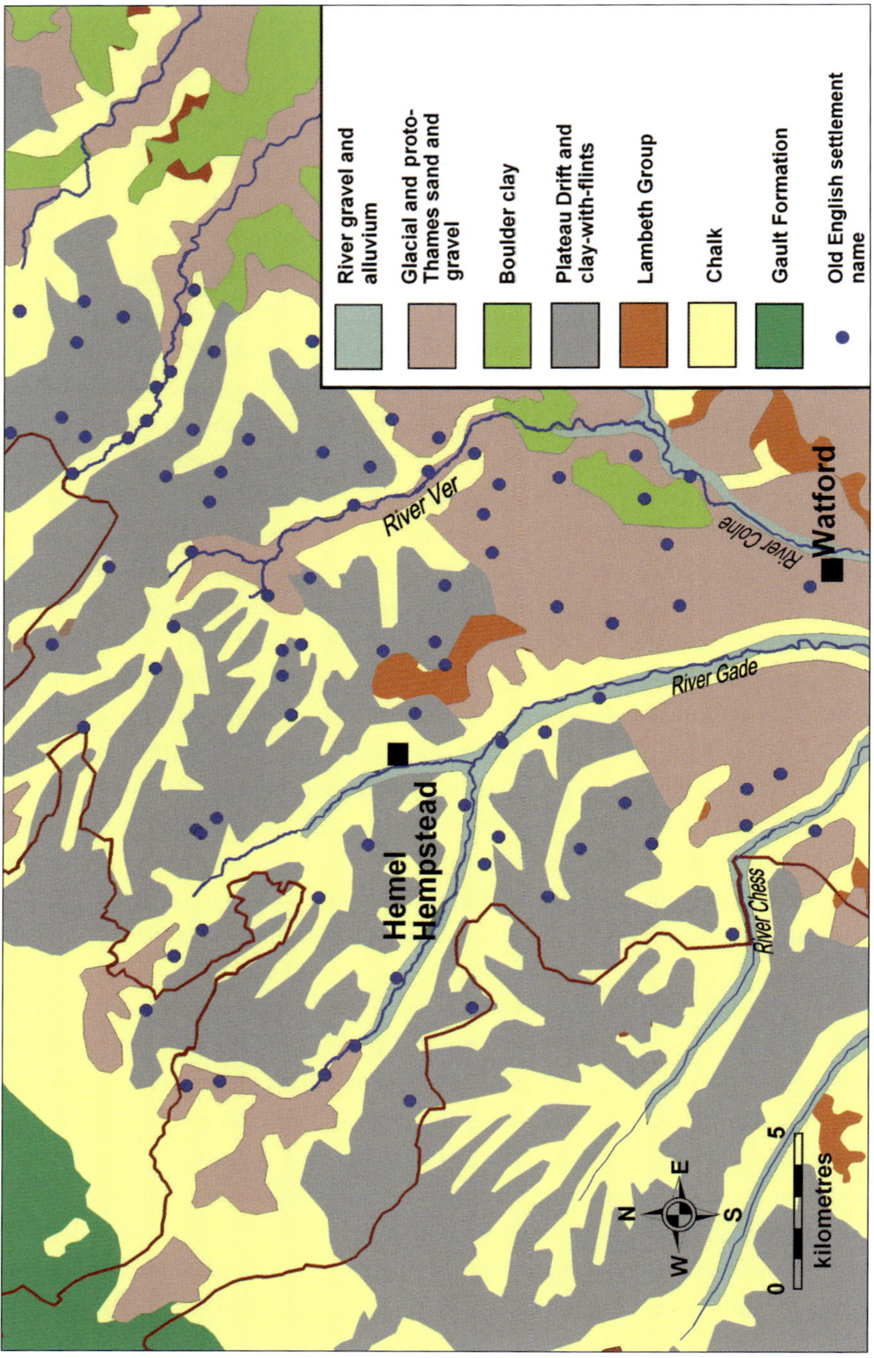

Figure 4.2. The distribution of minor settlements bearing Old English names in the Hertfordshire Chilterns. Some occur beside major watercourses on the floors of the principal valleys; rather more are found on the Plateau Drift of the uplands, although almost always towards the edges of that formation, close to the exposures of chalk.

away from the main river valleys, had individual names in Saxon times likewise implies that the landscape was not clothed in dense and continuous tree cover. One example, running roughly parallel to and to the west of the Gade, was called *Whippa's valley*, and connects Whippendell Wood with Whippendell Bottom, some five kilometres to the north. Another, which ran into it from the west, is the 'boundary valley' mentioned in the pre-Conquest charter for Cashio, an estate embracing what is now Watford.[13] In this context we should also note the paucity of major or minor place-names in the district which contain elements relating to woodland or its clearance, a surprising circumstance if this had, indeed, been densely and continuously wooded country.

What is also noticeable, however, is that the overwhelming majority of settlements with Old English names lying away from the main river valleys are either located on the sands and gravels in the south of the district or are closely associated with the principal dry valleys, where the Chalk is exposed or the clay soils better drained and more calcareous in character. The more remote parts of the drift-covered uplands appear to have been devoid of Anglo-Saxon settlements. There were other subtle variations in the character of the pre-Conquest landscape. In the north-east of the county, as we have seen, there was a tendency for medieval woodland to be located towards the margins of the clay plateaux, with greens and commons (often clearly the remnants of once more extensive tracts of common grazing) towards the interiors of the clay masses: the contrast between open 'cores' and wooded 'peripheries' was ancient, for field names indicating the assarting of woodland are closely associated with the marginal woods rather than the central greens and commons. Such a contrast is less evident on the Chiltern dipslope. Field names referencing the clearance of woodland can be found in the central parts of the uplands, as, for example, in the area between Leverstock Green and Potters Crouch, and in association with commons as much as woods. In part, perhaps, this difference reflects the fact that the Plateau Drift was a less water-retentive formation, ensuring that bringing wood and timber along unsurfaced lanes posed less of a problem. This said, some aspects of this important contrast between 'cores' and 'peripheries' are again evident. While some ancient woods can be found in the central parts of the interfluves the largest areas and biggest concentrations were generally on their margins, extending down onto lower ground, such as Whippendell Wood, Lees Wood, Harrocks Wood and Cartpath Wood near Watford, or Prae Wood, Park Wood, Birch Wood and Appspond Wood to the west of St Albans. Conversely, while there were, and still are, some large commons on the valley sides, such as Sheethanger or Roughdown in Hemel Hempstead, for the most part the larger tracts of common land were, by post-medieval times and still today, located in the centre of the interfluves, such as Harpenden Common or Chipperfield Common. This association, it should be emphasised, is topographic, rather than geologically determined, for while some of these commons overlie pockets of particularly acidic sandy Plateau Drift, unsuitable for cultivation, others are on the clay-rich Plateau Drift, while that at Harpenden occupies, in part, Chalk exposed in a dry valley (the *Herpe denu* from which the place takes its name).[14]

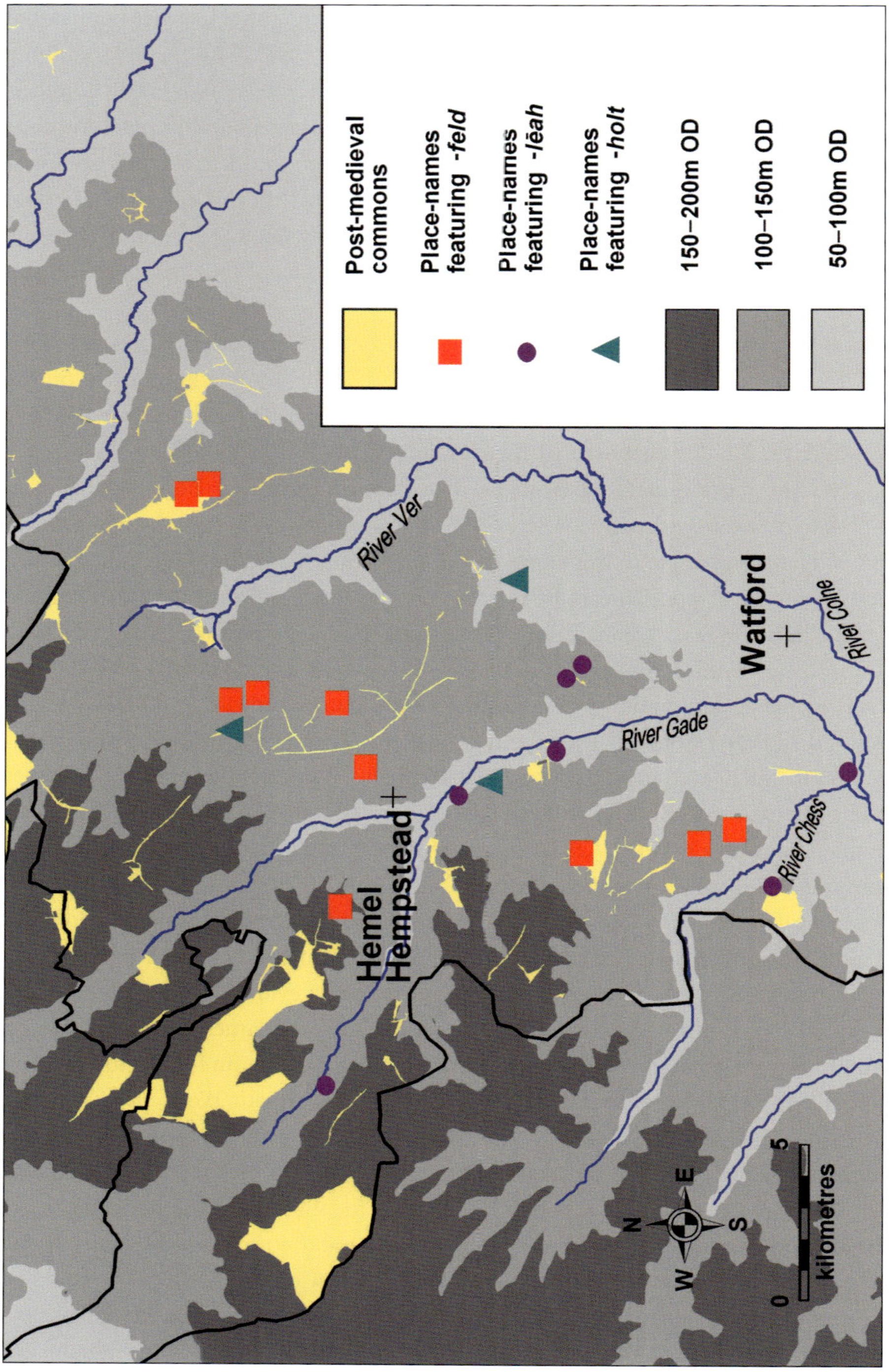

Figure 4.3. Place-names containing Old English terms for woodland on the Chiltern dipslope. Note how those incorporating the element *lēah* tend to occur towards the edges of the clay uplands, while those incorporating *feld* are found towards the centres of interfluves; *holt*, 'a wood', displays a mixed pattern.

Elsewhere, towards the centres of the interfluves, networks of interconnected linear 'greens' exist which appear to represent the tattered fragments of once continuous tracts of grazing, as, in particular, in the area to the north-east of Hemel Hempstead.

An ancient contrast between wooded peripheries and more open interiors is also suggested by place-names. The surprisingly few names featuring the element *lēah*, meaning a wood, wood-pasture or clearing, thus occur in or beside the major river valleys: Apsley, King's Langley, Abbot's Langley, Chorleywood, Ley Green, Croxley and Trowley (in Abbot's Langley). The Old English element *feld*, 'an open area', is found in contrast in minor place-names located toward the *centres* of the upland plateaux, and often in close association with surviving commons. Chipperfield in King's Langley, the 'open area used by traders', still has one of the largest tracts of unenclosed common in the district, while Bamville Wood Farm on the edge of the even larger Harpenden Common (although within Wheathampstead parish) derives its name from *beam feld*, 'the open space marked by a tree', perhaps the original name for the common itself. Micklefield, 'the great *feld*', in the north of Rickmansworth parish, was once associated with the now lost Micklefield Green, while Great and Little Revel End farms in the west of Redbourn – originally *Rutherfeld*, 'the open space for cattle' – lie close to the network of interconnected linear commons already noted which occupies the high ground to the north-east of Hemel Hempstead, as do the settlements of Buncefield and Adeyfield, names for which we have no early spellings, but which may in origin describe *felds* which were by a hillock (*benc*) and dirty (*adel*) respectively. In a similar way, the hamlet of Fields End in the west of Hemel Hempstead, lying as it does at the eastern end of the great interconnected commons to the north of Berkhamsted, may also incorporate the element *feld* (Figure 4.3).

While it is thus possible that, as in the north-east of the county, the central sections of clay uplands were grazed in common by wider communities than the margins, and thus tended to degenerate at an early date to more open ground, it is also evident that many of the larger commons in the area continued to carry a measure of tree cover into the Middle Ages and sometimes beyond. Chipperfield may have become relatively open ground at an early date but Commonwood Common, immediately to the south, clearly did not. Gustardwood Common may have as its first element the OE term *gorst*, 'gorse', a plant of open heath, but the second clearly indicates a measure of tree cover. As we shall see, some of these dipslope commons continued to carry some tree cover well into the nineteenth century, although seldom on the scale of the great wood-pasture commons found on the high Chiltern ridge.

The picture that emerges from an examination of charters and place-names is thus that the landscape of the clay-covered uplands of the Chiltern dipslope, far from comprising a continuous area of unsettled, unexploited and wooded land in Saxon times, was varied in character, with some clearings and open *felds* – especially towards the interiors of the uplands – and with a scatter of small settlements, some perhaps with associated plots of arable land attached, where soils were lighter. This said, there is no doubt that the eleventh,

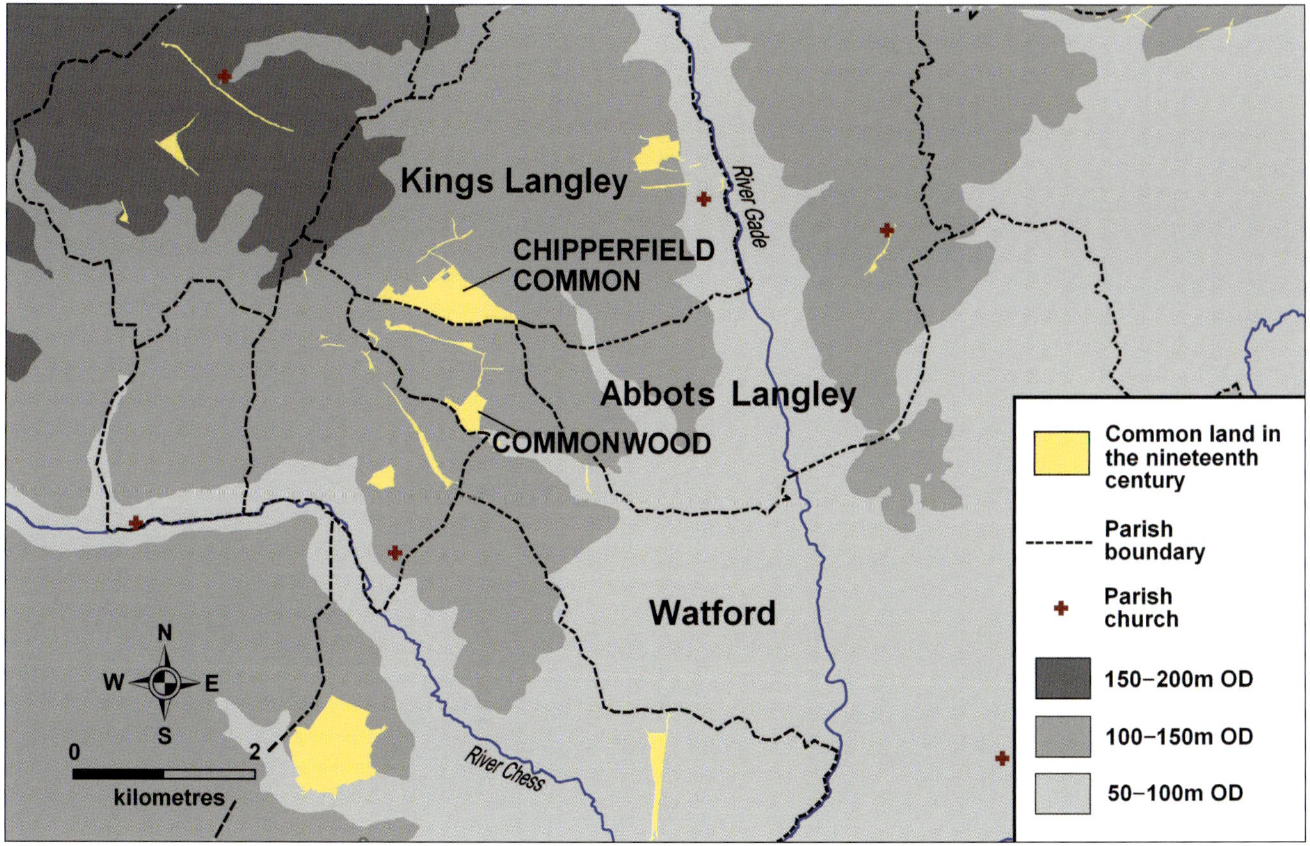

Figure 4.4. The parish boundaries of Watford and Abbot's Langley seem to stretch and distort in order to reach the complex of commons, evidently the shattered remnants of a once more extensive area of 'waste', around Commonwood and Chipperfield. The boundaries were probably fixed on their present line by the end of the twelfth century, suggesting that the contraction of the upland woods and pastures must have occurred fairly rapidly in the period after the Norman Conquest.

twelfth and thirteenth centuries did see an extraordinary expansion of settlement and cultivation, so that woods and common pastures shrank dramatically, although many areas of woodland – by now mainly enclosed and coppiced – and some areas of common land remained. Expansion was probably rapid in the period after the Conquest. It is noteworthy that the parish boundaries of Watford and Abbot's Langley, probably finally fixed on their present lines some time in the twelfth or early thirteenth centuries, appear to stretch and distort in order to reach, and thus obtain a share of, the block of commons around Chipperfield and Commonwood, suggesting that grazing was already in short supply in the district (Figure 4.4). By the thirteenth century the scattered pattern of settlement which today characterises the interfluves was in place, with dwellings sometimes standing alone, on public roads or a little back from them; in small hamlets, usually referred to as 'Ends'; in loose rows along lanes, called 'Rows'; or scattered around the margins of greens and commons. Small, triangular greens of the kind often encountered in the north-east of the county, connected by a maze of winding lanes, are less in evidence in this district. Instead, the commons took the form either – as already intimated – of extensive blocks, as at Harpenden, Chorleywood, Chipperfield or Gustardwood in Wheathampstead, which still cover 1.0, 0.75, 0.45 and 0.43 square kilometres respectively; or of long, narrow, linear 'greens', in effect wide roads. One striking example of the latter ran, and still largely runs, for some four kilometres along the watershed between the Gade and the Ver. Its southern section is called Leverstock Green and its northern High Street Green, the entire road originally being known, on account of its topographic location, as the High Street.

Where particularly extensive areas of common land survived into the twelfth century and beyond they often formed major magnets for settlement, suppressing the growth of villages around churches located beside the major valleys. Sarratt church is thus associated with only a small group of houses, the main focus of settlement being around Sarratt Green, a kilometre away to the north, while Little Gaddesden church stands quite alone in the fields, isolated from the village itself, which is strung around the margins of common land some 750 metres to the south-west. The most important of these upland common-edge settlements is the town of Harpenden. Originally part of Wheathampstead, the size of the place clustered around this vast common was such that, in the course of the twelfth century, it was provided with its own chapel, erected at one end of the common, although it continued to be part of Wheathampstead parish until the nineteenth century.[15]

It is difficult to ascertain the extent or distribution of open fields in west Hertfordshire in the Middle Ages owing to a paucity of pre-nineteenth-century maps and the fact that most appear to have disappeared through informal, piecemeal enclosure rather earlier than was usual in the east of the county. A 1768 survey of the Gorhambury estate, covering a vast acreage in Hemel Hempstead, St Albans, Redbourn and Wheathampstead, shows virtually no open arable; surveys of farms in Hemel Hempstead made in the 1750s and 1760s show only occasional strips; a survey of the manor of Abbot's Langley dating to 1604 records none at all; while as early as 1556 there were in King's Langley only around 30 acres.[16] However,

Figure 4.5. Remnants of the once extensive Chiltern heaths survive around the golf course on Berkhamsted Common.

documents suggest that open fields had been much more extensive in the Middle Ages. A survey of Hemel Hempstead, dated 1523 but apparently based on one of late medieval date, thus describes around 7,000 acres of land, of which 3,000 lay in open fields.[17] As in the east of the county, open arable tended to be most extensive and continuous on the lighter land in the principal valleys which had been brought into cultivation before the later twelfth century. In Sarratt, for example, a glebe terrier of 1610 describes a number of strips all of which appear to have been located in fields immediately above the river Chess, where today one of the finest series of strip lynchets surviving in the county provides striking archaeological evidence for open-field cultivation on the steeply sloping land to the west of the church.[18] Where exposures of light land were extensive open fields sometimes survived on a large scale into the seventeenth century: over a third of King's Walden still comprised open arable in *c.*1600. Here, typically, there were numerous separate areas of open fields of very varying size. The largest and most continuous lay on the slopes of wide chalk valleys and included fields with names directly associating them with early hamlets in the parish – Flexmore and Wandon, both separate vills in Domesday; others, occupying heavier soils on the narrow plateau between valleys, were probably created in the late eleventh or twelfth century and sometimes have names suggesting woodland clearance, such as Redding Field (from the Old English *ryding*, 'cleared land').[19] In all there were probably 30 separate 'fields' in the parish.

The Chiltern ridge

The highest ground of the Chiltern ridge, above *c.*150 metres OD, appears to have formed in early medieval times an almost continuous tract of woodland pasture. This began in the west with Tring common, which merged eastwards into Wigginton common. Beyond the Bulbourne valley lay the conjoined commons of Aldbury and Berkhamsted and, continuing north-eastwards, these were followed in turn by the commons of Hudnall, Studham, Buckwood, Kensworth and Caddington. These vast tracts of unenclosed ground continued across the county boundary into Buckinghamshire, with Cholesbury and Hawridge forming continuations of Tring common and Pitstone and Ivinghoe commons adjoining the north side of Aldbury common. These large areas of common land mostly survived until the late eighteenth century before falling victim to agricultural 'improvement', but parts of Aldbury, Berkhamsted and Hudnall commons remain today, largely in the care of the National Trust (Figure 4.5).

Earthworks of Iron Age and Roman enclosures, fields and lanes have been discovered by Angus Wainwright on Aldbury and Berkhamsted Commons, although these appear to have formed limited 'islands' of enclosed ground within undivided pastures or woodland.[20] Such upland ranches appear to have been abandoned at the end of the Roman period and evidence for early or even middle Saxon settlement on the higher ground is non-existent. The general paucity of Old English names across this high ground, away from the principal valleys, suggests that settlement was sparse and that the uplands were by this time largely exploited in common by communities living on lower ground, in the principal valleys to the south or on the wide plains to the north of the escarpment: the name of Mayne Wood, a former common wood at the south end of Tring parish, may be derived from Old English *gemæne*, 'common' and the same word may also be preserved in Mangrove Green, east of the Lea valley near Cockernhoe.[21]

In Saxon times the upland woods were probably exploited by numerous communities living in the surrounding area but increasing population pressure led to closer definition and, in all probability, some limitation of common rights. One early stage in the parcelling up of the commons can be traced in the configuration of both parochial and county boundaries, which display numerous anomalies and complexities created as distant communities acquired their own defined area of woodland. Mangrove Green, for example, lay in a small detached part of the parish of Lilley within the parish of Great Offley, and no fewer than five parishes staked a claim to the high and steeply sloping pastures between the Gade and the Bulbourne around Potten End, where the boundaries of Berkhamsted, Northchurch, Hemel Hempstead, Great Gaddesden and Pitstone in Buckinghamshire all converge. Yet, while the trend was towards increasing definition, some areas remained common to numerous communities. Even in the mid-sixteenth century rights to use Berkhamsted common were claimed by 2,000 people living not only in Berkhamsted but also in Northchurch, Aldbury, Little Gaddesden, Hemel Hempstead, Bovingdon and Flaunden, as well as in the Buckinghamshire townships of Frithsden, Nettleden, Pitstone and Cheddington.[22] Some Hertfordshire residents had rights of common in woods at Pitstone in Buckinghamshire, and in Buckwood, which lay largely in Bedfordshire.[23] Most of the complex boundaries in the border area between Hertfordshire and its neighbouring counties were simplified in the late nineteenth century. The hamlet of Hudnall had, up until 1885, formed a detached portion of the Buckinghamshire parish of Edlesborough, but was now annexed to Little Gaddesden; ten years later Nettleden, originally a chapelry in the Buckinghamshire parish of Pitstone, became a parish within Hertfordshire; while in 1897 a detached part of the Bedfordshire parish of Whipsnade at Ballingdon Bottom was added to the new parish of Markyate and the parish of Studham, which straddled the county boundary, was transferred wholly into Bedfordshire.[24]

In medieval times commoners enjoyed rights to graze livestock on the commons (herbage), to allow their pigs to forage in the woods in the autumn (pannage) and to collect wood for fuel (firebote).[25] It is likely that the lords of the various manors owned the trees although most were probably managed as pollards, their branches removed for use as fuel by the tenants every few years. The fifteenth- and sixteenth-century grantees of the manors of Champneys and Fosters (which included lands in the parishes of Tring, Aldbury and Wigginton) were entitled to take ten cartloads of wood per annum from the common wood of Tring, and a sixteenth-century lease of lands in Studham included an entitlement to a load and a half of wood yearly from the common wood of Studham.[26] In Saxon times the uplands may have formed one continuous tract of woodland, but by the high Middle

Ages grazing pressure had produced a mosaic of wood-pasture, grassland and heath, which formed once the acidic soils of the Plateau Drift had been cleared of trees. It would have been hard to regulate the exploitation of such remote areas, especially when demand for firewood soared in the twelfth and thirteenth centuries. Many manorial lords, perhaps in response to the dwindling resources, enclosed parts of the commons and managed them as coppiced woodlands, as at Frithsden, where the Great and Little Frithsden Copses were enclosed from the 'common woods or coppices called Southwoods',[27] and at The Leith south-west of Tring, where the tenants retained their grazing rights in the woods but their cattle were excluded for seven years every time the trees were coppiced to protect the regrowing shoots.[28]

Other enclosures were made in the course of the Middle Ages as new farms were carved out of the woods and pastures, and in some areas it is possible to trace stages in this process in the configuration of the local landscape. Hedgeswood Common in the far north of Great Gaddesden is thus clearly a fragment of what was once a much larger area of unenclosed woodland and pasture, approached from the south-east by the ancient lane called Gaddesden Row and encircled on its south and east sides by equally ancient-looking lanes. Hatches Farm, on the edge of the plateau south-west of the common, perhaps marks the site of a gate, or *hœcc*, onto the common land. The area of the former common can be traced to the north, where it was evidently shared by the parishes of Studham, Flamstead and Whipsnade, the latter's share being fossilised, until the nineteenth century, as a detached portion of the parish covering around 160 acres.

This area of land appears to have been held by the priory of St Giles in the Wood from the twelfth to the sixteenth centuries. It was one of many areas granted to medieval religious houses in this relatively empty landscape. The manors of Kensworth and Caddington were, by the eleventh century, the property of the Dean and Chapter of St Paul's, London. Dunstable Priory, founded *c.*1132 by Henry I, held the manor of Studham, as well as land called Bradwin Fields lying near Studham common in Flamstead, land in 'Bradden Wood' (Bradwin's Wood in Flamstead) and Buckwood, and rights of common pasture in both Kensworth and Caddington.[29] The manor of Tring was granted to the monks of Faversham Abbey in the twelfth century and the priory of St Giles in the Wood was founded in the parish of Flamstead, with lands extending into Whipsnade and Studham in Bedfordshire.[30] In the late thirteenth century the manor of Ashridge, with land in the parishes of Berkhamsted, Little Gaddesden and Pitstone, was granted to a religious foundation which became known as the Bonhommes.[31] Sales of timber from the extensive woodlands on these monastic estates provided a valuable source of income and a 'keeper of the woods' was recorded at Tring in 1387.[32]

Encroachments and assarts steadily reduced the extent of the commons while over-grazing and poor regulation ensured that they became increasingly open in character, and by 1538 the antiquarian John Leland, travelling from Markyate to Berkhamsted, described the 'Chilterne-hilles' as 'baren, woody and ferne [ferny] grounde for the most parte, the soile waxing chalky and flinty, as al Chilterne ys'. Beyond Berkhamsted he 'passid by hilly, woody, and much baren ground from Hertfordshire into Buckinghamshire'.[33] Further losses of trees on commons were recorded in the later sixteenth century. 'Divers parcels of wood' – said to be at least 80 years old – were felled and sold for £160 from the 200-acre common 'woodgrounde' called Mayne Wood in the parishes of Tring and Wigginton in 1584. The unauthorised felling of substantial amounts of timber and underwood from the 300-acre 'great wood', another common wood, was recorded in the court rolls of the manor of Caddington in the 1580s and 1590s, resulting in the complete clearance of trees from the area.[34] Significantly, most of the known records of Mayne Wood at Tring relate to grazing rights, not to wood production, and at the beginning of the seventeenth century the common was described as the 'waste or pasture called "The Mayne Wood"'.[35] In the south-west of Tring parish another common wood, the West Wood, covered 300 acres, but here too there were few trees by the middle of the seventeenth century. Indeed, a survey in 1650 valued the wood and trees growing on *all* the waste of the manor of Tring at just £10 and the customary tenants, in addition to the herbage and pannage on the wastes, were entitled to gather not wood but small shrubs and bushes growing on West Wood, for which they had to pay four pennies to the lord of the manor for every load felled and carried away.[36]

As the wood and timber were removed the vegetation on the commons became more heathy in character, with an abundance of gorse, broom, heather and ferns or bracken. By the mid-seventeenth century the southern end of Berkhamsted Frith had lost its tree cover and was marked on a map of 1638 as 'Barkhamsteed Heathe with out trees'.[37] Court rolls for the manor of Berkhamsted contain orders for preserving the 'Furze [gorse] and Heath [heather]' which, by then, were gathered by the commoners for fuel in place of wood: tenants were fined if they sold furze gathered from The Frith to any 'Forrainer' (someone outside the manor) and also if they cut the furze with a bill with a double blade, or with a handle that was more than 12 inches long, although exceptions were made for those over 60, under 14, or infirm.[38] In 1748 Pehr Kalm, a visitor from Sweden, recorded that much of Berkhamsted common 'was overgrown with *Genista spinosa* [gorse], which was here not much over a hand's breadth high because the poor people are continually cutting it down to the ground and taking it home as fuel'.[39] Ferns were also cut for fuel on the commons in Aldbury, Berkhamsted and Tring: Kalm reported that one local considered dried bracken 'among the best of fuels'. It burned with a 'more intense heat than many kinds of wood' and was used for baking bread and in the brick kilns on the adjacent Ashridge estate.[40] As these resources in turn became scarcer in the seventeenth and eighteenth centuries manor courts sought to regulate the quantities that people could take: at Tring it was ordered that 'no person [was] to carry any scrubbs [*sic*] or heath [heather] out of the commons upon carts carriages or on horse back' but only as much as 'they can carry upon their backs'.[41] Aldbury manor court in 1764 ordered that no one could cut furze or fern on the commons of Aldbury and Berkhamsted between 31 July and 1 September each year.[42] The rights of tenants to gather the bushes on Tring common were still being guarded in 1783, when the Tring Vestry had the task of dealing with people from outside the parish who were unlawfully cutting them.[43]

The great tracts of common land provided other resources. On payment of a licence fee, tenants of the manor of Tring were allowed to dig clay, gravel, sand or chalk on the commons, providing that they did not 'sell the same to any foreyner living out of the manor'. Clay and sand were dug in the Mayne Wood and chalk in Allowes Down (Oddy Hill today) in the early seventeenth century,[44] while the West Wood was also dug for clay, sand and chalk in the 1640s by three men, each of whom had a kiln and made bricks and lime.[45] A series of pits shown on the first edition Ordnance Survey map in Shrubbs Wood and the farm called Kiln Farm just to the east on modern Ordnance Survey maps may mark the site of these kilns, and also indicate the former extent of the West Wood. Another Kiln Farm lay at the north end of Wigginton Common in the nineteenth century and an area of pits labelled 'Brickfield' further south on the first edition Ordnance Survey map may indicate the northern extent of the Mayne Wood.

According to David Roden, enclosure in the Chilterns appears to have been widespread by the mid-sixteenth century, reaching a peak in the seventeenth.[46] Enclosing wastes and common woods, however, was usually more difficult than enclosing strips in common arable fields as there were often numerous commoners unwilling to relinquish their rights and attempts to enclose in the Hertfordshire Chilterns met with strong resistance. At Tring 460 acres of the commons called Mayne Wood and West Wood were inclosed into fields early in the seventeenth century but the enclosures were broken down during the Civil War and were not reinstated until the end of the eighteenth century.[47] At Caddington the Dean and Chapter of St Paul's attempted to enclose 400 acres of the common in 1635 after the former wood-pasture had degenerated to 'open scrub and semi-heath' of little economic value. But determined resistance from local yeomen farmers led to three years of legal proceedings and 'violent disturbances in the enclosures'.[48] Despite losing the legal battle, the Caddington yeomen won the war and the common was preserved for a further century and a half. Commoners were not always successful, however. The inhabitants of Berkhamsted failed to prevent Prince Charles from enclosing 300 acres of The Frith when he enlarged Berkhamsted Park in 1619, while inhabitants of Flamstead, Studham and Kensworth lost their rights of common pasture in Buckwood Stubbs and Buckwood Deane in 1634 after a 15-year legal battle against their enclosure by Sir John Ferrers of Markyate.[49]

Co-axial landscapes

The most striking feature of the landscape of west Hertfordshire is its distinctive field pattern. Over large areas of the Chiltern dipslope this still displays a marked, and often rather regular, 'co-axial' form: that is, the landscape has a dominant 'grain' defined by axes which run for longer distances in one direction than another, so that the pattern of roads and fields resembles in plan rather wavy and irregular brickwork (Figure 4.6).[50] These long axes run from the valley floors and up onto the clay-covered interfluves, where they

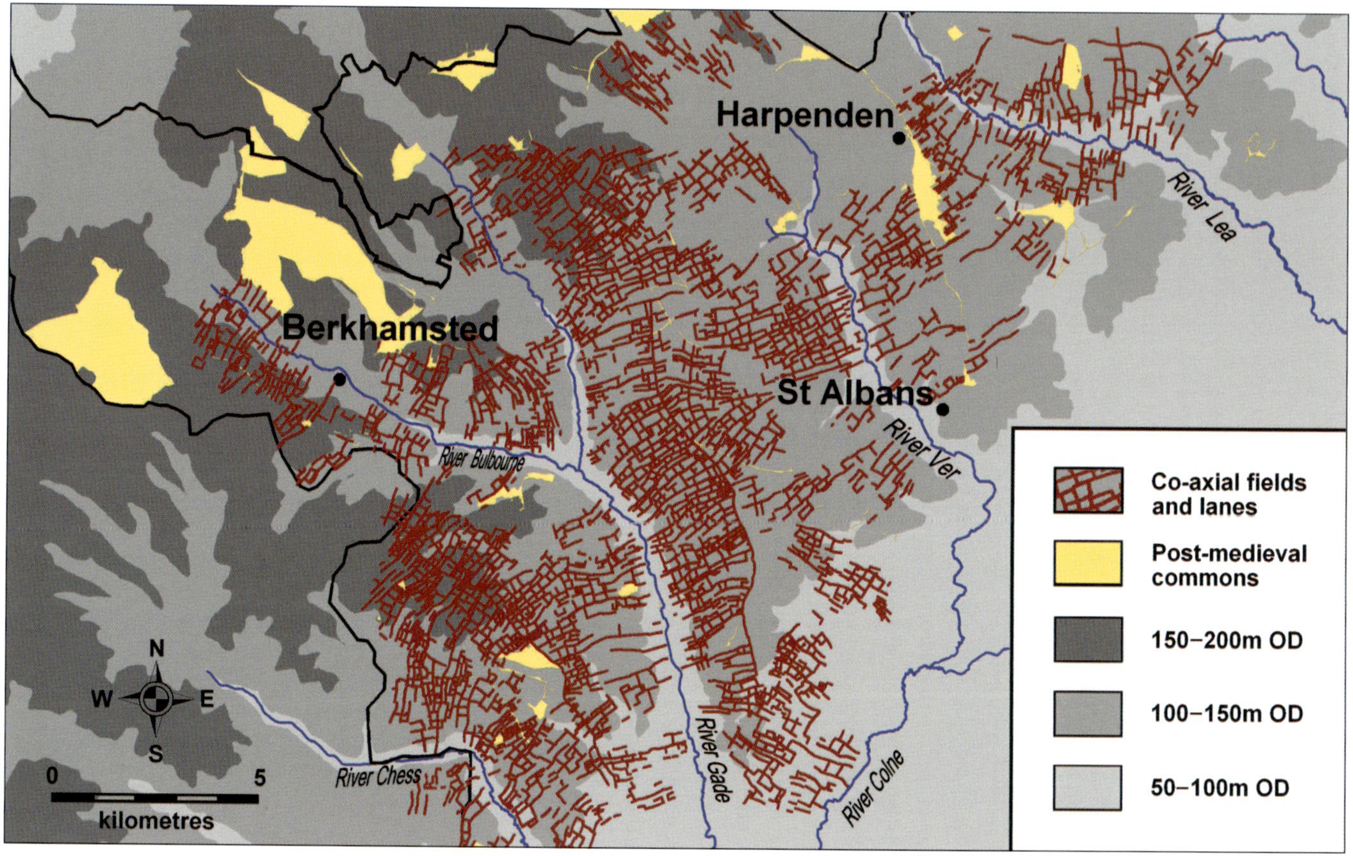

Figure 4.6. The Chiltern dipslope is characterised by distinctive 'co-axial' field patterns. Roughly parallel axes formed by long, slightly sinuous boundaries and lanes run up from the valley floors onto the drift-covered plateaux above: transverse boundaries are generally much shorter, giving the landscape the appearance, in plan, of slightly wavy brickwork. The axes defining each 'system' terminate at continuous lanes which run along the watersheds between the principal valleys. Based on the first edition 6-inch Ordnance Survey, with additions from earlier maps.

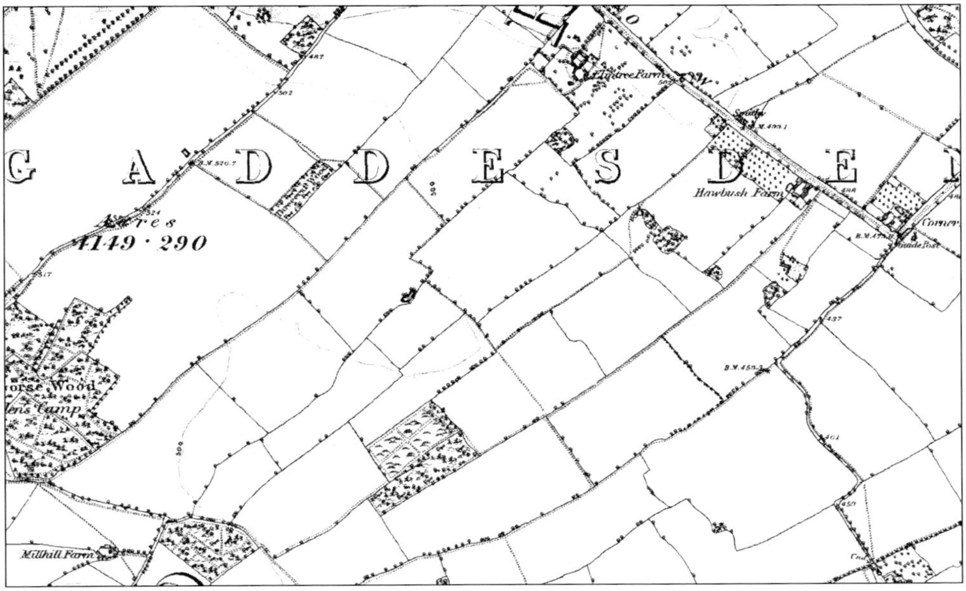

Figure 4.7. Some of the co-axial field patterns on the Chiltern dipslope display a high degree of regularity, such as this example to the south-west of the hamlet of Gaddesden Row, shown on the first edition 6-inch Ordnance Survey, 1884. These fields demonstrate the typical 'wavy brickwork' appearance of such landscapes, with long boundaries trending south-west to north-east, linked by much shorter boundaries at right angles.

appear to terminate at long trackways running along the watersheds between valleys. Co-axial field patterns running up from the Gade and the Chess thus meet at a long trackway, sometimes a linear common, which runs as a continuous line for some eight kilometres from Chandlers Cross to Bovingdon, and with signs that it formerly extended further in each direction; while similar patterns running up from the Gade and the Ver meet at a feature which begins as the linear common already described, running from High Street Green to Leverstock Green, and which continues to the south as a lane – now Bedmond Road, Tibbs Hill Road and Langley Lane – as far as Leavesden, giving a total length of around 10 kilometres. This route may originally have continued further to the north of High Street Green, linking up with the stretch of lane along which the hamlet of Gaddesden Row is scattered.

On the interfluves between the Chess and Gade and the Gade and Ver these arrangements of co-axial fields and watershed tracks are clear, dominant and, in some places, as to the south-west of Gaddesden Row, highly regular in character (Figure 4.7). Between the Ver and Lea and the Lea and Mimram, they become less clear, continuous and regular, but the landscape still displays a marked parallelism and there are, again, traces of the long watershed tracks. Moving towards the east and south such patterns rapidly disappear – they are a particular feature of the Plateau Drift and Chalk countryside. They also become less apparent to the north-west, towards the higher ground of the Chiltern ridge, tending to disappear around the sources of the rivers, even though their valleys continue, in 'dry' form, beyond this. There are, however, in some places signs of co-axial arrangements lying parallel to the valleys floors, running up towards the high ridge to the north-west.

Similar co-axial field patterns are known from many parts of England, including Norfolk, Suffolk, Essex, Cambridgeshire and Herefordshire.[51] Field systems of prehistoric date, surviving as cropmarks or in the form of low earthworks or tumbled walls, also often take this distinctive form – most notably the 'reave systems' of Bronze Age date on Dartmoor, which each cover several square kilometres of ground.[52] It has been suggested that some of the examples existing as hedged fields in the modern landscape may, in fact, represent much-altered survivors from prehistory. In parts of Norfolk and Essex early Roman military roads appear to slice through such arrangements in a way analogous to a railway line or bypass, leaving awkward corners, in such a manner as to suggest that the fields were earlier, and thus of prehistoric date.[53] Looked at in this way, the distinctive Chiltern field patterns superficially appear to represent the remains of a series of huge planned field systems laid out in remote prehistory.

There are, however, problems with such an interpretation. The extremely low levels of population and plough teams evident in Domesday Book and the extensive areas of woodland it records clearly suggest that the area over which these field patterns extend cannot have been cleared in prehistory and continuously cultivated since. Indeed, they sometimes include fields with names indicating woodland clearance in the early medieval period. An undated early eighteenth-century map of a farm in Flaunden, for example, shows two fields in part of a co-axial 'system' called 'Reading', a map of the Gorhambury estate surveyed in 1768 shows 'Upper and Lower Readings' with another near Gustard Wood in Wheathampstead, while examples west of St Albans include 'Long Readings' near Westwick Row and Ninnings (from Old English '*inning*') a little to the north of Bedmond.[54] It is possible, as John Hunter suggested of similar field patterns on the heavy clays of south Essex, that boundaries laid out in remote prehistory, but abandoned to woodland at the end of the Roman period, may have been re-employed as land was returned to cultivation in the early Middle Ages because it was easier to dig out existing silted ditches than to create new ones from scratch.[55] But there is another and more plausible explanation for these distinctive landscapes.

We have already noted the contrast between the river valleys, with good supplies of water and ribbons of light soil, and the extensive clay- and gravel-covered uplands. The major valleys appear to have been continuously occupied and cultivated, albeit with varying degrees of intensity, from prehistoric times; the uplands, in contrast, were mainly exploited as woodland and grazing before the high Middle Ages. Many of the main axes of these co-axial systems are formed by tracks and lanes, such as Bunkers Lane in Hemel Hempstead or Chipperfield Lane in King's Langley; early maps, such as a survey of land at Fields End in Hemel Hempstead made in 1764, generally suggest that more of them took this form even in the relatively recent past.[56] Systems of parallel tracks probably developed because there was a recurrent direction of movement between lower and higher ground, as wood and timber were brought from uplands down to valley settlements, and as livestock – including pigs – were repeatedly driven to the upland woods and pastures and back again. This said, some of the long axes may not have functioned as tracks or lanes, but only as boundaries, dividing resources between different communities living within the valleys. Only from the eleventh century, as more intensive forms of land use were adopted and more and more land was brought into cultivation, did these loose networks of lanes and boundaries gradually become infilled with fields, the boundaries of which replicated the dominant 'grain' of the surrounding features. Most, if not all, examples of 'co-axial landscapes' in England probably developed in this way, for all are found in similar topographic contexts, where valleys form ribbons of light land separated by wide clay uplands, or where escarpments of light soils are capped by drift-covered plateaux.

The Chiltern co-axial landscapes are thus not planned systems of prehistoric land division but organic developments structured by topography, the early forms of land use which this engendered and the gradual intensification of exploitation over time. Nevertheless, given that they derive from patterns of land use and movement which may be of extreme antiquity, it remains possible that some of their elements may have originated before the Middle Ages. Angus Wainwright surveyed a complex of late prehistoric and Roman earthworks – lanes, boundaries and enclosures – preserved on Ashridge and Berkhamsted Commons and in the adjacent woodland of the Ashridge estate, and many of these appear to represent continuations of the surviving co-axial patterns in the surrounding countryside (Figure 4.8).[57] The widely spaced tracks and small enclosures separated by undivided ground indicate the kind of landscapes out of which the denser mesh of the extant 'field systems' must have developed. Charles Le Quesne's excavations at The Grove, in the Gade valley just to the north of Watford – which involved the systematic stripping of around four hectares prior to development – revealed lost elements of the local co-axial field pattern which had been removed when the landscape park was laid out here in the 1760s. Several of these ditches contained, in their basal fills, sherds of Iron Age pottery apparently deposited when still relatively new.[58] Moreover, some small sections of these landscapes are so regular in character that they may have been laid out in planned form in the past, within a wider framework of sub-parallel lanes, such as the area to the south-west of Gaddesden Row.

The date and origin of the continuous watershed tracks which appear to form terminations of the Chiltern 'systems' present more of a problem. One interpretation is that these originated not simply as transport routes but as boundaries which divided the woods and pastures of the 'uplands' between those dwelling in adjacent valleys: early systems of land division often included features which served both as roads and as boundaries, such as the late prehistoric linear patterns on the Yorkshire Wolds, some of which survive as modern roads.[59] Such a dual function might arise because a boundary line provided a neutral passage between territories occupied by different social groups. Elsewhere

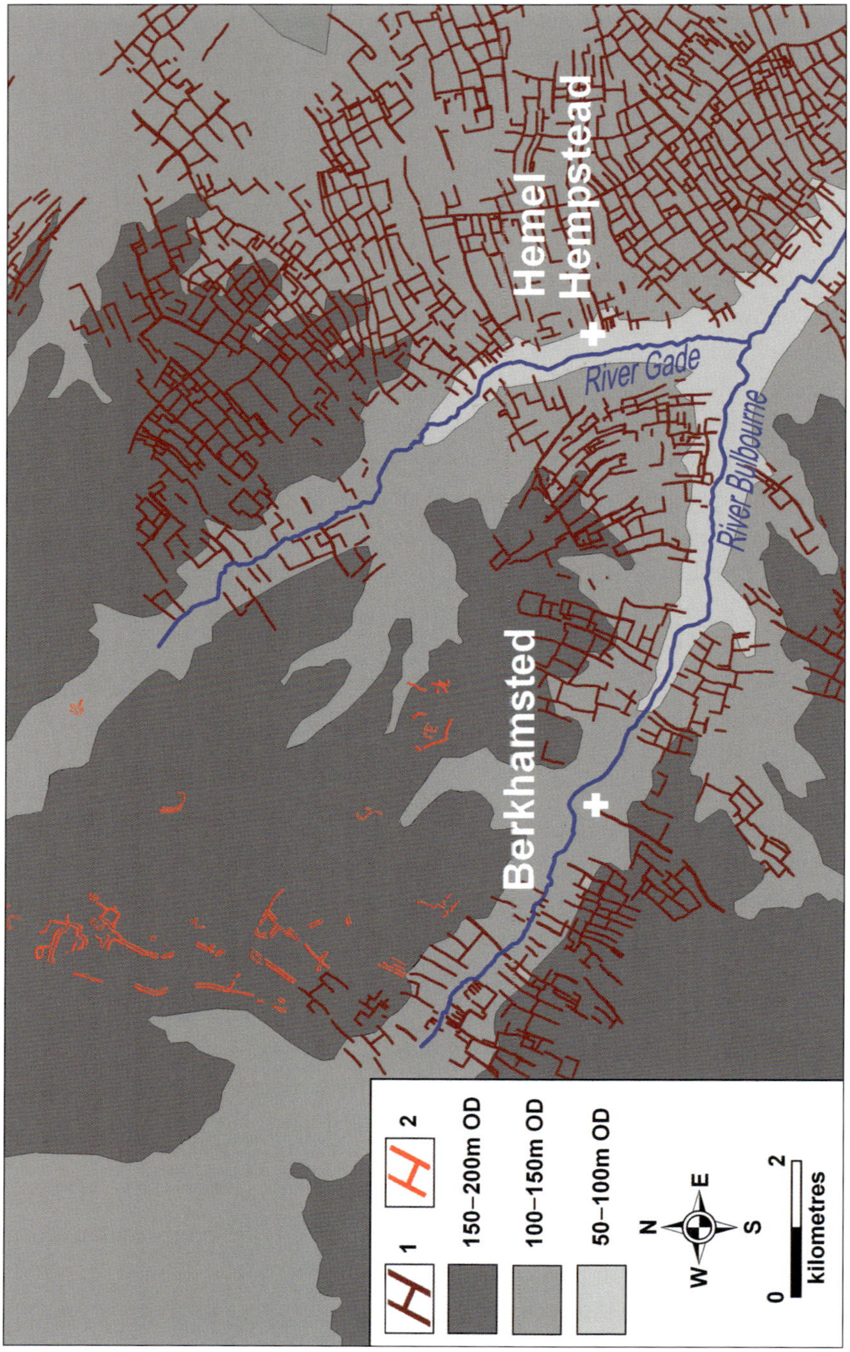

Figure 4.8. Detail from Figure 4.6, showing co-axial field patterns, defined by modern lanes and boundaries, in the area around Berkhamsted and Hemel Hempstead (1). Angus Wainwright discovered a number of earthworks, representing lanes and enclosures of Iron Age and Roman date, buried in woodland on the Ashridge estate (2). These appear to share the same general alignment as the surviving co-axial boundary patterns, suggesting that elements of the latter may likewise have late prehistoric origins.

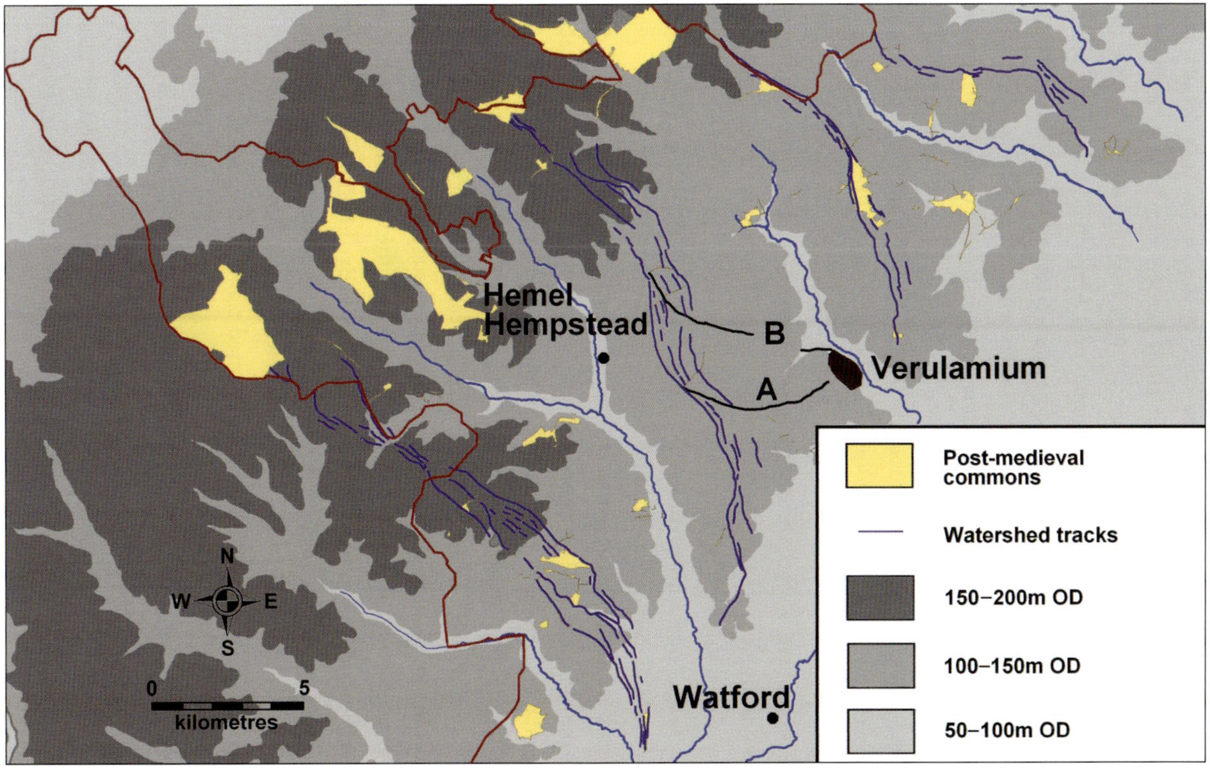

Figure 4.9. Careful inspection suggests that the single roads which now run along the watersheds, such a striking feature of the landscapes of the Chiltern dipslope, were originally bundles of tracks and ribbons of waste providing access to the woods and pastures on the high Chiltern ridge. Many of the principal upland commons lie within these bands of former trackways and presumably represent vestigial traces of these broad corridors of movement. The features marked 'A' and 'B' are long, continuous roads which still run from the watershed between the Ver and the Gade down towards the Roman town of Verulamium. They appear to have originated as routes along which livestock, especially pigs, could be driven from the site of the old Roman town – the Saxon settlement of Kingsbury – up onto the watershed tracks, and thence northwards to the woods and pastures of the Chiltern ridge.

in Yorkshire, on the Tabular Hills on the southern edge of the North York Moors, late prehistoric dyke systems seem to have developed from earlier, Bronze Age territorial divisions which were marked only by round barrows.[60] It is striking, therefore, that the few surviving round barrows on the Chiltern dipslope, on Chipperfield Common and at Highfield to the east of Hemel Hempstead, all stand beside watershed tracks. But there are signs that the history of these distinctive upland routes may be more complex, and that they too developed gradually and organically over time. It is noteworthy that they sometimes (as at Chipperfield or Harpenden) widen into large commons, and often themselves take the form of linear greens, suggesting that as the woods on the interfluves gradually degenerated through pressure of exploitation, and as the frontiers of enclosure and cultivation expanded and the networks of co-axial fields spread up the valley sides, filling the spaces between parallel trackways, the upland wastes shrank in stages to bands and ribbons, and only finally, in most places, to single tracks. At the same time, access to the remaining commons and pannage woods several kilometres away on the high Chiltern ridge to the north-west would have become increasingly important to the valley-based communities, so that these upland ribbons of pasture became long-distance routes for moving stock, especially pigs – animals difficult to herd through the more crowded landscapes, with crops and fields and settlements, in the valleys. In this context it is noteworthy that the watershed tracks often have long boundaries, or lanes, lying parallel to them, which together form a broad band within which the co-axial patterns of fields, lying at right angles are less marked, or absent altogether (Figure 4.9). Some of these parallel features, such as East Lane in Bedmond, run for more than a kilometre without interruption, suggesting that there were originally *bundles* of tracks running along the watersheds, which the continued expansion of farmland up the valley sides eventually reduced to just one. By the thirteenth century, when local documents become available, the Chiltern commons were generally exploited only by the villages lying relatively close to them. But, as we have already seen, even in the sixteenth century it was said that 2,000 individuals claimed rights in the commons to the north of Berkhamsted, including some dwelling as far away as Bovingdon and Flaunden.[61] The landscape evidence suggests that in Saxon times, when the ridge-top commons were more extensive, they may have been more widely shared.

What is also interesting is that the ridgeway tracks are in places joined by other prominent, continuous routeways which sweep up from the valleys, cutting at an awkward angle through the co-axial patterns on the valley sides. These appear to be later additions to the landscape, as they were clearly only ever single features, rather than bundles: and they make their way unerringly to the final, single surviving watershed track. One – the modern A4147 Hemel Hempstead Road – leaves the High Street at Leverstock Green and runs in a great bend – south then east – directly to the west gate of Verulamium ('A' on Figure 4.9); another, less continuous, leaves the High Street further north and may originally, before the creation of Gorhambury Park, have led to the north gate ('B' on Figure 4.9). It is possible that the walls of the Roman town were used, in late Saxon times, in part to corral stock, especially pigs, and that these two routes were established only after the abbey acquired Gaddesden (to which the watershed track directly leads) from Æthelgifu in *c.*980.[62]

Although cleared and settled at a relatively late date, many aspects of the local landscape were thus established in pre-Conquest times. In particular, the movement of livestock by Saxon farmers is deeply etched in the modern pattern of roads and boundaries. The bundles of watershed tracks have had a major impact on the pattern of settlement, as they have decided the location of most of the major population centres not positioned beside the major rivers, such as Bovingdon, Abbot's Langley and Harpenden. And people living in houses erected in the 1960s on Chambersbury Lane or St Albans Hill in Hemel Hempstead New Town walk out each morning onto trackways first established, in all probability, centuries before the Norman Conquest.

Recent history

As in the east of the county, the basic pattern of settlement, roads and boundaries was firmly in place by the end of the Middle Ages: in fact, this was even more true in the west, as the enclosure of open fields came earlier here, in part because of the possibilities this offered for improving this relatively infertile land, stimulated by the proximity of the London market. Seventeenth-century surveys show that arable land use predominated in the district, although there was also much pasture and meadow. On the demesne of Abbot's Langley in 1604, for example, around 500 acres were arable out of a total of 800, the rest being woodland and grass, while wheat, oats, peas and vetches were the principal crops.[63] But in the eighteenth century barley became more important, the arable acreage increased and the new crops of the 'agricultural revolution' were widely planted.[64] Indeed, sainfoin and clover were being cultivated in the locality by the 1670s, turnips by the 1690s (William Ellis, the eighteenth-century agricultural writer from Gaddesden, described in 1736 how they had been grown for forty years even in the open fields).[65] By the end of the century 'improved' rotations, featuring regular courses of cereals, clover and turnips combined in various ways, were widely adopted. This allowed more stock to be kept, especially young sheep and cattle destined for the London market, and this in turn allowed more manure to be produced, which increased arable yields on these poor soils. But fertility was also, on the interfluves, limited by the acidic nature of the soils formed in the drift deposits, and the eighteenth and nineteenth centuries appear to have witnessed an increase in the scale of 'chalking', the practice of spreading the underlying Chalk on the surface of the fields in order to neutralise acidity. Pehr Kalm observed in 1748 how 'In many places between St Albans and Hempstead we saw chalk being spread on the land as a fertiliser. The arable plots were mostly chalky white with it.'[66] The district is liberally peppered with 'dells' from which the chalk was extracted, although Arthur Young also described how chalk was obtained by digging vertical shafts, leading to networks of tunnels locally referred to as 'dene holes'.[67]

New rotations and chalking were both easier to practise where land lay in hedged fields in individual occupancy, and the seventeenth and eighteenth centuries accordingly saw the piecemeal, gradual enclosure of almost all remaining open fields, so that only small areas were enclosed by parliamentary act, such as the 63 acres of Watford Fields removed in 1855.[68] Some of the great commons on the high Chiltern ridge were enclosed by act, most notably Tring Common in 1804, Kensworth in 1798 and Caddington in 1800;[69] but most of those which existed on the dipslope in 1700 were unaffected, although some experienced extensive encroachment. Hudnall, Aldbury and Berkhamsted Commons also survived, although the last of these only by a whisker. In 1865 the second Earl Brownlow enclosed around a third of it with iron fences five feet high. It was an illegal act, for he had bought out most, but not all, of the commoners: Augustus Smith, a wealthy local resident (and Lord of the Scilly Isles), had not parted with his common rights and on the night of 7 March 1866 brought 120 labourers by train from London to break down the fence. In a subsequent court case, in which the newly formed Commons Preservation Society was much involved, the earl was defeated.[70] The commons of west Hertfordshire were beginning a new career, as an important recreational resource for an increasingly suburbanised district.

In the course of the eighteenth and nineteenth centuries the proximity to London influenced the development of local agriculture to an increasing extent. The continued expansion of the capital and its suburbs, coupled with increasing levels of groundwater pollution, led to the outward migration of watercress production, which had become a specialised activity at the end of the eighteenth century. West Hertfordshire was ideal for cress growing because of the purity of its chalk streams, the waters of which – fed in part from springs in the valley sides – were comparatively warm in the winter months.

Figure 4.10. Watercress beds on the river Mimram at Whitwell, one of only three such businesses still functioning in the Hertfordshire Chilterns.

The Sanson family established the first commercial beds in the county at West Hyde near Rickmansworth in the 1830s. By the end of the century watercress beds, with varying arrangements of channels and lagoons, were densely scattered along the rivers draining the dipslope. The second half of the twentieth century, however, saw a sharp decline in the industry, caused in part by foreign competition and in part by increasingly strict legislation concerning water quality. Now only the beds at Sarratt in the valley of the Chess and those at Kimpton and Whitwell on the Mimram remain in production (Figure 4.10).[71]

The great agricultural depression which began in the late 1870s affected west Hertfordshire more than the east. The Plateau Drift soils were more agriculturally marginal and the climate was slightly, but significantly, wetter. In 1941 Cameron was able to describe how 'less than half a century ago' the entire area, other than woods and commons, was under cultivation, but 'the amount of tillage on the Clay-with-Flints has now declined so much as to be negligible', although cereals continued to be grown on the lighter land.[72] There was a significant increase in dairying, providing milk for London, and in poultry-keeping and market gardening, stimulated in part by the activities of the County Council's smallholding schemes, which were established under a series of government acts from the late nineteenth century, although largely perhaps by the markets provided by London and its suburbs.[73] The most dramatic expression of this pattern of early twentieth-century diversification is, perhaps, the Ovaltine Egg Farm, built in King's Langley in the 1930s, although this owes its existence primarily to the proximity of the Ovaltine factory in the valley of the Gade below. Indeed, while changes in agriculture – including a re-expansion of the area under cereal crops after 1940, associated to some extent (as in the east of the county) with hedge removal – continued to shape the landscape in the twentieth century, the growth of towns and industry and the spread of suburbia were now more important influences, as we shall see.

Notes

1. Williamson, *Origins of Hertfordshire*, pp. 141, 152–3, 205; Page (ed.), *VCH Hertfordshire*, vol. 2 (London, 1908), pp. 296, 216, 187.
2. Hodge *et al.*, *Soils and their use in eastern England.*
3. Catt and Cheshire, 'Quaternary deposits', pp. 118–82; 130–36, 146, 152.
4. Bailey, 'Introduction', pp. xxi–ii.
5. A.E. Levett, *Studies in manorial history* (Oxford, 1983), p. 180.
6. Gover *et al.*, *Place-names of Hertfordshire*, p. 29.
7. *Ibid.*, p. 102.
8. H.T. Riley (ed.), *Gesta Abbatum Monasterii S. Albani*, vols I and II, Rolls Series (London, 1867), p. 63.
9. Mainly based on Gover *et al.*, *Place-names of Hertfordshire.*
10. Gover *et al.*, *Place-names of Hertfordshire*, p. 19; Crick, *Charters of St Albans*, no. 7.
11. Gover *et al.*, *Place-names of Hertfordshire*, p. 92; Crick, *Charters of St Albans*, nos 7 and 9; Riley, *Gesta Abbatum* I, p. 54.
12. Crick, *Charters of St Albans*, nos 1 and 7.
13. Gover *et al.*, *Place-names of Hertfordshire*, p. 46; Crick, *Charters of St Albans*, no. 1.
14. Crick, *Charters of St Albans*, no. 11; Gover *et al.*, *Place-names of Hertfordshire*, pp. 37–8.
15. Page (ed.), *VCH Hertfordshire*, vol. 2, p. 296; N. Doggett, 'The medieval chapels of the Chilterns: the evidence from south-west Hertfordshire', in Holgate (ed.), *Chiltern archaeology*, pp. 113–17.
16. HALS DE/V/P2 a survey of the estates of The Right Honourable James Lord Viscount Grimston by J.L. Davis, 1768; 76831 plan of Haxten-end, Little Heath and Hill-end Farms, Northchurch and Hemel Hempstead, 1758; 79703 copy plan of Agnells Manor, Hemel Hempstead, 1645; AH/681 plan of Fields End Farm, Hemel Hempstead, 1764; 80251 Corner Farm, Hemel Hempstead, 1772; Valor of the manor of Abbot's Langley, *c.*1604, TNA:PRO SP14/11/22, quoted in G. Longman, *A corner of England's garden: an agrarian history of south west Hertfordshire 1600–1850*, vol. 2 (Bushey, 1977), p. 5; King's Langley survey 1556 discussed in L.M. Munby (ed.), *The history of King's Langley* (King's Langley, 1963), pp. 41–2, 57.
17. Munby, *Hertfordshire landscape*, p. 165.
18. Longman, *Corner of England's garden*, 2, p. 6.
19. Roden, 'Field systems of the Chiltern Hills', pp. 325–76, 331.
20. Morris and Wainwright, 'Iron Age and Romano-British settlement and economy', pp. 68–75.
21. For an interpretation of the name 'Mens' see O. Rackham, *Ancient woodland: its history, vegetation and uses in England* (Dalbeattie, 2003), p. 175; M. Gelling and A. Cole, *The landscape of place-names* (Stamford, 2000), pp. 226–9. The *grāf* element of Mangrove may indicate that the wood was coppiced.
22. Page (ed.), *VCH Hertfordshire*, vol. 2, p. 162.
23. HALS DE/FL/15719 Mortgage 'of land formerly part of a wood called Buckwood', 1631.
24. W. Page (ed.), *VCH Buckinghamshire*, vol. 3, pp. 350–51; Page (ed.), *VCH Hertfordshire*, vol. 2, p. 317; W. Page (ed.), *VCH Bedfordshire*, vol. 3, pp. 455–7; Page (ed.), *VCH Hertfordshire*, vol. 2, p. 274.
25. *Calendar of Charter Rolls, vol. 2, Henry III, 1257–1300* (London, 1906), p. 324. Among the rights granted by the Earl of Cornwall to the brothers of the religious foundation at Ashridge were the right of common pasture for their cattle, together with free pannage for their pigs, in Berkhamsted Frith; S. Hindle, 'Persuasion and protest in the Caddington common enclosure dispute 1635–1639', *Past & Present*, 158 (1998), pp. 48–9 citing D. Roden, 'Woodland and its management in the medieval Chilterns', *Forestry*, 41 (1968), p. 68. 'Customary tenants [of the manor of Caddington] had claimed since at least 1297 "to take freely in the woods according to the nature of their holdings"'.
26. HALS 44831 grant of the manors of Champneys and Fosters, 1427; 46126 grant of the manors of Champneys and Fosters, 1546; DE/FL/17075 lease of lands in Studham, 1578.
27. HALS AH/1163 re Manor of Frithesden, 25 Sept 1672 reciting Articles of Agreement (to Inclose); South Woode Corner was marked on a map of the area of *c.*1638 (HALS 1985).
28. TNA:PRO E134/43Eliz/East3 legal depositions concerning a warren and ground called the Leith, 1601.
29. Page (ed.), *VCH Bedfordshire*, vol. 3, pp. 371–7; Page (ed.), *VCH Hertfordshire*, vol. 2, p. 231; HALS DE/FL/17246 grant of 'Bradwynfelders, Flamstead, formerly the property of Dunstable Monastery', 1545; DE/FL/18159 mortgage of 'property in Flamstead, near Studham Common, including the two Home Closes, formerly part of Bradwin Fields', 1756.
30. Page (ed.), *VCH Hertfordshire*, vol. 2, pp. 282–3; HALS DE/FL/17248 lease of the 'manor house of Beechwood, late called the Priory of St Giles in the Wood', 1549. This probably relates to the detached portion of Whipsnade occupying the south-west corner of the nineteenth-century Beechwood Park.

31. *Calendar of Charter Rolls, vol. 2*, p. 324.
32. S. Richards, *A history of Tring* (Tring UDC, 1974), p. 19.
33. M. Tomkins, *So that was Hertfordshire: travellers' jottings 1322–1887* (Hertford, 1998), pp. 5–6 quoting from L. Toulmin Smith (ed.), *The itinerary of John Leland in or about the years 1535–43* (London, 1907), pp. 104–5.
34. TNA:PRO E134/26Eliz/Hil7 and E134/MISC/2415 legal depositions concerning a wood called Mayne Wood, in the parish of Tringe, parcel of the manor, 1583–4; Hindle, 'Caddington common enclosure dispute', pp. 48–9.
35. TNA:PRO E134/4Jas1/Mich19 legal depositions concerning damage to the pasture on the common caused by rabbits from a neighbouring warren, 1606–7. Several deponents refer to 'the wood called Mayne Wood'.
36. TNA:PRO E317/Herts/29 parliamentary survey of the Manor of Tring, 1650, p. 18.
37. HALS 1985 photograph of a map of 'Barkhampsted Parke', *c.*1638 (original at The National Archives).
38. HALS DE/Hn/Z5 notes about Berkhamsted Common including extracts from the manor court rolls in April 1666 and October 1726.
39. W.R. Mead, *Pehr Kalm: A Finnish Visitor to the Chilterns in 1748* (Aston Clinton, 2003), p. 84. Although of Finnish descent, Kalm was born and resident in Sweden.
40. *Ibid.*, pp. 51–2.
41. HALS DE/X1010/M1 Tring manor – court book with Tring Rectory, 1691–1708.
42. HALS AH/2790 extracts from Court Rolls as to Tenants and Inhabitants of the Manor turning on the Common and cutting of Furze and Fern.
43. HALS DP/111/8/19 vestry minutes 1782–1815, see pp. 6, 15.
44. Richards, *History of Tring*, pp. 60–61 citing TNA:PRO SC2/178/70 Tring estreats 1615–16 & 1621–3.
45. TNA:PRO E317/Herts/29, pp. 15–16.
46. Hindle, 'Caddington common enclosure dispute', p. 44, citing D. Roden, 'Enclosure in the Chiltern Hills', *Geografiska Annaler*, 52B (1969), p. 119.
47. TNA:PRO E317/Herts/29, pp. 17–18.
48. For a full account of the dispute see Hindle, 'Caddington common enclosure dispute', pp. 37–78.
49. H. Falvey, 'Crown policy and local economic context in the Berkhamsted Common enclosure dispute, 1618–42', *Rural History*, 12 (2001), pp. 131–42; Hindle, 'Caddington common enclosure dispute', p. 60, citing TNA:PRO, SP14/124/65; Bedfordshire Record Office, DDBS 18,974–19,012 and DDBS 19,058.
50. Based on the Ordnance Survey first edition 6-inch maps, the various tithe maps of the district and the following archive maps: HALS DE/V/P2; 76831; 79703; AH/681; 80251; AH/680, plan of Boxted farm, Hemel Hempstead, 1735; 7568 plan of Langley Lodge and Balls Pond Farms, King's Langley, 1797; DE/X905/P1 map of farm in Flaunden, early eighteenth century; DE/Fn/P1 plan of Redheath estate in Watford, 1760; 57310 map of Little Gaddesden, 1802.
51. W. Rodwell, 'Relict landscapes in Essex', in H.C. Bowen and P.J. Fowler (eds), *Early land allotment* (London, 1978), pp. 89–98; P. Drury and W. Rodwell, 'Settlement in the later Iron Age and Roman periods', in D.G. Buckley (ed.), *The archaeology of Essex to AD 1500*, Council for British Archaeology Research Report 34 (London, 1980), pp. 59–75; T. Williamson, 'Early co-axial field systems on the East Anglian boulder clays', *Proceedings of the Prehistoric Society*, 53 (1987), pp. 419–31; T. Williamson, 'The "Scole-Dickleburgh Field System" revisited', *Landscape History*, 20 (1998), pp. 19–28; A. Davison, *The evolution of settlement in three parishes in south east Norfolk*, East Anglian Archaeology 49 (Dereham, 1990), pp. 73–4; M. Hesse, 'Fields, tracks and boundaries in the Creakes, south Norfolk', *Norfolk Archaeology*, 41 (1992), pp. 305–24; S. Oosthuizen, 'Prehistoric fields into medieval furlongs? Evidence from Caxton, south Cambridgeshire', *Proceedings of the Cambridgeshire Antiquarian Society*, 86 (1998), pp. 145–52; S. Oosthuizen, 'The roots of common fields: linking prehistoric and medieval field systems in west Cambridgeshire', *Landscapes*, 4 (2003), pp. 40–64; S. Harrison, 'Open fields and earlier landscapes: six parishes in south east Cambridgeshire', *Landscapes*, 3 (2002), pp. 35–54; S. Bryant, B. Perry and T. Williamson, 'A "relict landscape" in south east Hertfordshire: archaeological and topographic investigations in the Wormley area', *Landscape History*, 27 (2005), pp. 5–15; P. White, *The Arrow Valley, Herefordshire: archaeology, landscape change and conservation* (Hereford, 2003), pp. 37–47, 73–5.
52. A. Fleming, 'The prehistoric landscape of Dartmoor: Part 1, south Dartmoor', *Proceedings of the Prehistoric Society*, 44, pp. 97–123; A. Fleming, *The Dartmoor Reaves* (London, 1988).
53. Rodwell, 'Relict landscapes in Essex'; Drury and Rodwell, 'Settlement in the later Iron Age and Roman periods'.
54. HALS DE/X905/P1; DE/V/P2; J. Field, *English field names: a dictionary* (Newton Abbot, 1972), p. 150.

55. J. Hunter, *Field systems in Essex* (Colchester, 2003), p. 35.
56. HALS AH/681.
57. Morris and Wainwright, 'Iron Age and Romano-British settlement and economy', pp. 68–75.
58. Le Quesne *et al.*, 'Grove Estate'.
59. C. Stoertz, *Ancient landscapes of the Yorkshire Wolds: aerial photographic transcription and analysis*, English Heritage (Swindon, 1997).
60. D. Spratt, *The linear earthworks of the Tabular Hills, North Yorks*, Department of Archaeology and Prehistory, University of Sheffield (Sheffield, 1989).
61. Page (ed.), *VCH Hertfordshire*, vol. 2, p. 162.
62. Crick, *Charters of St Albans*, no. 7.
63. Valor of the manor of Abbot's Langley, *c.*1604, TNA:PRO SP14/11/22, quoted in G. Longman, *Corner of England's garden*, p. 5.
64. Longman, *Corner of England's garden*, pp. 28–64.
65. W. Ellis, *New experiments in husbandry* (London, 1736), pp. 12–13.
66. Mead, *Pehr Kalm*, p. 44.
67. Young, *Agriculture*, p. 159.
68. Page (ed.), *VCH Hertfordshire*, vol. 2, p. 449; HALS QS/E70A–B Watford enclosure award and map, 1854.
69. Page (ed.), *VCH Hertfordshire*, vol. 2, pp. 281 and 187; HALS DE/X437/Z13 Kensworth Inclosure Act, 38 George III c67, 1798.
70. B. Cowell, 'The Commons Preservation Society and the campaign for Berkhamsted Common, 1866–70', *Rural History: economy, society, culture*, 13 (2002), pp. 145–61.
71. S. Fletcher, 'Watercress growing', in Short (ed.), *Historical atlas*, pp. 94–5.
72. Cameron, *Land of Britain*, p. 321.
73. For information about the county council smallholdings, see M. Bowyer, *We have to deal with the farmers. Episodes in the history of North Hertfordshire in the 19th and 20th centuries* (Girton, 2010).

– 5 –

The landscape of south Hertfordshire

Introduction

The area lying to the south and east of the river Colne and to the south and west of the Lea is today the most suburbanised and built-up part of Hertfordshire, largely a consequence of its proximity to London. Indeed, the southern extremities of the ancient county are now physically part of outer London, and some parishes here were transferred to Greater London by local government reorganisations in the later twentieth century. In this latter district much of the land is so densely packed with housing that it is difficult to appreciate the character of the local topography and, in particular, the fact that the ground rises in places to a height of over 150 metres above sea level, although both residents and visitors are aware of the often steep inclines, for the numerous streams feeding the Colne and the Lea, or flowing south to the Thames, have carved steep-sided valleys. In places, however, it is still possible to enjoy extensive views, especially northwards across the Vale of St Albans. Further north there are more tracts of open ground between the suburbs and housing estates, some occupied by golf courses and playing fields but others remaining as farmland, usually under pasture. Here it is often possible to experience something of the rural landscape that the suburbs replaced, although always with the sound of near and distant traffic coming from the many main roads, or from the M25, which slices its way through much of this open land. Further east, however, beyond Potters Bar and north of the M25, the growth of suburbia has been much more localised and largely confined to the Lea valley, leaving extensive tracts of farmland – an attractive countryside of relatively small, hedged fields and large areas of woodland.

As always, the key to understanding the landscape lies in the character of geology, soils and topography. Most of the district comprises a dissected plateau of London Clay, a formation which gives rise to heavy, poorly draining, relatively infertile and acidic soils. On the highest ground the clay is capped in many places by spreads of Pebble Gravels, giving rise to soils even more acidic and even less fertile in character – Arthur Young considered them the 'most unfertile' to be found in the south of England[1] – and here extensive tracts

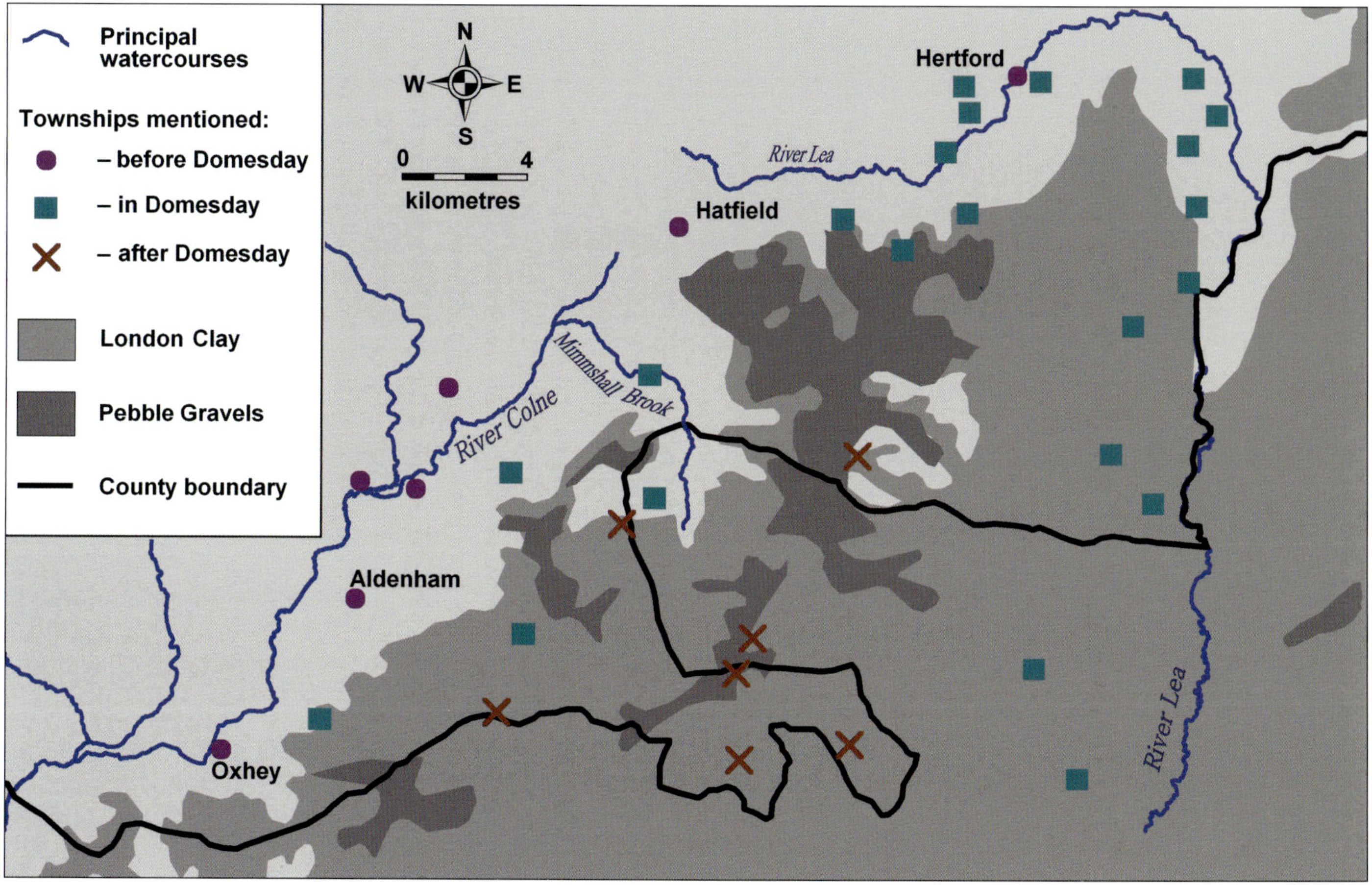

Figure 5.1. The chronology of settlements in south Hertfordshire. The locations shown are those of parish churches, or of manors now bearing the name of the places in question.

of heathland often survived into the nineteenth century. Only on the lower ground, in and beside the Vale of St Albans and the valleys of the rivers Colne and Lea, are lighter and more amenable soils to be found, formed in exposures of Chalk, in glacial sands and gravels or in the clays and sands of the Lambeth Group. As we shall see, this contrast, between this lower ground and the higher, less fertile land, is deeply etched in the fabric of the landscape.

Early settlement

Not surprisingly, it is from the low-lying areas in Vale and valleys that most of the evidence for prehistoric and Roman settlement has been found, including a number of Roman villas, such as the excavated examples at Netherwilde Farm and Park Street.[2] Although some Roman settlements are also known from the broad clay 'uplands' many of these appear, significantly, to have been involved in pottery production, exploiting the local clays and presumably using the local woodlands as fuel.[3] Minor settlements – farms and hamlets – with Old English names, some referred to in pre-Conquest documents, are likewise quite densely packed on the lower ground, close to the rivers. Spaced at intervals of two or three kilometres along each bank of the Colne we thus have places such as Brightwells Farm and Oxhey Hall in Watford, 'bright spring' and 'enclosure for oxen' respectively; Bushey Hall, the original focus for Bushey parish; Wall Hall near Aldenham (although actually in the parish of St Stephen's, St Albans), 'the valley of the slaves or Britons'; *Achurst* in Shenley, 'the oak wood', mentioned in a will of *c.*980; and Broad and London Colney, 'the island in the Colne'.[4] Some at least of these places may have developed directly from significant Roman settlements: some, certainly, are found in close proximity to them. The probable sites of *Achurst*, Brightwells Farm and Munden all lie less than 200 metres from known villa sites. Even more striking is the fact that all the settlements which bear the names of estates described in pre-Conquest charters or other documents, including Oxhey in Watford, Hanstead in St Albans, Aldenham, Napsbury, Hatfield and, of course, Hertford, are likewise located in close proximity to the two rivers (Figure 5.1).[5] The vills mentioned by Domesday were also mainly located within the major valleys, although their distribution is slightly different, as many – such as Shenley or Wormley – lie a kilometre or so back from the rivers, on the rising ground overlooking them, often at the junction of the Lambeth Group sands and clays and the underlying Chalk, and close to the margins of the London Clay.

Away from these settled margins, in contrast, the difficult and unrewarding claylands are a blank on the map of Domesday settlements. Only one Domesday township – *Titeberst*, the modern Theobald Street in Aldenham – was located more than a kilometre from exposures of light valley soils, and this was a small place with only a small area under cultivation.[6] Medieval settlements in the area almost all have names with strong woodland associations, such as Northaw, the 'northern woodland enclosure', or Barnet, the 'place cleared by burning'. Many feature the Old English term *lēah* – Shenley, the 'bright *lēah*', Barnetley (an earlier name for Chipping Barnet) and Wormley, the 'snake-infested *lēah*'.[7] Place-name scholars often translate *lēah* as either 'wood' or 'clearing' but Della Hooke has suggested that it was used to describe 'open woodland or wood-pasture'.[8] Yet it is clear that the uplands were already being used, and exploited, on some scale in later and probably in middle Saxon times, and it is likely that some isolated farms already existed at the time of Domesday, even in the more remote parts of the clay uplands, subsumed within the descriptions of the manors located on lower ground. Settlements such as Elstree and Northaw, that is, may well have existed from a very early date as isolated ranches, perhaps seasonally occupied. Indeed, although there are fewer minor settlements – farms and hamlets – on the uplands than in the neighbouring river valleys that bear names featuring Old English elements, such places do nevertheless exist. The majority again incorporate elements with woodland associations. The manor of Weld in Shenley and Ridge, for example, is derived from the Old English *Wald*, 'tract of upland forest'; Borehamwood in Elstree is the 'wood by the hill homestead'; Rowley Green in Shenley is the 'rough *lēah*'; Cuffley is 'Cuffa's *lēah*', while Arkley, originally a hamlet in Barnet but created as a distinct parish in 1894, has a name of uncertain significance although again evidently incorporating the element *lēah*.[9] Monken Hadley, on the county boundary with Middlesex, is referred to in a charter of 1005 as the *hæð lēah*, the 'heathy *lēah*': it lay on the northern boundary of a great tract of outlying woodland and pasture attached to Kingsbury in St Albans, granted by King Æthelred to the monks of St Albans. This embraced, as Taylor has shown, most of what is now Barnet, between the parishes of Edmonton in the east and Totteridge in the west.[10] The heathy '*lēah*' was probably not an actual settlement at this time – it is likely that most of the Old English names attached to minor settlements scattered across the uplands may have been coined originally to describe topographic features, rather than farms or hamlets. Nevertheless, the naming of ridges, separate portions of woodland and even – in the case of Elstree, *Tiðulfes treow* ('Tithwulf's tree'), individual trees – indicates that this was not, in Saxon times, an untenanted, unvisited wilderness.[11]

More detail about the character of the middle and later Saxon landscape of the uplands is provided by the boundary clauses of early charters. The markers on the charter of 1005 relating to the tract of woodland attached to Kingsbury included a burial mound, a 'forest gate', several other gates, a stile, a drainage ditch and a wood-bank. Where the bounds of Aldenham, described in a charter of uncertain but evidently pre-Conquest date, ran through the remote clay uplands near the county boundary with Middlesex they passed the 'gate to the hazel wood', the *fulen gate*, or 'dirty gate', and a hedge.[12] Those of Oxhey, a few kilometres to the west, set out in a charter of 1007,[13] similarly included – again within the high, unsettled country near the county boundary – a *wudwyrðe* ('enclosure in the wood') and *mapuldorgeat* ('deer gate made of maple wood', suggesting perhaps some feature like a deer leap: *geat* often means something like 'gap in a fence or bank through which only particular animals could pass').[14] The presence of gates, stiles, banks, hedges and the rest indicate that subdivision of the uplands was necessary both to assign particular areas to different individuals and groups, and perhaps to allow parts to be managed in particular ways, as woodland, grazing and so on. Indeed, woodland was not simply a valueless resource, waiting to be turned into farmland, and it retained an economic importance throughout the Middle Ages, particularly so given the proximity of London, with its insatiable demands for fuel. In the thirteenth century tenants of the manor of Barnet held their land by the service of carrying firewood from the demesne wood in Barnet to the hospice of the abbot of St Albans in London when requested to do so by the bailiff.[15]

As in the Chilterns – although on a larger scale and at an earlier date – religious houses often acquired extensive tracts of this marginal and under-exploited land. St Albans thus held not only the area around Barnet already described, including the woods of 'Suthawe' and 'Huzeheog', confirmed to the monks in a charter of *c.*1176,[16] but also Northaw, the *sylvam quae dicitur North Haga*, which the monks managed to retain against the claims of Robert Fitzwalter in the late eleventh century.[17] Domesday records that they held land in 'Shenley' but this was probably what became the parish of Elstree, which was confirmed to the monks by a papal grant of 1188, together with the 'wood of Boreham' to be used as swine pasture: this may have been granted to the abbey in the later tenth or early eleventh century,[18] likewise remaining a chapelry until promoted to full parochial status in 1424. Domesday records that the abbey held part of *Titeberst*, the modern Theobald Street, while the village of Ridge, first mentioned in 1248, was also one of its possessions.[19] It formed part of the manor of 'Ridge or Tyttenhanger', the latter a hamlet to the north of the river Colne, in St Peter's parish in St Albans, and may have been granted to St Albans at the same time as the woods around Barnet, attached to Kingsbury;[20] it similarly remained a chapelry annexed to St Albans until the early fourteenth century.[21] The abbey of Ely also held land in these remote uplands. The manor of Hatfield, granted to the abbey in the late tenth century, extended for some distance eastwards onto the upland clays, while the abbey had also acquired, by the time of Domesday, the manor of Totteridge on the boundary with the county of Middlesex, some 15 kilometres to the south of Hatfield. It remained a detached chapelry of Hatfield parish until 1892, the chapel originally dedicated – like the church at Hatfield itself – to St Etheldreda, the founder of Ely Abbey.[22] Waltham Abbey held the manors of Wormley and Brickendon, both granted to the canons of Waltham Holy Cross before the Conquest; the abbey later acquired the manors of Little Amwell, perhaps in the twelfth century, and Cullings, in Cheshunt, in 1387.[23] Hertford Priory, a cell of St Albans Abbey established around the end of the eleventh century, held lands in Cheshunt and Hoddesdon and woodlands at Hertford Heath and in Broxbourne.

One of the striking features of the district is the way that the parishes attached to settlements in the Colne or Lea valleys tend to have long, attenuated shapes, reaching far up onto the clays, as if to stake a claim to the varied resources which these had to offer.[24] This striking parallelism is shared more widely by the pattern of roads and field boundaries. As in the Chilterns, the movement of timber, wood and livestock to and from the upland woods and grazing grounds evidently generated in many places a 'co-axial' pattern, the antiquity of which is demonstrated by the way in which the parish

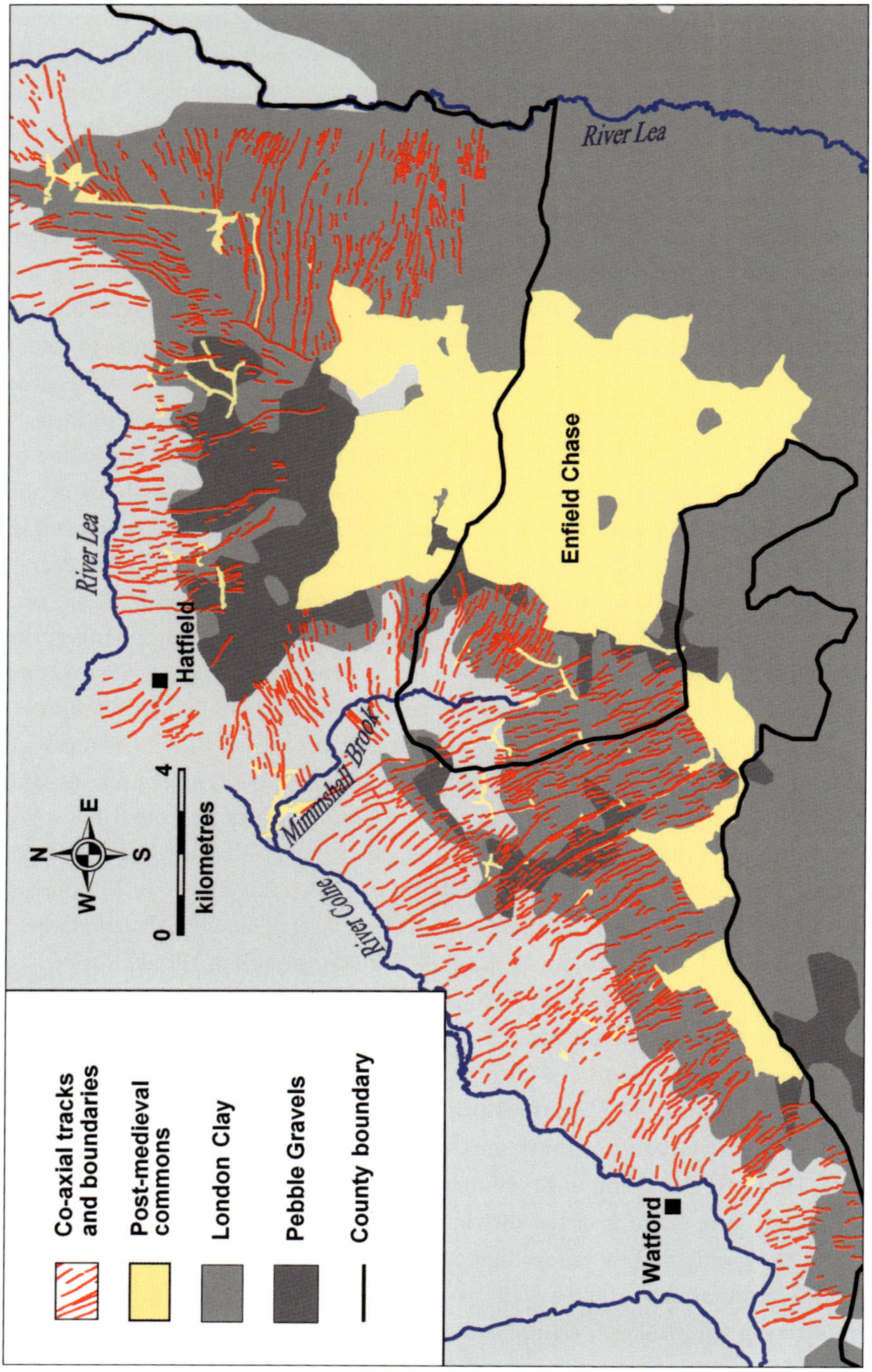

Figure 5.2. Co-axial and sub-parallel lanes and boundaries, and the principal areas of common land in the eighteenth century, in south Hertfordshire. Some boundaries, lanes and areas of common land in the adjacent parts of Middlesex (most notably the extensive commons of Enfield Chase) are also shown.

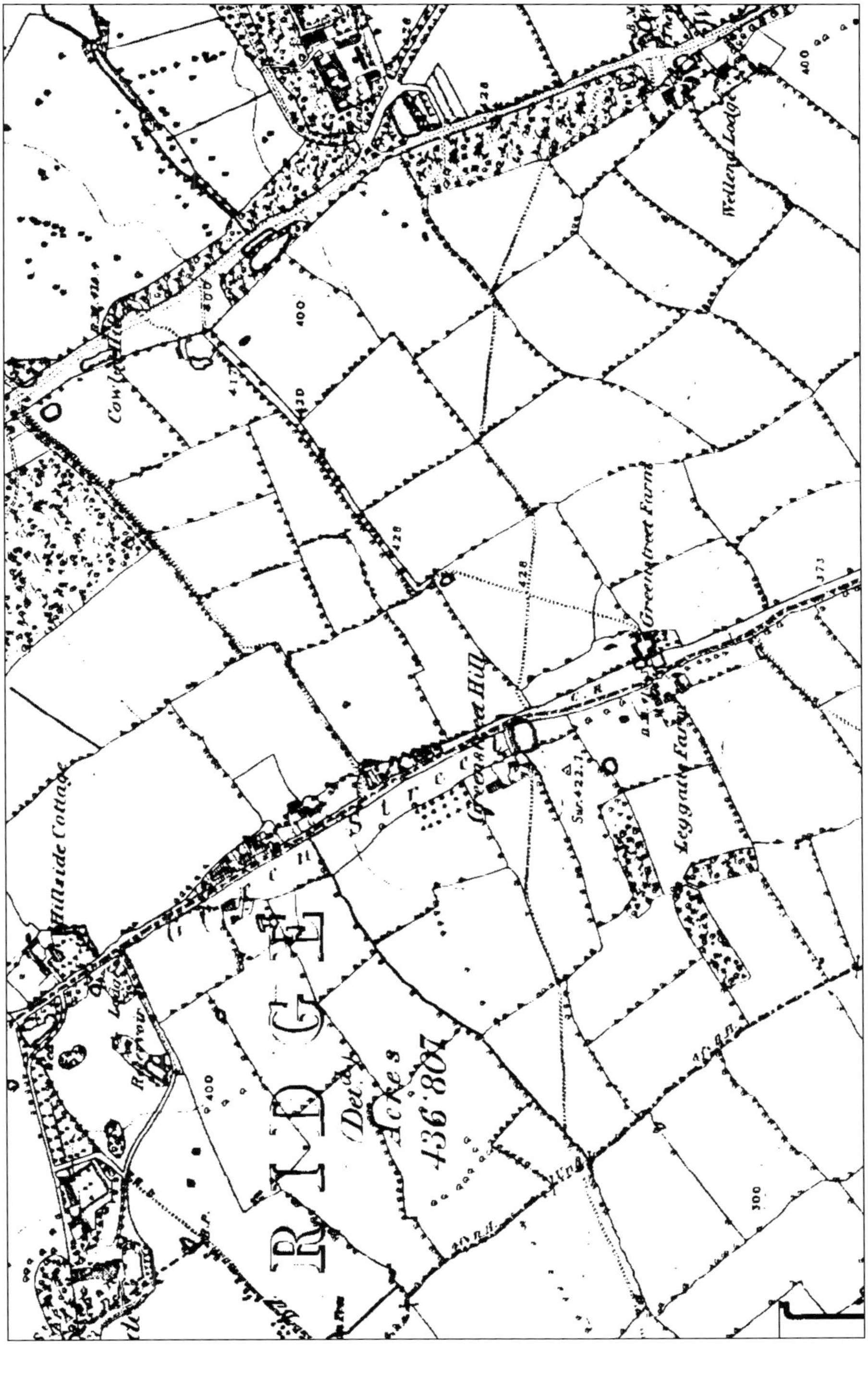

Figure 5.3. Parallel boundaries and lanes in Ridge and Shenley. Note how Green Street is flanked for much of its length by narrow, strip-like fields, showing how it was once a much wider drove way.

boundaries appear to join and leave major axes in such a manner as to leave no doubt that the latter were already established features of the landscape when the former came into existence, relatively early in the Middle Ages (Figure 5.2). However, the parallelism appears in general less regular than in the Chilterns, the lanes and boundaries less striking in plan. Moreover, while in the Chilterns linear greens and commons tend to be found in the centres of interfluves, often forming the terminations for blocks of 'co-axial' fields, here in the south of the county many of the main axes running up from the river valleys themselves took, and sometimes still take, the form of wide droves or linear commons. The configuration of field boundaries suggests that many others originally did so, but have been reduced to single, relatively narrow lanes through progressive encroachment (Figure 5.3). These differences are probably to be explained by topography and by the extremely uninviting character of the soils in the south of the county. The London Clay uplands were both more extensive and less attractive to farmers, compared with the adjacent valleys, than the Plateau Drift interfluves on the Chiltern dipslope. Tracks meandered across the heavy clays for a greater distance from the valley settlements, becoming less evenly spaced as they did so. In addition, as the areas between them gradually became filled with a network of enclosed fields, drainage was a much more important consideration for farmers on this particularly impermeable ground than it was on the Plateau Drift of the Chilterns. Field boundaries, defined by deep drainage ditches as well as by hedges, were thus more likely to be aligned in such a manner as to take maximum advantage of the natural gradients than to be laid out parallel with the dominant 'grain' of the adjoining lanes. In addition, the topography is more complex than in the Chilterns, featuring a number of stream valleys, including those of the Mimmshall Brook and the St Catherine Bourne, which have carved deep valleys running often at an awkward angle to the broader sweeps of topography, thus obscuring and confusing the simplicity of a landscape developing from a contrast between river valleys and uplands. Subsidiary co-axial 'systems' thus run out of the valley of the Mimmshall Brook, particularly to the north-east and east, occupying much of the parishes of North Mymms and South Mimms, and confusing the overall 'grain' of the landscape, which otherwise in this area runs at right angles to the Colne valley to the north.

This said, there are places on the Southern Uplands where highly regular and extensive co-axial patterns did develop, as, for example, in the area of Ridge discussed by Jonathan Hunn, or on the western side of the Lea valley, in the parishes of Cheshunt, Wormley and Broxbourne.[25] And, as in the Chilterns, there are hints that some elements of these parallel landscapes may be of extreme antiquity. Excavations of a Roman settlement beside the course of Ermine Street in the middle of Cheshunt Park, for example, revealed ditches which shared the orientation of the surrounding co-axial landscape, and which contained first- and second-century pottery in their basal infill: they had apparently been filled in during the construction or expansion of the settlement.[26] Portions of this particular 'field system' are so regular that they may

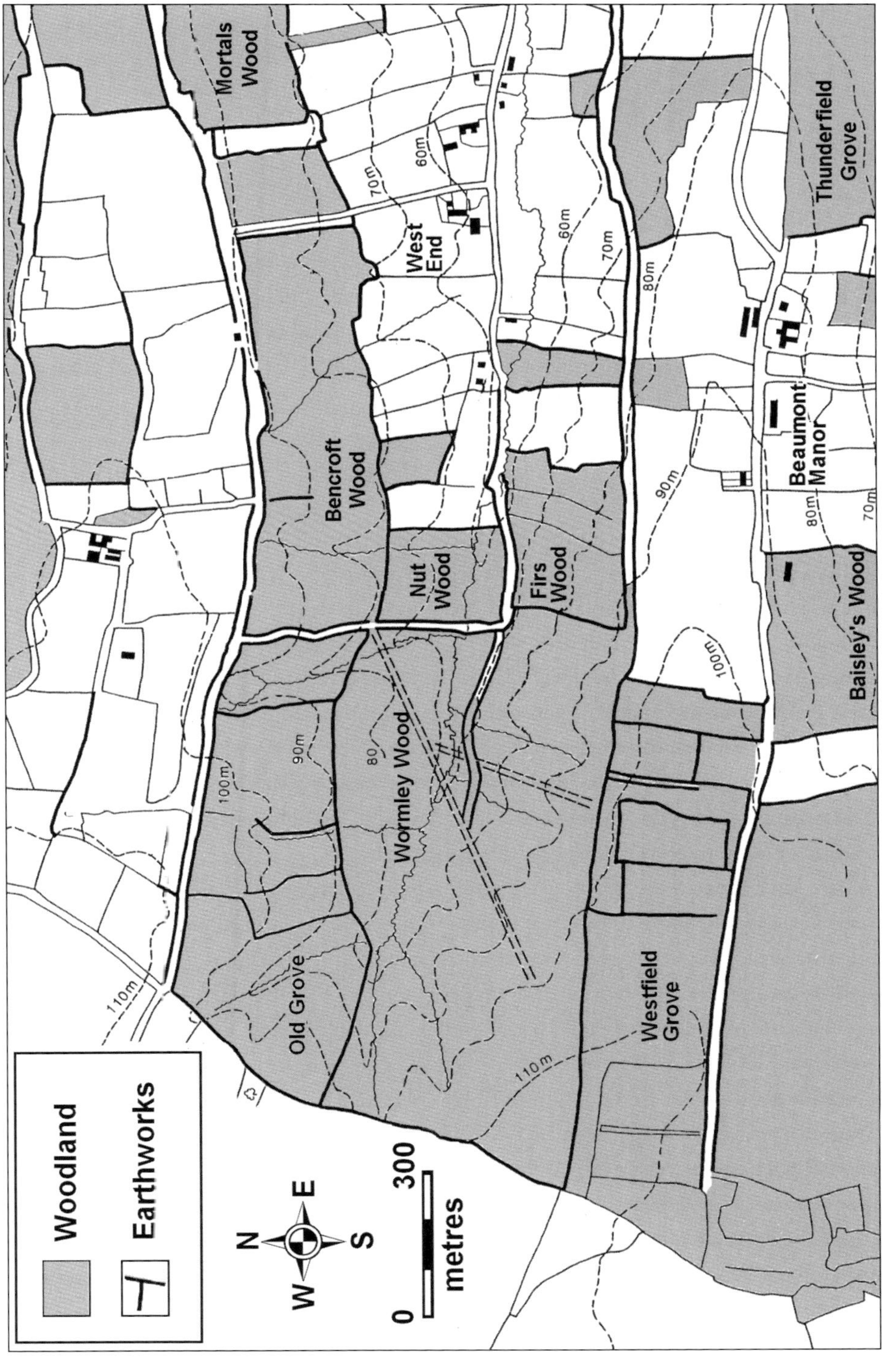

Figure 5.4. Co-axial lanes, field boundaries and earthworks in the west of Broxbourne, Wormley and Cheshunt.

well have been planned in the remote past. In the area between Beaumont Manor and Westfield Grove, for example, two principal axes maintain, almost exactly, a separation of 275 metres for nearly two kilometres (Figure 5.4). There is a great deal that we still do not understand about these striking field patterns: they offer ample scope for further research by Hertfordshire's local and landscape historians and archaeologists.

Patterns of boundaries and lanes in the district have other stories to tell. In the area immediately to the south and west of the river Lea a string of parishes – Essendon, Little Berkhamsted, Bayford, Hertford All Saints, Hertford St John, Broxbourne, Wormley and Cheshunt – all have roughly parallel boundaries which run up onto the clay uplands, apparently towards a block of woodland which, originally shared by all, was annexed at some time in the remote past by just one parish – Hatfield – which extends, in elongated fashion, far to the east. It is possible that this annexation occurred very early indeed, for the *Liber Eliensis* describes how, when King Edgar gave the 40 hides of Hatfield to the Abbey of Ely in *c.*970, he did so 'because, since the country is wooded, the brethren can find timber for the fabric of their church, and wood for other purposes'.[27] The main section of Hatfield comprises fairly well-drained gravel soils within the Vale of St Albans, land which had probably been cleared and settled for centuries. The woods referred to may well have been on these clay uplands, and taken away from the other communities in the area which had formerly exploited them. This entire area, essentially comprising the land extending from Millwards Park in the west to Newgate Street in the east, remained heavily wooded throughout the Middle Ages. Once the abbey's holdings had been transferred to the bishopric of Ely in 1109, this easterly extension of the parish was enclosed by a fence and made into a huge deer park covering over 750 hectares. The tenants of the manor of Hatfield retained the right to graze their livestock and to collect firewood there but no such rights were enjoyed by those living on adjoining manors.[28]

As we have seen in the previous chapter, when upland wood-pastures were divided between various lowland manors properties often became confused and intermingled, a complexity which was often perpetuated in the later configuration of parish boundaries. Ridge, for example, which remained a chapelry annexed to the parish of St Peter until the early fourteenth century, had to the west a small detached section from which it was separated by Shenley parish; beyond *this* was a detached section of Shenley. At the northern end of the detached portion of Ridge there was a manorial site called Weld Hall (derived, as previously noted, from *wald*, 'upland wood-pasture'), located within a moat which still partly survives beside Wild Farm, a kilometre to the south-west of Shenley Church. It was, according to Chauncy, 'compassed with a Mote, having a Park adjoyning to it, at the Centre of four Parishes, St Peter's in St Albans, St Stephen's, Aldenham and Ridge'; its land also extended into Shenley.[29] According to the first edition Ordnance Survey map, most of the moated platform lay in the parish of Aldenham; the eastern part lay in the detached part of Ridge and a small area on the north side lay in the parish of St Stephen's: the boundary with Shenley parish lay some 150 metres to the east.

The medieval expansion

Although there must have been some arable farms on the uplands by the time of Domesday, as the case of *Titeberst* suggests, the twelfth and thirteenth centuries appear to have witnessed a rapid expansion of clearance, cultivation and settlement out from the major river valleys and up onto the damp and difficult clays. Numerous permanent settlements had appeared at the furthest extremities of parishes by the middle of the thirteenth century. Little Bushey, for example, clustered on the edge of Bushey Heath, appears in documents as early as 1245, by which time remote Chipping Barnet had grown into a substantial urban centre. Settlement appears to have spread along the edge of wide parallel droves, carving up the land between into enclosed fields, occasionally leaving small patches of poorly draining or acidic ground to form small greens and commons. By the time the earliest maps were made farms were widely scattered along the margins of what were still wide droves, or linear commons, such as Organ Hall or Brooklands in the south of Aldenham. More usually, droves had themselves eventually been filled in so that narrow strips of field flanked the road on both sides, some of which had been built on, while the older, larger farms stood a little way back from the road, as at Green Street in Ridge parish (Figure 5.3). This is a district in which scattered, isolated farms and 'halls' are a major part of the settlement pattern. Some were moated, as we would expect in this area of heavy, poorly draining soils, but it is noteworthy that most examples of these appear to have been manors, in contrast to the situation in the more prosperous north-east, where a significant minority of moats appear to have surrounded the homes of wealthy freeholders. This was not, however, only a landscape of scattered, isolated farms. Larger concentrations of settlement could also be found on the clays. Some, such as the hamlet of Wormley West End or the village of Shenley, appear to have developed through the infilling of wide droves with houses. Others clustered around (and sometimes partly filled) the kinds of more irregular, often triangular greens of the kind familiar from the east of the county, such as Brickendon Green. Many of the largest concentrations of settlement, however, were associated with the margins of the great upland commons: places like Little Bushey, Elstree, Borehamwood or Potters Bar.

Until the nineteenth century the heaths and commons occupying the highest ground would have been the most striking features of the landscape. Smaller greens and droves were widely distributed across the claylands and smaller areas of heath could be found on the lower ground, on areas of glacial gravels in the major valleys, such as Colney Heath. But all would have been dwarfed by the sheer extent of the wastes on the Pebble Gravels, which formed an almost continuous chain rambling across the highest ground in the district (Figure 5.5). Bushey Heath, which still extended across a square kilometre of ground when enclosed in the early nineteenth century, was joined directly on the east by Aldenham Common, which covered at enclosure just under two square kilometres. A small gap, occupied by Elstree village, separated this from the common of Boreham Wood (2.6 square kilometres), which was directly joined to Barnet Common (2.3 square kilometres). A further narrow gap, occupied by Chipping Barnet, was followed – over

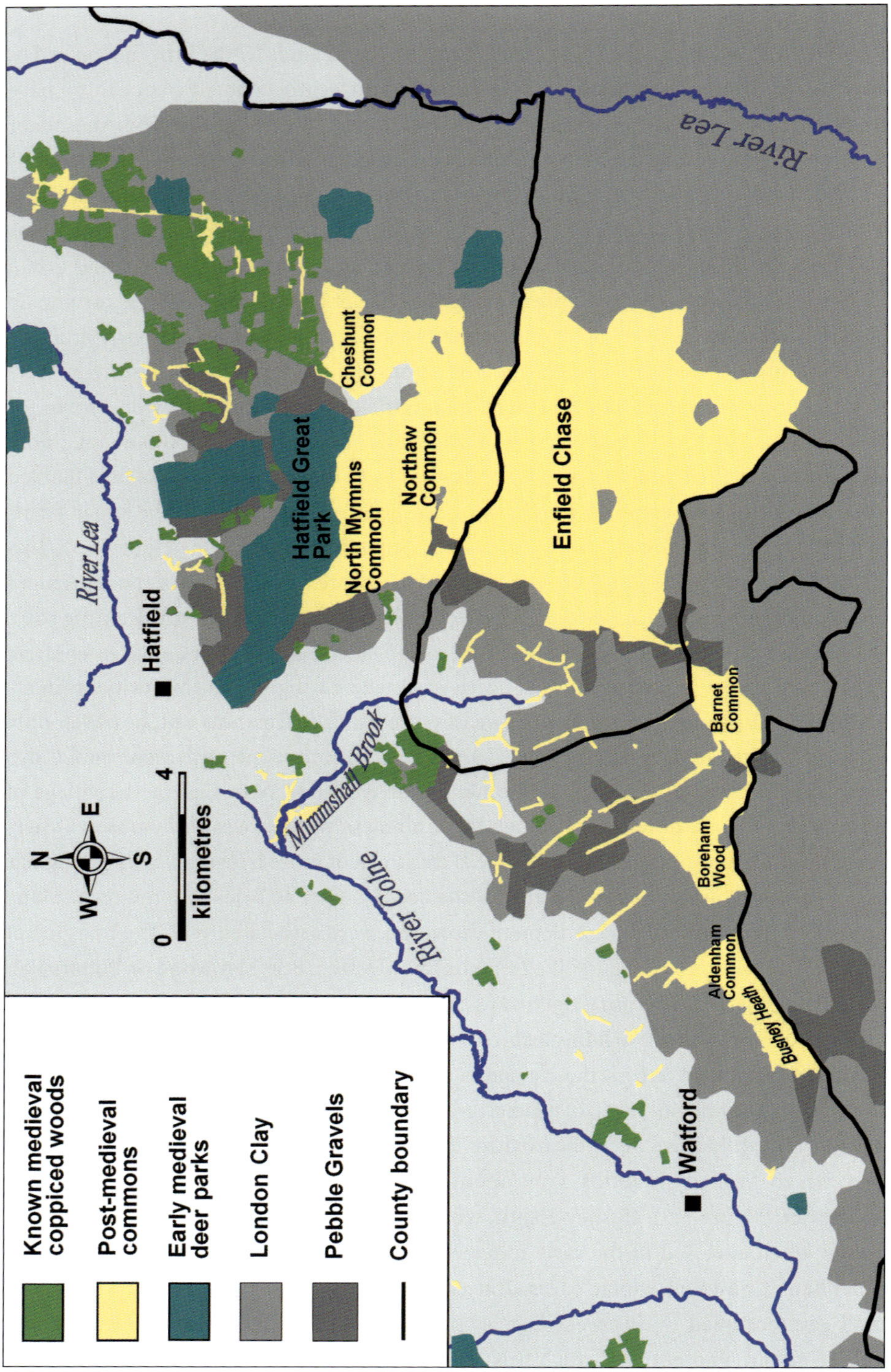

Figure 5.5. The distribution of commons, woods and deer parks in south Hertfordshire.

the county boundary – by Monken Hadley Common and Bentley Heath, and by Enfield Chase, which covered some 33.8 square kilometres (8,350 acres) when enclosed in 1777; a further 1.9 square kilometres (475 acres) had been incorporated within Theobalds Park in the early seventeenth century. These in turn were succeeded, back on the Hertfordshire side of the county boundary, by the conjoined commons of Northaw, North Mymms and Cheshunt, which together covered no less than 19 square kilometres even in the late eighteenth century. Large though these commons were, their often irregular, indented boundaries showed where encroachments had continued into the late medieval and even post-medieval periods.

Most of these large commons coincide with areas mapped by the Geological Survey as Pebble Gravels: the coincidence is probably greater than Figure 5.5 suggests, as geological surveying tends to be more restricted in heavily built-up districts. The early medieval parks in the area are also clustered on the same formation, and Hatfield great park, in particular, appears to be a direct extension of the Northaw/Cheshunt commons, a portion of a once continuous tract of waste which was (as noted above) taken under the control of the abbey and bishops of Ely. It is noteworthy that there was a scatter of farms strung along the northern margins of the park, including those in the hamlet of Woodside, Brewhouse Farm and Camfield, identical to the girdles of settlement found around the margins of the great upland commons. This clearly suggests that farms and cottages had already arrived on the margins of this great tract of 'waste' before its northern portion was emparked by the bishop, before 1222. Evidently, as in the Chiltern Hills, the expansion of settlement onto the upland soils must have occurred very rapidly in the later eleventh and twelfth centuries.

The expansion of farmland thus generally petered out as the largest areas of Pebble Gravels capping the highest ground were reached. The patterns of parallel boundaries, droves and tracks reach up to and terminate on the margins of these last fragments of the 'waste', some of which continued to be occupied by common wood-pastures well into the post-medieval period. Yet it was not on this highest ground that the majority of private, enclosed woods, managed as coppice-with-standards, were to be found. As in the Chilterns, although perhaps to a less marked degree, these tended to cluster towards the margins of the uplands, where they were preserved by manorial lords to facilitate the removal of wood and timber. Good examples include the complex of interconnected woods in the Broxbourne and Hoddesdon area, mostly on sloping ground above the river Lea, and Berrygrove Wood in Aldenham, which almost touches the river Colne (Figure 5.5).

We saw how, in other areas of Hertfordshire, the expansion of settlement from ancient sites located in the principal valleys far out onto clay-covered uplands sometimes led to a profusion of isolated churches, left alone in the fields – or accompanied only by a manorial 'bury' and small cluster of houses. Not surprisingly, in this district many churches likewise stand in isolated or marginal positions. Shenley church thus stands on light soils, formed in glacial gravel overlying Chalk, about a kilometre to the north of the main village, which is mainly located on London Clay and appears originally to have comprised clusters of houses around a linear green and other areas of common grazing. The churches at Aldenham, Wormley and Bayford likewise stand beside relatively light valley soils, accompanied by only small clusters of houses, and with the majority of dwellings scattered around greens and linear commons located further onto the clay uplands. At Essendon the church stands at one end of a village which has clearly expanded away from it, spreading into a narrow strip of common grazing left on the western margins of Bedwell Park; while at Bayford, to quote the *Victoria County History*, the church 'stands in an isolated position about a quarter of a mile north-west of the village'. But there are places where this pattern of development did not occur. Bushey is an interesting case, for the site of the main manor in the parish is beside the river Colne, in an area of light, freely draining soils, while the church itself is placed some three kilometres to the south-east, on the London Clay. Here it is probable that the church was a late foundation, for Bushey seems to have originally formed part of Watford parish. It was apparently established in the middle of the twelfth century, possibly as part of a plan to develop what is now Bushey village as a small town on the road to London (below, pp. 249–51). The church was established, that is, *after* settlement had expanded out of the valley and onto the heavy soils of the uplands, and it was thus conveniently placed to serve the medieval rather than the Saxon arrangement of settlement. The churches at Elstree, Totteridge and East Barnet are all likewise closely associated with the main focus of settlement within their respective parishes. All lie towards the centre of the 'uplands', on London Clay or Pebble Gravels, and far from light, valley soils. In at least two of these cases the church appears to have been established at an early date, probably during the twelfth century. East Barnet church thus includes Norman fabric and the chapel at Elstree is referred to in 1188: these places were outlying possessions of the abbeys of St Albans and Ely, and their churches were perhaps established as soon as the villages with which they are associated had developed through the expansion of smaller and perhaps seasonally occupied settlements. Monkish owners may have been particularly keen to provide a religious focus for, and perhaps ensure a measure of social control over, their tenants in these remote places.

Open fields and enclosure

As far as we can tell the bulk of the land reclaimed from woodland and waste in the course of the twelfth and thirteenth centuries was from the start enclosed into hedged fields. But on the older-settled, lighter soils on lower ground, in and around the valleys of the Colne and the Lea, open fields appear to have been extensive in medieval times, and small areas of intermixed arable could sometimes be found on the clays themselves. In the valley of the river Lea, in the parishes of Broxbourne and Cheshunt, quite large areas of open-field land still remained in the eighteenth and nineteenth centuries, but elsewhere only fragments usually survived. Their former extent, however, can often be reconstructed from references in early documents, the distribution of particular kinds of field name or the configuration of field boundaries.

The distinction between lighter valley land, characterised by open fields, and heavier and more acid soils, mainly occupied by enclosures made direct from woodland, is manifested in the traditional division of Aldenham, in the valley of the Colne, as expressed in a variety of local documents, into two parts: the 'Field Side' and the 'Wood Side'. Yet the earliest surviving maps – a group of farm surveys from 1765 and a map of the whole parish made in 1803 – show only small areas of open-field land in the parish: Dee Common, covering 29.5 acres (12 hectares); The Great Common, of 43 acres (17.4 hectares); and the Little Common, covering 16.4 acres (6.6 hectares).[30] Many of the neighbouring fields, however, have the kind of slightly sinuous boundary – mirroring those of the arable strips in open fields – characteristic of early piecemeal enclosure. In the neighbouring parish of Bushey, immediately to the south-west, the evidence for open-field farming is clearer. A document drawn up as late as 1825 lists a number of unhedged parcels lying within the fields called Bournehall Bottom, Church Bottom, Great Grove Field, Little Grove Field, Skippetts, Upper and Lower Wardells, Little Broadfield, Upper Rye Hill and Barn Croft Field, all in the north of the parish, some of which are described as 'dispersed lands in the Common Field of Bushey'.[31] Most of these fields appear to have contained the intermixed properties of the two main manors in the parish, Bournehall and Bushey, and this probably explains why some memory of ancient arrangements persisted, for a map of the parish drawn up in 1799 leaves no doubt that these distinct, unhedged parcels were entirely notional in character, for no open fields survived in the parish at this time.[32] The field called Wardells can be traced back into the Middle Ages: in 1461 two pieces of land in '*campo voc Greteweredell*' were held of Bushey manor, while in 1322 53 cropped acres were described in 'Muchele Wyrdellefeld'.[33] The phrase '*in campo*' – 'in the field' – occurs many times in late medieval and early modern documents relating to the parish, apparently indicating the existence of unhedged parcels in open fields.[34] In addition, various medieval and early post-medieval documents refer to 'acres' or 'parcels' lying 'in' various fields, many of which can be identified, as well as to the presence of 'baulkes' – that is, narrow unploughed strips lying between the individual arable lands.[35] Other areas of former open arable are indicated by fields with boundaries displaying the distinctive forms characteristic of piecemeal enclosure, gently sinuous and with marked 'kinks' where bundles of strips were enclosed in adjoining furlongs, but not in direct alignment; and by the field name 'shot' or 'shott', the Hertfordshire term for a furlong.

When all these names and boundary types are mapped it becomes clear that in these two adjacent parishes, and in the adjacent parts of Watford, extensive and in places continuous areas of open arable once existed in a broad band on the lighter soils formed in the sands and light clays of the Lambeth Group, in patches of gravel and, to some extent, in the Chalk (Figure 5.6). On the higher ground and heavier soils to the south and east, however, there is only scattered evidence for small areas of open fields, interspersed with enclosed land. There is slightly less evidence for open fields in the extreme north-west of both parishes, where the best soils, overlying exposures of Chalk, are found on the lower ground close to the river

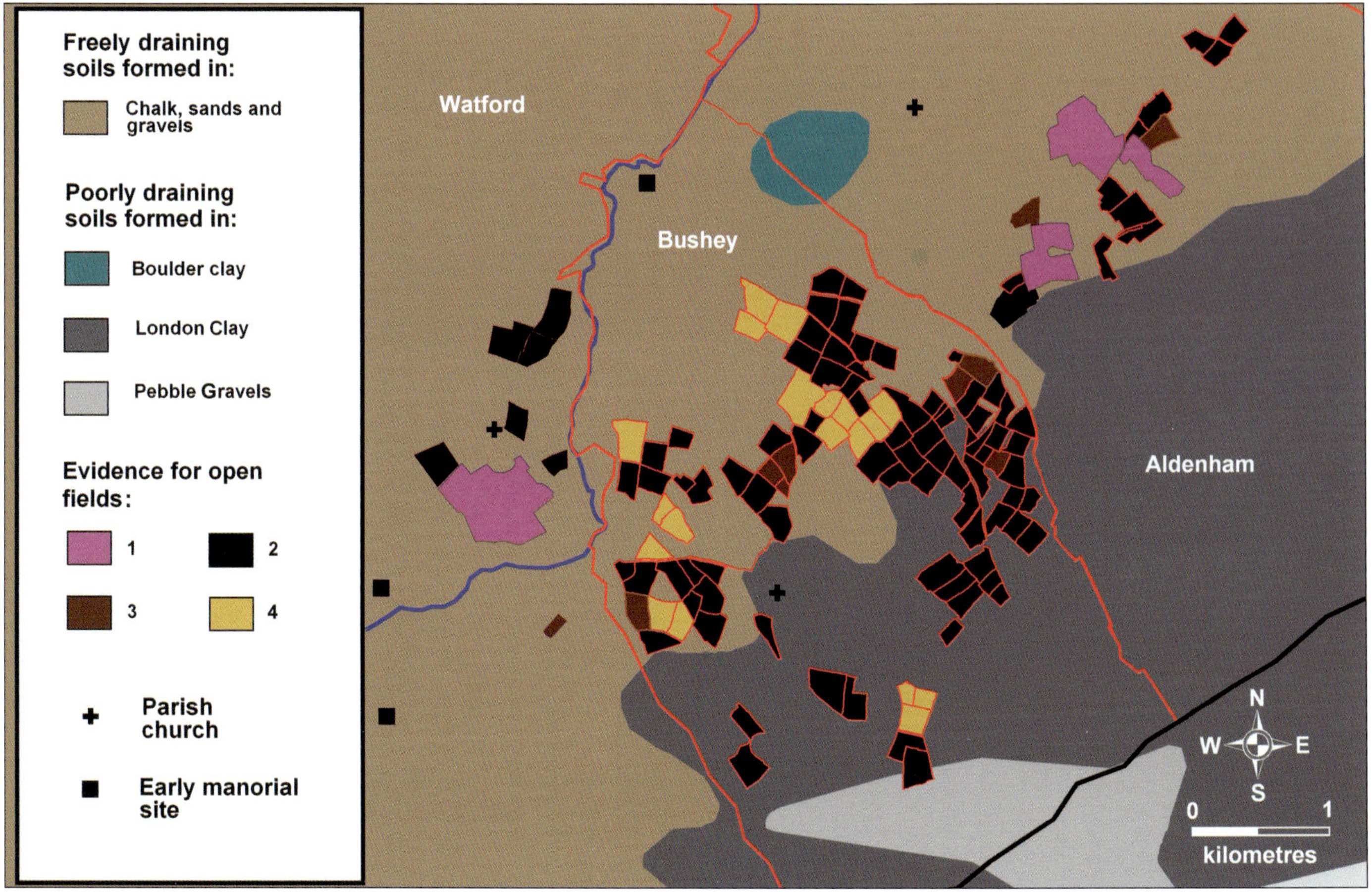

Figure 5.6. Evidence for the former existence of open fields in Bushey, Watford and Aldenham. 1: Open fields surviving in nineteenth century. 2: Fields with sinuous and 'reversed S' boundaries. 3: Field names featuring the element 'shott'. 4: Probable location of open fields mentioned in early documents.

Colne. The main areas of manorial demesne were located here, in areas which had probably been continuously settled since the Roman period – lands which had been held quite separate from the peasant holdings from earliest times. It is possible, however, that the lack of evidence for open-field cultivation (in the form of the layout of field boundaries and field names) is also due to the fact that field patterns in these locations were often reorganised on more 'rational' lines in the course of the nineteenth century (see below, p. 139). Evidence from neighbouring parishes certainly suggests that in some cases the demesne arable was interspersed in open fields with the land of tenants. In Shenley, for example, an extent of 1276 describes how much of the demesne lay in three fields on the light land in the vicinity of the parish church and manor house – 136½ acres in Adgranosfeld, 134½ in 'Shenlefeld' and 118½ in 'Westfeld' – suggesting the scattering of strips evenly through three great open fields.[36] Elsewhere the contrast between open arable on the light valley soils and a predominance of enclosed fields on the higher and heavier ground is repeated. An undated early eighteenth-century map of the Gobions estate in North Mymms, for example, shows a landscape largely dominated by hedged closes on the clays but with surviving patches of open-field land on soils formed in the formations of the Lambeth Group.[37]

Whatever the relative proportion of large demesne closes and open fields on the very best soils in the district, there is thus no doubt that the latter were widespread on the second-best soils, formed in the sands of the Lambeth Group. Conversely, on the heavier and higher ground occupied by the London Clay most of the land always lay in enclosed fields. This does not mean, however, that the clays were necessarily exploited any differently from the lighter land, as pasture rather than as arable. Such was the pressure of population by the thirteenth century, coupled in this district with the proximity of London and the ready market provided by its growing population, that most land appears to have been under crops except the commons, which – as already noted – were particularly extensive on the highest ground. Most appear to have carried at least some woodland even in the high Middle Ages. Barnet Common, the remains of the Southaw recorded in twelfth-century documents, was still a wood – or, perhaps more accurately, a wood-pasture – covering some 200 acres at the end of the fifteenth century.[38] As with all commons, those in south Hertfordshire remained the property of the lords of the manors within which they lay, and some profit could be derived from the timber, and wood from pollards, growing on them, the former always and the latter sometimes the property of the lord. Various manorial rights also yielded an income, although the situation was made complicated by the fact that the larger tracts of 'waste' were shared between two or more manors or vills: the inhabitants of both Chipping and East Barnet thus both had rights to Barnet Common.[39] Complex arrangements were often made to allocate income to neighbouring lordships. The manor of North Mymms had the right every third year to round up animals on 'Cheshunt Common, North Hall (Northaw) Common and Mymwood Common' (a practice known as 'the drift') and to collect fines for those with no grazing rights; in other years the right was enjoyed by manors in the neighbouring vills.[40]

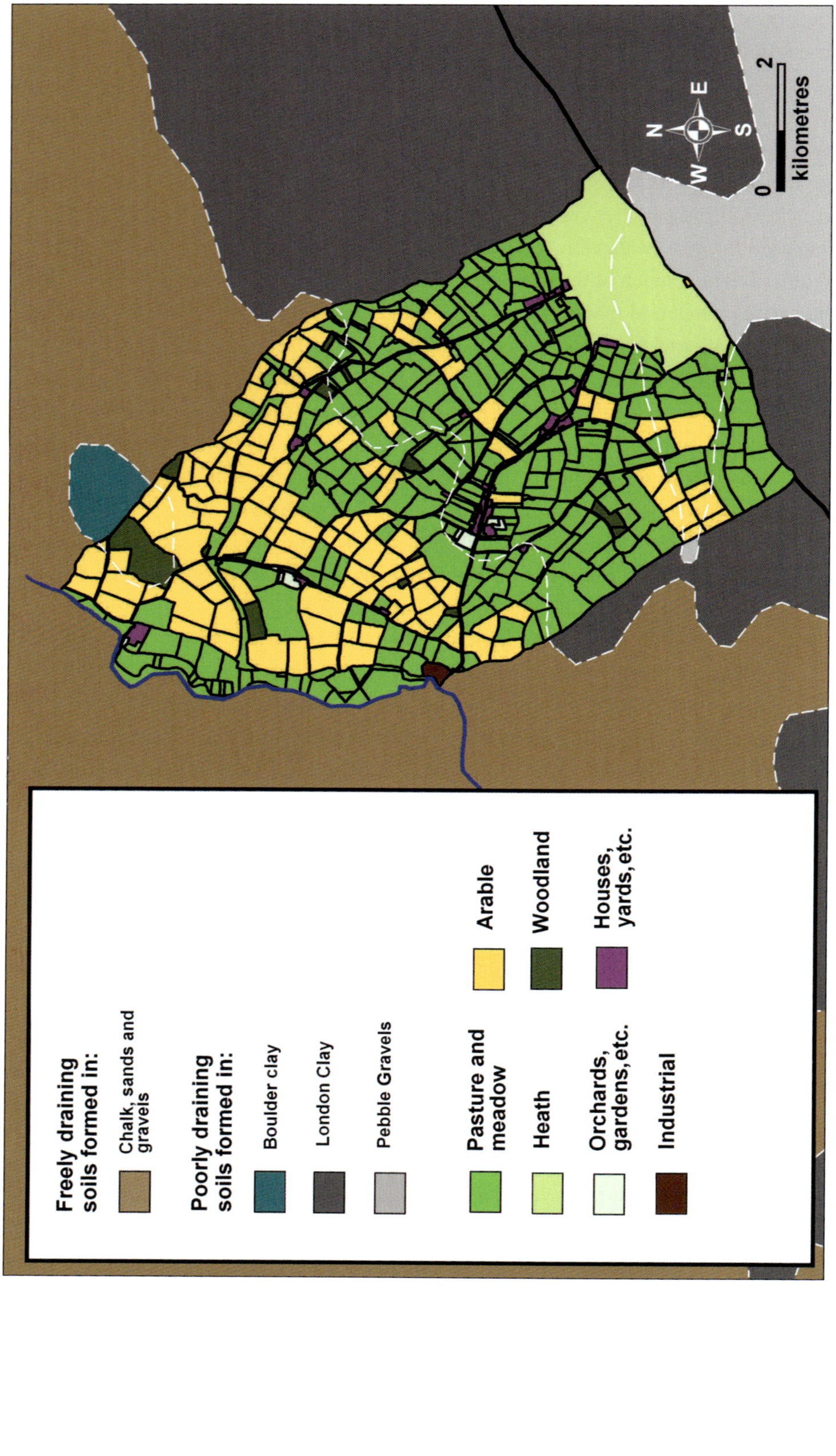

Figure 5.7. Land use in Bushey, south Hertfordshire, in 1840, as shown on the tithe map.

Post-medieval changes

The amount of land cultivated as arable on the heavier soils of the district almost certainly declined as population fell during the late Middle Ages, but by the sixteenth century it was again extensive almost everywhere and probate inventories show that crops were grown on some scale even in parishes such as Elstree, entirely occupied by London Clay soils.[41] One of the key developments of the seventeenth, eighteenth and nineteenth centuries, however, was the steady contraction of the cultivated acreage on the clays and the concomitant expansion of the area under grass. A survey of the demesne lands in the manors of Baas, Hoddesdon and Geddings, in the parish of Broxbourne, for example, shows that in 1692 more than half of the fields on the heavy clays here were being cultivated, but by the nineteenth century almost all were under grass.[42] An *inquisition post mortem* drawn up on the death of Henry Hickman in 1594 estimated that his two manors in the parish of Bushey, one of which extended into Leavesden in Watford, comprised 2,000 acres of 'land' – that is, arable – but only 800 of meadow and pasture, together with 300 acres of wood and 600 of furze and heath.[43] These figures are rounded and rough but confirmed by a document of 1632, which records 1,833 of arable in the parish but only 347 acres of grass (the 1,000 or more acres unaccounted for were partly common heath, woodland and land covered by yards, gardens and orchards).[44] By 1772, to judge from the journal kept by the rector, the area of cropped land had dropped to 1,780 acres, around 55 per cent of the parish, and by 1800 a detailed parish map shows that it covered only 1,128 acres, a mere 35 per cent, almost entirely on the lighter soils in the north-west of the parish. The tithe map shows a similar pattern, which was shared by the adjacent parishes (Figure 5.7).[45] The shift to pasture clearly occurred at different rates in different places, reflecting in part local environmental circumstances, in part the policies of land owners. Already, in 1694, the 84 acres described in a lease for Medburn Farm in Aldenham consisted entirely of 'meadow and pasture land', although the lessee was permitted to plough up 12 acres of this.[46] Elsewhere, in contrast, patches of arable survived on the London Clay even in the nineteenth century. Yet, gradual or rapid, a shift to permanent pasture occurred everywhere on the clays, and had a number of causes.

On the lower, lighter land of the district farmers were able, in the course of the seventeenth and eighteenth centuries, to adopt the new crops such as turnips and clover. But the former were hard or impossible to grow successfully on the sticky London Clay and William Ellis described in 1736 how 'Between Watford and London few, or none, of their farmers sow Turnips, and but little Barley because, they say, their yellow clayey and wet loamy Ground is rather too wet for either of them'.[47] Arable farming thus became increasingly uncompetitive on these difficult soils as the new rotations were widely adopted. Yet at the same time the inexorable expansion of London created a ready market for other produce. Farmers on the clays thus concentrated on the fattening of beef, lamb and mutton for the city, and grew hay to feed the ever-increasing numbers of draft horses kept there. The meadows and pastures within this 'hay belt' were, in turn, fertilised with muck brought

Figure 5.8. Narrow, post-medieval 'ridge and furrow' surviving on King George VI Recreation Ground, Bushey.

out in carts from the capital. William Ellis described in 1742 how 'From Watford to London the yellow Butter-flower grows so thick that there seemed, in many Fields at a Distance little else but this pernicious weed'.[48] Its growth may have been encouraged by the intensive management of the meadows, some of which (according to the Finnish visitor, Pehr Kalm) were mown three times a year (in May, July and September).[49]

Where these pastures have survived to the present, and not been ploughed up or built over in the course of the last century, areas of narrow ridge-and-furrow with a 'wavelength' of less than four metres (of the kind already noted in our discussion of north Hertfordshire) are occasionally found, sometimes in unlikely places, as in the middle of the King George VI Recreation Ground in Bushey (Figure 5.8); in fields in Wormley, now engulfed by woodland but shown as arable in 1692;[50] or in the middle of Sandy Lodge golf course near Oxhey. They clearly attest to the drift from arable to pasture which was the central feature of the district's agrarian history in the post-medieval period.

The seventeenth and eighteenth centuries also saw a significant reduction in the area of woodland in the district, which affected both the enclosed woods largely located towards the margins of the clays and the wooded commons clustered towards the interior. In Bushey, for example, Hickman's *inquisition post mortem* of 1594 records 300 acres of woodland on his two manors, but a parish map of 1799 shows only 92, and even this includes belts and clumps in local parkland, almost certainly newly planted in the course of the eighteenth century.[51] A marked cluster of woodland in the centre of the parish (Bourn Hall Grove, Knights Grove and Curburds Grove) shown on an undated mid-eighteenth-century survey had entirely disappeared by 1799.[52] Other losses are referred to in documents. A lease from the 1630s thus describes a 'close of land called Great Oatfield with the ground thereunto adjoining which lately was woodground & now is grubbed & stocked'.[53] In addition to such deliberate clearance of private woods, moreover, there is clear evidence that many of the upland commons became increasingly open landscapes, continuing a process which had begun in the Middle Ages. As trees were felled, died or were brought down by gales they were difficult to replace in the face of grazing pressure. In the late seventeenth century it was said that Barnet Common 'was formerly a wood, but had been of recent years laid waste, and used as a common'.[54] Steady degeneration was encouraged by the fact that in many cases there was little control over the numbers of livestock which tenants could pasture on a common. The inhabitants of Northaw, Cheshunt, North Mymms and the manor of Brookmans (in North Mymms parish) thus all claimed common of pasture without number for all their cattle and beasts, at all times of the year, in the three great wastes of Northaw, Cheshunt and North Mymms; tenants of Enfield also claimed pasture rights in Northaw Common.[55] Northaw did, however, retain significant numbers of trees until enclosure in the early nineteenth century, as did Borehamwood Common and Broxbourne Common. Other upland wastes, in contrast, slowly degenerated to open heathland dominated by gorse (*furze*), but with some heather and broom and with areas of grass and bracken. Such rough vegetation made poor grazing for cattle but could sustain sheep, although Ellis described in 1742 how they were vulnerable to foot-rot because the soils on the heaths were 'wet, springy, rough sands or Gravels'.[56] Goats were a prominent feature of local agriculture in the early modern period and were largely grazed on the heaths, as Glennie noted at Northaw.[57] But, as in other parts of Hertfordshire, the materials cut from heathy commons were probably as important in the local economy as any grazing they could offer, and stringent attempts had to be made to control the intensity of exploitation. In October 1600 the manor court in Bushey forbade the selling of furze cut from the waste to anyone living outside the manor, while in 1707 it ruled that 'every person that cuts Furze within the manor between May Day and Michaelmas [i.e., the summer growing season] shall forfeit to the Lord of ye manor 20d'.[58]

Where significant stands of trees survived, as on the commons of Broxbourne, sales of wood and timber could bring in a reasonable income for manorial lords. But where commons degenerated to open heaths they were of less value to the local gentry, especially as the money to be made from exercising manorial rights seems to have decreased with the passing centuries. By 1689 the income from the rights of 'drift' exercised by the manor of North Mymms every third year on the commons of Cheshunt, Northaw and Mymwood had dwindled to such an extent that it was sufficient to cover expenses only; in the same year it was said that there had been no 'drift' on Colney Heath for several years.[59] As population and food prices rose, in this area so close to London, local lords increasingly saw the advantages of enclosing all or part of their wastes and using them as improved pasture or arable, often against the wishes of their tenants, who sometimes mounted violent protests. In 1548 a rumour spread that Northaw Great Waste was about to be enclosed by Sir William Cavendish, lord of the manor. A crowd of 500 armed rioters, drawn not only from Northaw but also from Cheshunt, North Mymms and other local villages, laid siege to Cavendish's house. Ordered to disperse, they proceeded to Cavendish's rabbit warren, killed 1,000 rabbits and blew up the 'burrows' with gunpowder, in an expression of their 'great hate and dedlye dyspleasure'.[60] Similar disputes are recorded in the sixteenth and seventeenth centuries in Aldenham – where the tenants took their manorial lord to court in 1576 – and Shenley – where a group of rioters met at Blackheath in Shenley and threatened to 'lay all the said grounds common' which had been enclosed and sown with crops by Edward Wingate, MP for St Albans (and, at the time, a captive of the royalists).[61] Attempts at large-scale enclosures of the wastes were seldom successful in this period, in part perhaps because of the large numbers of people who had rights to use them; the main exception was the appropriation – in the face of stiff opposition from local commoners – by King James of 149 acres of Cheshunt Common, 336 acres of Northaw Common and 475 acres of Enfield Chase, and their incorporation into Theobalds Park.[62] Smaller encroachments, however, were frequent. Part of Barnet Heath was thus enclosed by the Duke of Chandos in the eighteenth century,[63] and much small-scale squatting occurred as cottages were established on areas of waste with manorial permission, especially in parishes in which the presence of major roads to London encouraged a measure of 'ribbon development'. In Bushey, for example, the properties lining both sides of the A411 on Clay Hill, leading up to Bushey Heath, are clearly taken in from the roadside waste and an inquisition held at the manorial court in 1630 alleged that five houses had been built on the 'waste' here within the previous 70 years. Significantly, their occupants were tradesmen rather than farmers, including a carpenter and a wheelwright.[64]

Most of the local heaths, however, survived into the late eighteenth or nineteenth century, when – as in the Chilterns – they were enclosed by a series of parliamentary acts which in some cases, as at Cheshunt and Aldenham, also removed residual pockets of open field. Borehamwood Common in Elstree was enclosed in 1777, North Mymms in 1782, Aldenham Heath in 1803, Cheshunt in 1804, Northaw in 1806 and Bushey in 1808.[65] The 60-acre wood-pasture of Broxbourne and Hoddesdon Common does not appear to have been affected by the enclosure act here in 1843, which dealt with common meadows and some open field. Instead it seems to have been enclosed 'unofficially' by the lord of the manor around 1870, the few tenants still exercising their grazing rights being compensated with a meagre 4.3 acres to which rights of public access were confirmed by the Bosanquet family in an act of parliament of 1890.[66]

Northaw Common, covering over 2,000 acres, was the largest of these enclosures: the work of division was under way when the Ordnance Survey draft map was made in 1805, the surveyor noting 'Clearing for Inclosure' across its still wooded northern section.[67] In fact, this part was subsequently planted up by its new owner as what is now Northaw Great Wood. Old hornbeam pollards and vestiges of heathland vegetation survive among nineteenth- and twentieth-century trees, and ongoing management by Welwyn-Hatfield Council, who now own the wood, is restoring more extensive areas to the state of heathy wood-pasture. Some other areas of enclosed common were planted with trees, such as the plantation established on the 56 acres of the former common at Broxbourne, but most were laid to pasture in straight-sided fields bounded by species-poor hedges and many – in the course of the nineteenth and early twentieth centuries – were built over, an option reflecting the low agricultural value of the land in question as well as proximity to London.

The former commons and heaths are not the only places in the south of the county where extensive areas of straight-sided fields were created in the nineteenth century. On the lower ground and better soils, where much land remained under cultivation and many of the largest landed properties (often developed from the demesnes of medieval manors) were located, the pattern of fields was often extensively reorganised in the later eighteenth and nineteenth centuries in the manner we have already described in the east of the county but sometimes to a greater extent. In Bushey, for example, the *majority* of boundaries shown on a parish map of 1799 in the northern third of the parish, on light soils beside the Colne, had been uprooted – realigned or removed altogether – by 1886. In the northern parts of Shenley, and to a lesser extent Ridge, changes were on a similar scale, and in places the field pattern by the start of the twentieth century resembled that found in the northern areas of the county, where extensive open fields were enclosed by parliamentary acts. The contrast between the landscape of the light valley soils, where arable farming continued to be significant, and that of the higher clay-covered ground, where pasture farming in small and often irregularly shaped fields was practised, thus remained sharp. The fact that local hedges were often composed largely of blackthorn – a shrub which tends to sucker sideways into the field, especially when under pasture, converting a hedge into a wide, unproductive thicket – doubtless encouraged a desire to 'tidy up' the landscape and provide 'modern' field boundaries composed of straight hawthorn hedges.

By the middle of the nineteenth century most of the upland clays were under grass, although a few ploughed fields could be found, as in Wormley in 1851.[68] However, most such survivors were laid to pasture as the agricultural depression began to bite after *c.*1880, local farmers by this stage becoming increasingly involved in dairy farming. In parishes which included extensive areas of light valley land, such as North Mymms or Ridge, the proportion of the land area under arable at the start of the twentieth century might reach a third but it was generally less than a fifth and in parishes largely occupied by London Clay soils it was usually much less than this. In Northaw, for example, only 11 per cent of the farmland was under crops, in Arkley 7 per cent, in Elstree 3 per cent and in Totteridge a mere 1.4 per cent.[69] By 1941 it was said of the district that 'practically every farmer is concerned with milk production', either selling directly to the neighbouring suburbs or to the London market, and the area was 'almost entirely in permanent grass', with the exception of a few fields planted with root crops for winter feed, usually on the better-draining sloping ground.[70]

The onset of agricultural depression on these rather marginal soils had other effects. In a few places land went out of cultivation altogether and was allowed to regenerate

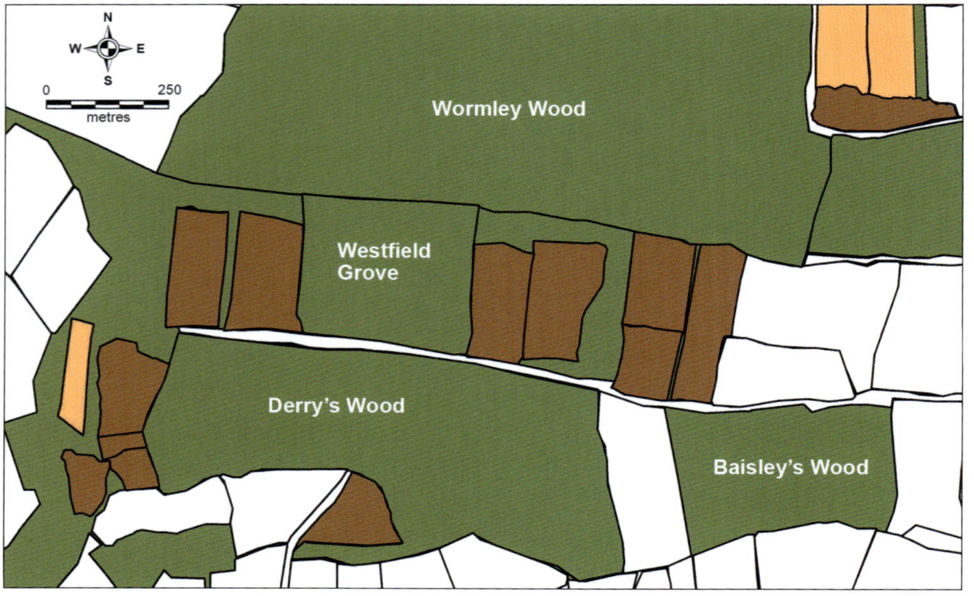

Figure 5.9. Fields abandoned in the area around Wormley Wood in the late nineteenth and first half of the twentieth century. Orange: abandoned by 1882. Brown: abandoned after 1882.

to woodland. The most striking case is in the west of Wormley and Cheshunt, where a number of fields – many of them arable in the seventeenth century but mostly pasture by the nineteenth – were progressively engulfed in woodland (Figure 5.9). Some were abandoned before 1882 but most between 1882 and 1940. The woodland also expanded over the adjoining lanes, part of an ancient co-axial layout: these and the field boundaries now survive as earthworks within the woods.[71] While land was going out of cultivation here, however, a few kilometres to the east, on the floor of the Lea valley, a new agricultural industry was developing, one which, until recently, had a major impact on the landscape. Beginning in 1883, when the firm of Rochford erected a single greenhouse, by the 1930s this was a 'region unique in Britain as the home of the largest glasshouse industry in the islands'.[72] Market gardeners were attracted to the area by the fertile soils of the flood plain, the easy availability of water from the saturated gravels of the valley floor and the expansion of London, which not only provided a ready market for tomatoes, cucumbers and other produce but also forced the migration northwards of growers who had been encouraged to sell land for development closer to the heart of the capital.[73] By the 1930s the availability of mains water encouraged the spread of glasshouses out of the valley, westwards onto the fringes of the London Clay uplands. The industry reached its peak around the 1960s. Since then, foreign competition coupled with a continuing rise in the demand for residential land has ensured a slow but steady decline and large areas of glass have, over the last few decades, been demolished and replaced with housing.[74]

Figure 5.10. View across the valley of the Cuffley Brook from Cheshunt towards Cuffley and Northaw, showing a mixture of farmland and settlement typical of south Hertfordshire. On the skyline top left is an office block in Potters Bar; top right is the parish church of Northaw; the southern edge of Cuffley is middle right; electricity pylons complete the scene.

The greatest change to the landscape of parts of south Hertfordshire since the later nineteenth century, as we have intimated on a number of occasions, was the inexorable expansion of suburbia, a subject to which we shall return at a later stage. Nevertheless, even in the south-west – where this process was particularly prevalent – pockets of farmland remain and, even where they do not, a surprising number of fragments of the rural landscape – especially ancient hedges and old farmland trees – survive amidst the houses, golf courses and playing fields. In the south-east the built-up areas are more localised and extensive areas of farmland and woodland survive (Figure 5.10).

Notes

1. Young, *Agriculture*, p. 11. As a farmer at North Mymms in the late eighteenth century he had personal experience of its agricultural value and described it as 'poor gravel'.
2. Niblett, *Roman Hertfordshire*, pp. 77–87.
3. Niblett, 'Roman settlements', in Short (ed.), *Historical atlas*, pp. 72–3.
4. Gover *et al.*, *Place-names of Hertfordshire*, pp. 99, 67–8; Crick, *Charters of St Albans*, no. 7.
5. Aldenham, Oxhey and Hanstead are mentioned in charters of pre-Conquest date: Gelling, *Early charters*, p. 80; W. De Gray Birch, *Cartularium Saxonicum: a collection of charters relating to Anglo-Saxon history* (London, 1885), vol. 1, pp. 339–40; Crick, *Charters of St Albans*, nos 1 and 12.
6. With arable sufficient for two and a half ploughs.
7. Gover *et al.*, *Place-names of Hertfordshire*, pp. 64, 67, 233.
8. D. Hooke, *Landscape of Anglo-Saxon England* (Leicester, 1998), pp. 147–8; D. Hooke, *Trees in Anglo-Saxon England* (Woodbridge, 2010), p. 126.
9. Gover *et al.*, *Place-names of Hertfordshire*, pp. 68, 69–70, 74–5, 114.
10. P. Taylor, 'Boundaries and margins: Barnet, Finchley and Totteridge', in M.J. Franklin and C. Harper-Bill (eds), *Medieval ecclesiastical studies in honour of Dorothy M. Owen* (Woodbridge, 1995), pp. 259–79; P. Taylor, 'Boundaries, margins and the delineation of the urban: the case of Barnet', in Slater and Goose (eds), *County of small towns*, pp. 249–75; Crick, *Charters of St Albans*, pp. 188–9.
11. The tree, as well as giving its name to Elstree, is mentioned as a boundary marker in an undated pre-Conquest charter: Gover *et al.*, *Place-names of Hertfordshire*, p. 60.
12. *Ibid.*
13. Crick, *Charters of St Albans*, no. 12.
14. A.H. Smith, *English place-name elements*, vol. I (Cambridge, 1956), p. 198.
15. Page (ed.), *VCH Hertfordshire,* vol. 2, pp. 329–37.
16. Taylor, 'Boundaries, margins', p. 257. The charter is undated but ascribed to 1176 by R.W. Eyton, *Court, household and itinerary of King Henry II* (London, 1878), p. 204. Suthawe or Southaw was also referred to as Southallwodd (1485–1509); Huzeheog had many variant spellings, including Huzeog, Hossegge, Owsage, Osnage and Osidge.
17. Page (ed.), *VCH Hertfordshire*, vol. 2, p. 358.
18. The charters for Kingsbury (1005) and for Burston (996) both have very confusing bounds, difficult to reconstruct, which include references to the 'valley of the Shenley people'. See Williamson, *Origins of Hertfordshire*, pp. 148–51; Crick, *Charters of St Albans*, nos 9 and 11.
19. Page (ed.), *VCH Hertfordshire*, vol. 2, p. 386; P. Barton, *Manorial economy and society in Shenley: a Hertfordshire manor in the 13th and 14th centuries* (Hertford, 1981).
20. Again, the bounds of Kingsbury may have extended this far south.
21. Page (ed.), *VCH Hertfordshire*, vol. 2, pp. 387, 391.
22. *Ibid.*, p. 148.
23. W. Page and J.H. Round, 'Houses of Austin canons: abbey of Waltham Holy Cross', in W. Page and J.H. Round (eds), *A history of the county of Essex, vol. 2* (1907), pp. 166–72; Page (ed.), *VCH Hertfordshire*, vol. 3, pp. 409–10.
24. Barton, *Manorial economy and society*, pp. 8–9; S.W. Wooldridge and G.F. Hutchings, *London's countryside* (London, 1957).
25. J.R. Hunn, *Tyttenhanger: excavation and survey in the parish of Ridge* (Oxford, 2004); Bryant *et al.*, 'A "relict landscape"', pp. 5–15.
26. K.D. Ely and K. Edwards, 'Cheshunt Park, Hertfordshire. An archaeological investigation by the Time Team', unpublished report (Bristol, 2003), p. 15; Bryant *et al.*, 'A "relict landscape"', pp. 12–14.
27. Page (ed.), *VCH Hertfordshire*, vol. 3, p. 91.
28. Rowe, *Medieval parks*, pp. 108–12.
29. H. Chauncy, *The historical antiquities of Hertfordshire* (London, 1700), p. 461. Chauncy believed the house to be in the parish of St Peter but the family attended Shenley parish church as it was nearest; Barton, *Manorial economy and society*, p. 20.
30. HALS 54804 plan of Tile-house Farm, Haydons Farm, Cressalls Farm, Theobald Street, 1765; QS/E3 Aldenham enclosure map, 1803.
31. BL Phillips MS 82742 holdings of Michael Bellis, 1825.
32. HALS DP/26/29/2 plan of the parish of Bushey, 1800.

33. HALS 37123 Bushey manor court roll, 1461; TNA:PRO SC/1145/10 Ministers' and Receivers' Accounts, Bushey, 1321–2.
34. Based on unpublished research by the late Grant Longman, based on HALS 37123–33 Bushey manor court roll, 1461–9; 46143 Bourn Hall court roll, 1603–23; and TNA:PRO E199/10/2 Extent of manor of Bushey, held by John le Latimer (full details, including stock and grain), 1314–16.
35. For example, HALS 54807 manor of Bournhall court roll, 1627–34.
36. Barton, *Manorial economy and society*, p. 27.
37. Gloucester Record Office D1245/FF75 map of an estate belonging to Jeremy Sambrooke Esq.
38. HALS DE/B2067B/E24 a list of woods appertaining to the monastery of St Albans, *c.*1485–1509.
39. D. Lysons, 'County of Hertfordshire: Chipping Barnet', in *The environs of London: volume 4: counties of Herts, Essex & Kent* (1796), pp. 1–8. <http://www.british-history.ac.uk/report.aspx?compid=45461&strquery=hertfordshire waste>, accessed 6 November 2011.
40. HALS DE/X1012/M1 Typed transcript of the Steward's account book for North Mymms Manor, 1691 (original document is in the BL, Egerton MS 3390 'Leeds papers').
41. Longman, *Corner of England's garden*, pp. 16–27.
42. Hatfield House Archives (hereafter HHA) CP 354 survey of manors in Broxbourne by John Gascoyne; HALS DSA4/25/2 Broxbourne tithe map, 1839.
43. TNA:PRO C142/240/40 inquisition post mortem, Henry Hickman, 1593/4.
44. HALS DP/26/10/1–2 Bushey parish rate book (and transcript), 1632–49.
45. Transcribed by the late Grant Longman from a transcript in possession of Montague Hall; original now lost. See Longman, *Corner of England's garden*; HALS DP/26/29/2; DSA4/27/2 Bushey tithe map, 1840.
46. HALS DE/Am/E3 lease of Medburn Farm, Aldenham, 1694.
47. Ellis, *New experiments in husbandry*, pp. 12–13.
48. W. Ellis, *The modern husbandman* (London, 1742), p. 106.
49. P. Kalm, *Kalm's account of his visit to England on his way to America in 1748*, trans. J. Lucas (London, 1892), pp. 81–2.
50. HHA CP 354.
51. TNA:PRO C142/240/40; HALS DP/26/29/2.
52. HALS 21579 map of the demesne of the manor of Bourn Hall, n.d.
53. HALS DE/Am/E11 lease of Hartsbourne Manor Farm, Bushey, 1632.
54. Page (ed.), *VCH Hertfordshire*, vol. 2, pp. 329–37 citing *Calendar of State Papers Domestic* 1689–90, p. 79.
55. A.C. Jones, '"Commotion time": The English Risings of 1549', PhD thesis (University of Warwick, 2003), pp. 43–4 and p. 259, citing TNA:PRO STAC 3/1/49, bill and answer 2; DL43/7/4.
56. Ellis, *Modern husbandman*, p. 122.
57. P. Glennie, 'Continuity and change in Hertfordshire agriculture, 1550–1700: I – patterns of agricultural production', *Agricultural History Review*, 36 (1988), p. 74.
58. HALS 6398 and 6442 Bushey manor court rolls, 1600 and 1707.
59. HALS DE/X1012/M1.
60. Jones, '"Commotion time"', p. 33 citing TNA:PRO STAC 3/1/49.
61. Page (ed.), *VCH Hertfordshire*, vol. 2, pp. 151–2; *Journal of the House of Commons*, 3, 1643–4 (London 1802), p. 94.
62. TNA:PRO E317/Herts/27 parliamentary survey of Theobalds Park, 1650.
63. Lysons, 'County of Hertfordshire: Chipping Barnet', pp. 1–8.
64. HALS 6424 Bushey manor court roll, 2 October, 1630.
65. W.E. Tate, *Domesday of English enclosure acts and awards* (Reading, 1978), pp. 137–40.
66. Countryside Management Service noticeboard on Broxbourne Common.
67. BL Ordnance Survey Drawing 149, Hatfield, 1805.
68. HALS 80804 map in Wormleybury estate papers, 1851.
69. Page (ed.), *VCH Hertfordshire*, vol. 2, pp. 358, 330, 349; vol. 3, p. 148.
70. Cameron, *Land of Britain*, p. 342–3.
71. Bryant *et al.*, 'A "relict landscape"'.
72. Cameron, *Land of Britain*, pp. 344.
73. *Ibid.*, pp. 344–5.
74. See also K. Banister, 'Salads and ornamentals: a short history of the Lea Valley nursery industry', in Spring (ed.), *Hertfordshire garden history II*, pp. 202–29.

– 6 –

Woods, parks and pastures

The landscapes we have so far discussed in this book were created by the gradual transformation of the natural environment, a process which began in early prehistoric times and which has been completed with such thoroughness that ecologists now debate what England might have looked like before the impact of man. Some believe that in its natural state the country would have been densely and continuously wooded while others – most notably Frans Vera – argue that it contained much open, park-like ground, kept free of trees by herds of grazing herbivores.[1] Either way, such evidence as there is suggests that in the early Saxon period there were very large areas of wooded 'waste' in Hertfordshire, grazed with varying degrees of intensity and exploited for wood and timber. By the high Middle Ages, in contrast, arable land covered the majority of the county, although diverse areas of unploughed ground – the much altered residue of these wooded tracts – still remained, especially on the higher and heavier land, away from the light soil of the major valleys. Such semi-natural environments were shaped by particular forms of management and exploitation, and their surviving fragments today still provide a range of crucial habitats for the county's wildlife.

Ancient woodland

Perhaps the most important of these environments are ancient woods. In the early Middle Ages, as we have seen, many local lords enclosed portions of the remaining 'wastes' in order to preserve wood and timber from indiscriminate felling and damage from grazing livestock, and to manage them more intensively. The majority of trees and bushes within these enclosed, private areas of woodland were cut down, or coppiced, at or near ground level on a rotation of between 7 and 15 years in order to provide a regular crop of 'poles' suitable for fuel, fencing and other domestic uses. The plants then regenerated vigorously from the stumps, or stools, or suckered from the rootstock. Such woods contained relatively few standard trees – that is, full-grown specimens – for if these had been numerous the canopy shade would have suppressed the growth of the underwood beneath.[2] They were normally felled at around 80–100 years of age, when their growth rate began to decline and they were of a suitable size to be used in the construction of buildings or – more rarely – ships. Ancient woods today thus seldom contain any very old trees. Coppice stools, in contrast, can be extremely ancient, for they can continue to regenerate almost indefinitely. Coppicing was impossible on open land, for livestock would browse off the tender new growth – the 'spring' – and kill it. They needed to be excluded, at least for the first four years or so of the rotation, and so substantial banks, flanked by a ditch and usually topped with a hedge or fence, were constructed to keep them out.[3] Many woods today are still bounded by these traditional wood-banks, sometimes topped with ancient pollards or stubbs, but the accompanying hedges rarely survive. Maintaining such boundaries was one of the inescapable costs of wood production and was regularly recorded in accounts. In 1545, for example, William Lowyn, custodian of Henry VIII's woods in the manor of Cheshunt, was paid 20s 'for kepyng the hedge & beastes & cattell from the spryng in [a wood called] Rough Cattall [and] for the safe inclosure therof', while Thomas Browne received 13s 4d for 'the kepyng of the hedges & savegarde of Dyvers Coppices' in the king's manor of Bedwell in Essendon.[4] Maintenance of the boundary hedge was generally undertaken during the winter, prior to a wood being coppiced, and the material cut from the hedge, like that removed when farmland hedges were maintained, was used as or sold for fuel. That from the hedges around Stocking Wood on the bishop of Ely's manor of Little Hadham, for example, was sold for 5s in 1379/80, while wood from the hedge on the west side of the Eastwood was sold for 20d in 1388/9.[5]

The embanking and enclosure of woodland was probably already occurring on some scale in Hertfordshire before the Norman Conquest, in the densely settled north-east of the county at least, for while Domesday records most woodland in terms of the number of swine that could be grazed there on acorns and nuts during the autumn pannage season, in this district many manors had only 'wood for fences' – suggesting that their woods were limited in extent and not exploited for pannage, but instead cut to produce poles. Hertfordshire has long been one of the more densely wooded counties of England, and even at the end of the twentieth century approximately 12,370 hectares (30,550 acres) of its land surface, around 8 per cent, was wooded. But of this only around 28 per cent comprises ancient coppiced woodland of the kind just described, much having been destroyed or replanted in the course of the nineteenth and, in particular, twentieth centuries.[6] The majority of modern woodland now comprises either post-medieval plantations, deliberately planted with timber trees and lacking a coppiced understorey, or results from the natural regeneration of trees over former farmland or, more usually, neglected commons.[7] Woods have never been evenly distributed within the county. Woodland of any kind had mostly disappeared from the 'champion' areas of the north by the time of the Conquest, and within other districts enclosed woods were often, as we have seen, concentrated close to the margins of the main clay-covered plateaux, rather than towards their centres. In general, the large number of separate lordships in the east of the county ensured that substantial numbers of woods existed here: there were fewer in the south and west, although they often extended over large areas or formed long interconnected chains, as in the area around Broxbourne and Hoddesdon.

Woods – the larger examples, at least – were usually divided into compartments which were felled sequentially in order to produce a continuous supply of poles. These, cut into lengths called roundwood or talwood, together with the smaller twigs and brushwood which were tied together in bundles called faggots, constituted one of the county's main sources of fuel before coal became more widely available from the later eighteenth century. The recommended time for cutting was either from 'January at the latest, till Mid-March or April; or from Mid-September, till the end of November'.[8] A document recording the sale of the 'crop' of several woods in Hertingfordbury in 1637 stipulated that the coppicing was to be completed by 25 March and that all the cut wood was to be removed by the end of October. The purchaser was to pay £4 10s per acre of coppice, each acre marked out on the ground by means of stakes placed in the hedges around the woods. He was to cut only the underwood that had been felled the last time the wood was coppiced, taking care to leave all the trees and 'standles' (young trees). He also undertook to 'make good and stronge hedges ditches and Fences … about the Coppice grounds'.[9]

Sometimes the timber and underwood was felled by the owner of the wood and then sold, as at Wheathampstead in 1539, when 115 loads of hardwood (worth 10d per load), 113 loads of small faggots, or bavins (at 7d per load) and oak, beech and ash timber from woods called Marchalles and Butler brought in a total of £16 16s 10d. The costs for felling and dressing the wood and timber, for hedging the wood and for making and hanging a new gate came to 64s 2d, leaving a profit of £13 12s 8d.[10] Alternatively, as at Hertingfordbury in 1637, the underwood could be sold unfelled to a single purchaser, who would then bear the costs of coppicing. But underwood was also commonly sold in portions covering an acre, a rood or even half a rood to numerous different purchasers, each of whom was responsible for felling his own particular allocation. This method of management is recorded at Weston and Knebworth in the fourteenth century; at Walkern and Broxbourne in the seventeenth century; and at Wain Wood near Ippollitts, on the Lamer estate in Wheathampstead and on the Broxbournebury estate in the eighteenth century.[11] The practice continued into the early nineteenth century in the woods around Hertford on the Marden and Panshanger estates and also at Bramfield, where the diarist John Carrington regularly recorded wood sales during the spring months, followed by 'wood feasts' held towards the end of the year, when the purchase money was handed over.[12] Records of a 'wood feast' on the Knebworth estate in 1840 suggest that the practice continued at least into the middle of the century.[13] Sometimes the price paid for the different sections of the underwood varied significantly, as at Walkern in 1612, where they ranged from 22s 6d to 50s per rood and from 9s 9d to 27s 6d per half rood, presumably reflecting the uneven quality of the underwood. Here the purchasers were all from the local area: 15 from Much (Great) Munden, 12 from Benington, 7 from Little Munden, 4 each from Walkern, Ardeley and Aspenden and 1 each from Standon and Westmill.[14] Surviving records suggest that, in some woods, the precise parcel of underwood to be purchased was determined by drawing lots; the price per lot was fixed, irrespective of variations in the quality of the underwood being sold. In March 1800 John Carrington, the tenant of Bacon's Farm between Tewin and Bramfield, recorded in his diary 'I Drawed No. 1. in Hooks bushes', a wood on the Marden Hill estate and adjacent to his farm; in March 1803 he 'Drawd No. 1 a Rood ...' also on the Marden Hill estate. In April 1804 a sharp increase in the price demanded for underwood at Bramfield resulted in a minor revolt: 'To Bramfield Wood Sale, the Hartford people met & we Drawd for it but went into the wood but the price as 3 pounds pr Rood or More, So all gave inn our Tickets & had none...'[15]

This method of selling could damage a wood's productivity because a host of individual buyers, many with little concern for the next crop, tended to spread the felling and dressing of the wood over a prolonged period of time, thereby hindering the regrowth of the 'spring'. Particularly greedy buyers might also damage the coppice stools by cutting them too low, or even by taking the roots, and they might be tempted to fell any young trees on their parcel which should have been preserved as future timber. Such problems were apparently rife in the woodlands belonging to Lord Burghley at Hoddesdon in the late sixteenth century, where the 473 acres of underwood had been sold, unfelled, to local people for many years in one-acre parcels each costing £3. In 1595 Burghley decided to lease the woods to just two men, John Thorowgood of Amwell and William Keeling of London, for the same price per acre, thereby maintaining his income but reducing the complexities of administration. During the 21 years of the lease the two men were to complete two cycles of a ten-year rotation, felling about 45 acres each year. The productivity of the woods varied from acre to acre, but they believed that on average they could clear a profit of £4 5s per year after deducting rent, tax and the costs of felling and making faggots, fencing and wages, as well as the liveries and board required for the two men overseeing operations. Inevitably, the cost to local people of purchasing material cut from the wood – faggots at 4s 4d per load and roundwood at 7s per load – was significantly higher than that of buying an acre of underwood uncut and processing it themselves, and the aggrieved inhabitants of Hoddesdon petitioned Lord Burghley for redress. In response, the lessees agreed to set aside 60 loads of faggots for sale to the poor of Hoddesdon at the knock-down price of a penny a faggot, but they were adamant that they were not making excessive profits and that the new arrangement would significantly improve the productivity of the wood.[16]

Larger poles, used for fencing, minor timbers in buildings, tools and fuelwood, were the main product of coppiced woods. But nothing was wasted and the smaller material, tied together as faggots, produced a short hot blaze when fired and was widely used for household baking and brewing, as well as for a variety of industrial processes such as brick- and tile-making. Faggots were sold on a large scale: 1,000 from the bishop of Ely's park in Little Hadham were sold for 25s in 1327/8 and 2,700 from Brantingshay park in Cheshunt for 108s in 1435.[17] They appear, together with larger pieces of wood, in many inventories of household goods produced in the sixteenth and seventeenth centuries: in John Johnson's kitchen in Hertford in 1673, for example, there were 15 loads of 'Faggotts, Blocks and Roundwood' valued at £11 5s; in the woodhouse of William Turner, a Hertford gentleman, in 1683 were three stacks of wood and half a load of faggots worth £2 9s; and in the yard of John Bach, an Alderman and gentleman of Hertford, in 1699 were 'one thousand Jack Faggotts, Six Load Roundwood and blocks' as well as 12 bushels of coal (coal or charcoal).[18] In spite of the growing importance and easy availability of coal, faggots continued to be produced on a large scale into the nineteenth century. In the winter of 1834/5, for example, 44,517 were made on the Ashridge estate, of which about 70 per cent came from woods and copses.[19] Faggots loomed large in the lives of our ancestors, sometimes literally: the manor court in Bushey in 1738 had to resolve a dispute between neighbours relating to a 'pile of faggots [set] so near the house … as to darken the windows'.[20] Several court cases relating to faggot thefts were reported in the county newspapers into the late nineteenth century and a man was prosecuted for criminally setting fire to a faggot stack in Letchworth as late as 1888.[21]

Some of the wood cut from coppices was converted *in situ* into charcoal – the hornbeam which constituted the principal underwood species in many Hertfordshire woods was particularly suitable for this purpose. Charcoal was employed in a variety of industrial processes but also as a household fuel and, being lighter and with a higher calorific value than firewood, could be economically transported over long distances and especially to fuel-hungry London. Numerous place-names, such as Colliers End in the parish of Standon, Coles Green in Walkern, Cole Green in Hertingfordbury, Coles Heath in Little Hadham and Coles Hill in Northaw, testify to the prevalence of the practice in medieval times, although direct references in the documentary record are comparatively sparse. In 1475 Sir John Say's manors around Little Berkhamsted sent charcoal to London and in 1484/5 four carts of charcoal were taken there from Bedwell.[22] Large quantities were made in the bishop of Ely's great park at Hatfield at the end of the fourteenth century, making a profit of over £28, but whether destined for London or to supply the kilns of the south Hertfordshire pottery industry remains unclear.[23] Forty loads of 'carbon' made on the manor of Baas in Broxbourne in 1493 were delivered to the lord, Sir William Say, and may have been intended for sale in London or for his own use.[24] In the fifteenth century charcoal was also made in the deer parks at Knebworth, Bedwell and at Little Hadham where, in 1462, a ten-year lease for making charcoal was sold for £4 a year.[25] John Goffe, a charcoal burner from Cheshunt, took 12 sacks of charcoal to London in 1368, but his load was found to be deficient by two quarters and his sacks were burnt in punishment.[26] Charcoal continued to be made in the county in the seventeenth century. In 1610 William Brokett the elder, a gentleman in Essendon, left to his daughter Anne 'all the benefitt of the cole wood in Busshey Lepes, And Alsoe [the] croppe of wood next adioyninge to yt called Dringe hall grove, to be Felled next yeare'.[27] At the start of the century Sir Arthur Capel was paying about £12 18s each year to have charcoal made in his woods at Walkern in order to heat his home at Hadham Hall.[28] In 1611 his woodward was planning to coppice about eight acres of wood each year in an 11-year cycle and to make 35 loads of charcoal a year from the cut wood. The costs included felling and cutting the wood, peeling off the bark, making faggots of the brushwood, making the charcoal, carrying water to quench the 'hearth' and hedging and ditching the wood.[29]

Another important product of woodland was oak bark, stripped from timber trees or (less commonly) from coppice poles. Rich in tannin, it was added to the vats in which animal skins were soaked to make leather. Bark was regularly sold from the bishop of Ely's manor of Little Hadham in the fourteenth century: the bark from 60 trees fetched 19s 6d in 1332/3 alone.[30] Henry Gape paid 20s for bark from oaks felled in Wheathampstead in 1539:[31] Gape was a prominent industrialist who later became mayor of St Albans, leased properties in St Michael's from St Albans Abbey in 1538 and probably established the tannery excavated in the grounds of St Michael's manor in 1976.[32] Bark was a valuable commodity. That stored at a tannery described in the inventory of widow Judith Day of Hertford, taken after her death in 1688, represented 16.5 per cent of the total value of the business (£339 10s 6d) and was worth almost as much as all of her household goods (£58 12s 2d).[33] Once again, the trade continued well into the post-medieval period. In the late eighteenth century bark from Earl Cowper's Digswell estate was sold to Thomas Crawley of Welwyn and Daniel Green of Buntingford, both tanners;[34] while bark from Wain Wood near Ippollitts was sold in the years between 1755 and 1767 to George Mugliston, a tanner in Hitchin, in quantities ranging from 26 to 111 yards a year, the price rising steadily from 6d a yard in 1745 to 14d a yard in 1767.[35] The increasing scale of leather production, but static or even declining supplies of bark, ensured that in eastern England as a whole bark increased from a tenth to a third of the total value of an oak tree in the course of the eighteenth century.[36] The increasing use of bark, which ferments at high temperatures, in the 'pineries' (used for growing fashionable pineapples) and other hot-houses constructed in country house gardens of this period doubtless contributed to this price rise.

Because woods were private property their survival was dependent on the whims of particular owners, and major changes in patterns of ownership could have a significant impact on their extent. The great monastic houses owned large tracts of woodland, especially in the south of the county and, following the Dissolution, new lay owners were often tempted to fell and sell the timber, or even to grub out a wood altogether and turn land over to grass or arable, in order to reap the more immediate profits to be gained from crops or livestock and thus recoup the purchase price of their estate. Alarmed at the loss of woodland and the prospect of a shortage of 'timber for building, making, repairing and maintaining of houses and ships, and also for fewel and fire-wood', an act of parliament was passed in 1543 which decreed that on every acre of coppiced wood felled at over 24 years of growth, 12 'standrells or store' trees (preferably oak) were to be left standing. Coppices were to be fenced and protected from cattle and a substantial fine of 6s 8d was imposed per tree on anyone caught cutting trees on waste or common land.[37] Destruction of woodland nevertheless occurred on a substantial scale. Five centuries ago the area around Barnet was one of the most densely wooded in Hertfordshire but by 1905 the parish was 'quite destitute of woodland'.[38] The main losses were two large woods called Hossegge or Owsage and Monkenfrith, which by 1553 had passed to Thomas Savage, keeper of 'Marybone Park'.[39] They together contained '380 oaks of between 40 and 60 years' growth, but by 1556 Savage had felled 200 of the best trees, leaving 180 of the smallest specimens.[40] Monkenfrith had been converted to farmland by 1584 but there was still an 'Ouzage Wood' in East Barnet in 1615.[41] During the seventeenth century the latter was incorporated within the grounds of Ussage House, but very little woodland remained by the end of the eighteenth century.[42]

Around 30 per cent of Hertfordshire's woods were still being coppiced in 1905,[43] but the practice was already in steep decline owing to the widespread use of coal and the substitution of manufactured products for such things as wooden hurdles, formerly a major use of poles. By the end of the twentieth century only 5 per cent of the county's woods were thought to be actively coppiced.[44] This figure is probably now rising, owing mainly to a recognition of the value of well-managed woods for wildlife conservation, but also because, with increasing prices of fossil fuels and concerns about rising levels of carbon dioxide in the earth's atmosphere, managing woodlands as a valuable fuel resource is once again becoming a feasible and economically attractive option.

Deer parks

Some of the dwindling tracts of 'waste' enclosed by great lords in the course of the eleventh, twelfth and thirteenth centuries were managed not as coppice woods but as deer parks. These likewise provided a supply of pasture and wood, but also of pleasure and prestige: parks were private hunting grounds but also, and often more importantly, they were venison farms, providing food for feasts and a luxury commodity which could be bestowed as a gift on friends, neighbours and dependents, forging bonds of loyalty.[45] The earliest parks in Hertfordshire were probably established in the eleventh century, three being recorded in Domesday Book at Benington, Ware and St Albans. By 1300 there were around 40, with another 30 appearing by 1500. Many parks, especially the smaller ones, were relatively short-lived, but the overall number present in the county remained remarkably constant, at about 30, for most of the fourteenth and fifteenth centuries, the establishment of new parks compensating for the loss of old.

The earliest parks, from the eleventh and early twelfth centuries, were almost all created on the manors of major aristocrats or great religious houses. Benington, for example, was the power base of Peter de Valognes, nephew of William I; Cheshunt was granted to William's son-in-law, Count Alan, son of the Duke of Brittany; while Standon was the property of the powerful de Clare family. The largest early medieval park in Hertfordshire was at Hatfield, where the bishop of Ely enclosed around 1,860 acres of the undulating wood-pasture and heath between Lower Woodside and Newgate Street some time before 1222.[46] Unusually, the tenants of Hatfield manor retained their rights of common pasture, pannage and 'woderight' within the park but, in return, they were bound to maintain a section of the park boundary.[47] By 1285 the bishop also had a substantial park on his manor of Little Hadham, extending over some 260 acres, which had been enclosed from the waste in the north-west corner of the parish.[48] But it was the abbots of St Albans who had the largest number of parks in the county, although they were probably not all in use at the same time. The earliest were perhaps at Eywood, just south of St Albans, and at Derefold, a little to the west. Parks are also recorded on the abbey's manors of Childwick, Gorhambury (both in St Michael's parish), Redbourn, Tyttenhanger, Boreham Wood and Bramfield.[49] There were also large examples belonging to the Crown, at King's Langley, Berkhamsted and Hertingfordbury, but most medieval parks in the county, and especially those established after the middle of the twelfth century, were the property of minor manorial lords, men such as William de Say of Sawbridgeworth (whose park was first recorded in 1237), William de Ayot of Ayot St Lawrence (1268), Simon de Furneus of Furneux Pelham (1274) and Gerard de Furnival of Great Munden (1283).

Medieval parks were significantly more numerous in the eastern half of the county than in the west, reflecting the higher population levels in the former district and the greater numbers of manors whose lords might aspire to owning one. In the west, in contrast, a lower number of lordships, and the dominance of St Albans Abbey and other ecclesiastical landowners, militated against their proliferation (Figure 6.1). Most parks were relatively modest in size, and often located on the high ground towards the margins of a parish, where land sufficiently well-wooded to be worth enclosing for the purpose survived into the twelfth and thirteenth centuries. On the watershed between the valleys of the rivers Rib (and its tributary the Quin) and Ash, for example, there were, by 1336, parks belonging to the manors of Newsells (Barkway), Hormead, Furneux Pelham, Albury, Little Hadham and Standon; while on the interfluve between the Ash and Stort by 1347 there were examples belonging to the manors of Hadham Hall (Little Hadham), Bishop's Stortford, Much Hadham, Eastwick, Sayes and Hunsdon. Not all medieval parks were located in remote parts of the uplands, however: some, such as Ardeley, Standon, Hayley, Benington, Walkern and Great Munden, were placed towards the edges of the clay plateau, even extending down onto the lower ground of the valley sides. Some of these, such as Benington or Standon, were particularly early creations, established when large tracts of wooded waste survived even in locations close to settlements. Other parks, such as those at Little Hadham and Albury, occupied areas of particularly difficult soils, as where heavy London Clays rather than Chalk underlay the boulder clays. In the south-east of the county the parks at Cheshunt and Hoddesdon similarly lay on the fringe of the Southern Uplands above the Lea valley, in an area of peculiarly intractable soils.

Many early medieval parks lay some distance from the residence of their owner. Indeed, it was generally only at the highest social levels – at royal residences such as King's Langley palace or Berkhamsted castle – that park and mansion lay in close proximity. But towards the end of the Middle Ages, as we shall see, 'residential' parks proliferated among local landowners and new parks were laid out around manor houses

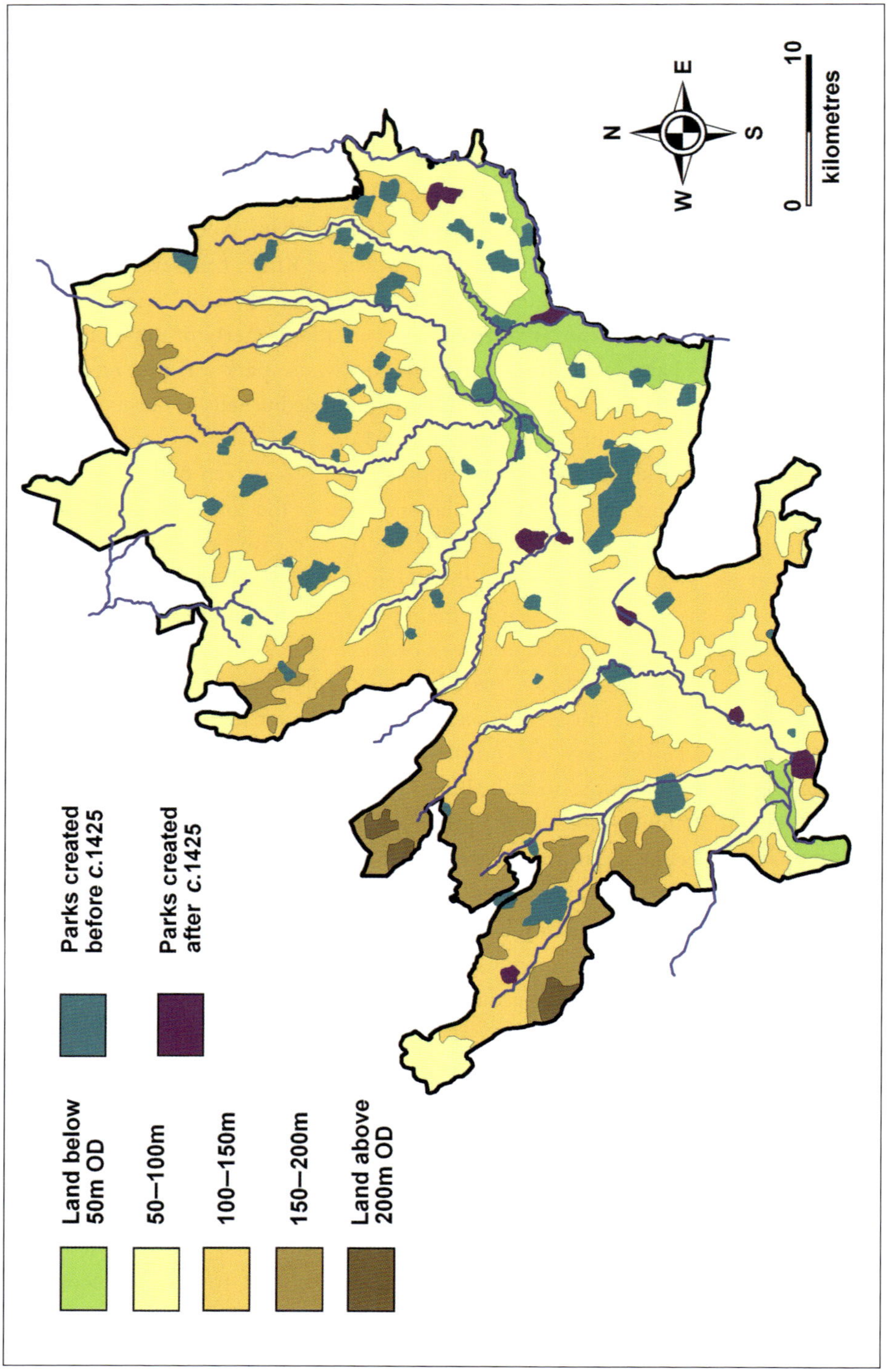

Figure 6.1. The distribution of medieval parks in Hertfordshire.

at the expense of farmland, such as those established on the banks of the Colne in the 1420s: at The More in Rickmansworth (1426), and at Bushey Hall and at Tyttenhanger (both *c.*1427). At least a third of the deer parks known to have existed in the fifteenth century continued in use into the sixteenth, mainly the larger examples (all but three were over 200 acres in size, and half covered more than 300 acres).[50] Indeed, almost all the medieval parks extending over 300 acres or more continued in use as deer parks into and beyond the sixteenth century.[51]

Although sometimes viewed primarily as landscapes of leisure and as essentially unproductive areas, most parks in fact were managed to provide a significant income for their owners. Grazing and pannage were leased or directly exploited by demesne farms, but parks were perhaps most highly prized for the timber that could be grown within them – either as free-standing trees or in areas of enclosed woodland. From medieval times onwards there are records of large quantities of timber being taken from the county's parks: 500 ash trees were sent from the park at King's Langley to the Tower of London in 1277, 66 oaks were sold from the park at Standon for £9 in 1328/9, 60 beech trees from Berkhamsted park in 1358, and beech worth £8 10s and £10 13s from Knebworth park in 1410/11 and 1412/13, respectively.[52] From the sixteenth century there was an increasing demand for timber for the navy – particularly for the curved timbers obtained from free-standing parkland oaks (standards in coppiced woods, in response to competition with the underwood, tended to grow in an upright form, without low branches). The survey of Theobalds Park made for the revolutionary government in 1650 recorded 15,608 trees 'marked for the use of the Navie ... besides what is already cutt Downe', which still left wood worth £7,259 in the 2,500-acre park.[53] At nearby Cheshunt Park 577 trees were considered suitable for the navy, leaving a further 713 trees valued at £77 7s.[54] Indeed, parks continued to be regarded as a major source of timber well into the post-medieval period: parkland trees represented vast reserves of capital which could be realised by the owner as and when needed. The timber trees growing in the 320-acre park at Broxbournebury in 1784 included 381 oaks, 172 elms, 115 chestnuts, 20 walnut, 8 ash and 3 beech, plus 1,325 young trees (spars) and 991 pollards.[55]

Wooded commons

Where the medieval 'wastes' remained unenclosed – where they were not converted into either managed parks or coppiced woods – they did not necessarily lose their tree cover rapidly. Wooded commons long remained a prominent feature of the landscape, at least in the less densely settled parts of the county and in places remote from large settlements – that is, on the broad interfluves in the Chiltern Hills and in the Southern Uplands. In most cases manorial lords were the owners of timber trees, but the grazing remained the shared right of the commoners. The larger commons were often 'intercommoned' – that is, rights to their use were shared between two or more manors and their tenants. As in deer parks, a high proportion of trees were pollarded – coppiced at a height of two metres or more,

Figure 6.2. Old beech pollards in Frithsden Beeches near Potten End: a surviving fragment of the once extensive wood-pasture commons of the Chiltern ridge.

out of range of browsing stock – in order to produce a regular crop of poles. The trunks (or *bollings*) were the property of the lord but the lopped wood usually belonged to the commoners. Well-spaced trees and regular pollarding enabled sufficient ground vegetation to flourish to provide grazing for the commoners' cattle. For reasons which remain unclear, many common woods in the county had names which incorporate the Old English term *fyrhð*, 'land grown over with scrub': thirteenth-century tenants of Westminster Abbey thus had rights of common pasture in the abbot's woods at Aldenham Frith; Monkenfrith was the name of a common wood in Barnet recorded in 1305; the 'Boreham Frith Farms', which lay near to the 'Common Wood called Boreham Wood', were recorded in 1607, as was Frith House near Bricket Wood at Smugoak Green (where a coppiced wood nearby was called the 'great Frithes'); Berkhamsted Frith and Frithsden Beeches were the names given to wooded portions of the extensive Berkhamsted Common (Figure 6.2).[56] Whatever their names, these common wood-pastures were, as already noted, inherently unstable environments. Once trees died, were felled or were brought down in gales they could not easily be replaced because of grazing pressure.

Some of the best documented wood-pasture commons were in the south-east of the county, on the Southern Uplands. Many commons here retained their tree cover longer than elsewhere, although they appear to have been losing it fast by the later seventeenth and eighteenth centuries. As we have already noted, in the late seventeenth century Barnet Common was described as 'formerly a wood, but … of recent years laid waste, and used

Figure 6.3. One of a handful of old hornbeam pollards surviving within a conifer plantation on the former Broxbourne Common. More than 3,000 trees were pollarded on this common every twelve years until the late eighteenth century.

as a common'.[57] The Finnish visitor Pehr Kalm described the great tract of undulating common land which existed 'between Cheshunt and Bell Bar', south of Hatfield, as follows:

> Common ling grew on it abundantly. It was covered with tufts of ling, between which bracken flourished and swamps abounded. But there was scarcely any grass. Sheep grazed here. In places *Carpinus* (hornbeam) grew fairly densely to a height of six feet, and the tops of it were cut for fuel. Otherwise it was of no particular use.[58]

Half a century earlier it seems that Cheshunt common would have been much more densely wooded. In 1695 there were said to be 24,000 hornbeam pollards growing on its 1,186 acres, around 20 trees per acre.[59] An average of around 3,800 of these were lopped each year over the 20-year period from 1658 to 1678, suggesting a short pollarding cycle of around six or seven years.[60] Unusually, the lord of the manor of Cheshunt owned not only the bollings of the pollards but also their 'heads': the wood was sold uncut at a set price, usually in batches of 100 or 50 heads, to local residents who lopped them themselves. The yearly income accruing to the estate ranged from £30 to £145 16s, averaging about £75 over the 20-year period. The same manorial lord also owned the much smaller Broxbourne Common, covering only around 60 acres, upon which in 1682–3 there were at least 4,420 pollards, an apparent density of 73 trees per acre.[61] The lops of these trees – likewise mainly hornbeam, to judge from the few surviving examples (Figure 6.3) – also belonged to the lord but the price charged per 100 heads was more variable, perhaps reflecting the variable quantity of wood produced by trees growing on different parts of the common. The pollards were also managed differently. Instead of a proportion being cut each year, after about six years' growth, as at Cheshunt, they were here cut in two phases in consecutive years as part of the 12-year rotation of coppicing and pollarding which extended to all the woods and hedgerows on the Broxbournebury estate. This, together with the different method of pricing the pollard heads, suggests that the pollards on the two commons had different management histories perhaps extending back into the Middle Ages. Despite sharing the same manorial lord, their management diverged even further from the end of the seventeenth century. The pollards on Cheshunt common were all sold and felled; those on Broxbourne common, in contrast, continued to be lopped every 12 years until at least 1777–8.[62]

Trees of woodland and wood-pasture

The extent to which the composition of coppiced woods and of private and common wood-pastures was primarily a consequence of environmental factors or of management practices remains unclear. Medieval documents suggest that oak, elm and ash were common as both timber trees and pollards throughout the county: on common land oak was more frequent than either ash or elm, although hornbeam, as we have seen, appears to have been the preferred choice on the commons in south Hertfordshire, while beech was dominant on many commons in the Chilterns. Ash made excellent firewood, and its somewhat elastic wood and timber were greatly valued for their load-bearing qualities. The wood matures early and poles of just three inches in diameter are sufficiently durable for a wide variety of uses, including axe-handles, ladders, shafts for carriages, hurdles and crates.[63] Elm had a range of specialist uses, especially for making water pipes, but was above all the main timber for floor boards. Oak, in contrast, was the most important structural timber, and trees from Hertfordshire were often transported to London for prestigious building projects, as for example in 1514 when 873 loads of timber from the abbot of St Albans' woods at Barvin near Northaw were purchased to build the Savoy Hospital.[64]

Beech (*Fagus sylvatica*) and hornbeam (*Carpinus betulus*) were also a familiar sight in the medieval county, and have interesting and slightly puzzling histories. Beech has long been a particular feature of woodlands on the Chilterns: the tree features prominently in the medieval timber records for Berkhamsted; the manor of Caddington contained a 'great beechwood of 300 acres' in the early thirteenth century; and the name of the former Buckwood near Markyate is derived from the Old English *boc*, 'beech'. It was also abundant on the Chiltern dipslope: there were 1,420 beech trees in the park at King's Langley in the mid-sixteenth century, far outnumbering the 300 ash and 180 oak,[65] while at Bovingdon in the first decade of the nineteenth century beech was the species most commonly recorded in the timber accounts.[66] Large quantities of beech were also sold from the lord's park at Knebworth in the early fifteenth century and there were still large beechwoods nearby, at Codicote and St Paul's Walden, at the start of the twentieth century.[67] What is perhaps surprising is that the available documentary evidence suggests that beech was also once abundant in the south-east of the county, where few now exist. Beech was thus a major component of the deer parks owned by the bishops of Ely to the south-east of Hatfield.[68] Beech timber worth 68s was sold from both the great and little parks at Hatfield in 1428 and a combined total of 12,000 oak and beech trees was recorded in a survey of the parks in 1538.[69] Another survey in 1626 recorded twice as many beech (520 trees) as oak (227) in Hatfield Great Wood, the former great park, and large amounts of beech timber were sold from the New Park (within the Great Wood), including 400 trees in 1629.[70] To the south of Hatfield, beech comprised about 20 per cent of the trees in woods on the manor of North Mymms in the early fifteenth century and it was a major component (together with oak and hornbeam) of the woods in Broxbourne and Hoddesdon.[71] It is unclear when, or why, beech declined in these districts but surveys show that this situation has to some extent been reversed once more in recent decades, and that 'on acidic gravels in the south of the county it has been until recently increasing steadily at the expense of oak'.[72]

Hornbeam was poorly represented in the prehistoric landscape, to judge from the evidence of pollen analysis, and it does not feature in documents relating to Hertfordshire before the second half of the fourteenth century, when it is mentioned growing in the deer parks at Little Hadham and Albury. In fifteenth-century manorial accounts for Great

Munden and Knebworth the bailiffs recorded receipts from 'trees called hornbeam', as though the name of the tree was new to them.[73] The earliest use of the tree in an English place-name is thought to be Hornbeamgate in Hatfield – 'Herinbenegatestrat' (street) was recorded in 1366 and the manor of 'Hermebemgate' first mentioned the following year:[74] the gate was an entrance into the bishop of Ely's great park at Hatfield. Most of the records relating to parks concern pollards but hornbeam was also managed as coppice, presumably within enclosed compartments, in Little Hadham park. It was recorded in mid-sixteenth-century surveys of the deer parks at Hunsdon and The More in Rickmansworth, but not as a timber tree: at Hunsdon hornbeam was confined to areas of 'wood ground' within the parks which were 'set with' maple, hornbeam, thorns and 'other fir[e]wood', while at The More a 261-acre wooded area within the park was said to be 'set with very little timber [but] well stored with pollards of maple and hornbeam'.[75] On the other hand, a survey of Hertingfordbury park in 1608 recorded '1,126 oaks, 3,311 hornbeams and other trees in the park' while in Hatfield Great Wood in 1626 there were 5,227 hornbeams 'of all sorts' – ten times the number of beech, the next most numerous species.[76]

It is hard to account for the apparent spread of the species through the county in the course of the fourteenth and fifteenth centuries, unless it was being actively planted or otherwise encouraged. The bishops of Ely and other landowners may have deliberately promoted it because of its value as firewood and for making charcoal, although according

Figure 6.4. Neglected hornbeam coppice in Broxbourne Woods. In some parts of the woods management by coppicing is being re-established under the auspices of the Countryside Management Service and the Woodland Trust, to the great benefit of wildlife.

to some sources hornbeam was also favoured as browse for deer, and it was certainly being used in this way at Little Hadham in 1435/6.[77] We should also note that the seventeenth-century Hertfordshire forester Moses Cook recommended planting hornbeam in parks because deer would not eat the bark.

Hornbeam also occurs as the main component of coppiced underwood in woods right across the centre and south-east of the county.[78] Indeed, Hertfordshire has some of the best examples of hornbeam coppice in England.[79] It is possible that its prominence in such woods may have increased in the century or so since coppicing was abandoned, for uncut stools have grown to canopy level and other species fail to thrive in the resultant deep shade (Figure 6.4). It is unfortunate that we know little of the species composition of these woods in medieval times: documents seldom describe the character of the coppiced understorey. Today this certainly displays much variation across the county, related in large measure to soils: hazel, ash and maple tend to characterise the understorey of woods on the more alkaline soils in the east, hornbeam, as noted, dominates in the south and centre, while in the Chilterns – where much coppice was converted to beech high forest in the eighteenth and nineteenth centuries – the underwood can include hazel, holly, birch, oak and hornbeam where the shade is not too dense.

Commons, warrens and meadow

Although a few commons, such as Broxbourne, continued to carry dense stands of pollards into the eighteenth century, they were unusual. By this time the majority of wood-pastures had degenerated to open grazing, a process which had continued, at rates that varied from place to place, for centuries. Indeed, many commons that were largely or entirely treeless by the time they first appear in the documentary or cartographic record have names which betray their wooded origins, including Commonwood Common in Rickmansworth, Mymwood Common in North Mymms and Gustardwood Common in Wheathampstead. The kinds of landscape that emerged as trees disappeared varied with environmental conditions. In the east of the county, on the boulder clay plateau, fragments of waste left after assarting gradually became the various open greens and commons which are so distinctive of the area: small or medium-sized areas of alkaline grassland (above, pp. 71–4). But in areas where soils were acidic, land cleared of trees developed into heath, with a vegetation characterised by heather (*Calluna vulgaris*), gorse and broom. Some areas of heathland may have existed naturally and others were created through clearance and grazing during prehistoric and Roman times, before the Anglo-Saxon period, as indicated by names incorporating the Old English element *hæð*, such as Hatfield (the 'heathy open land') or (Monken) Hadley (the 'heathy *lēah*').[80] But many were of medieval or even post-medieval origin.

Whenever they developed, by post-medieval times heaths were widespread in the county. They were prominent where Pebble Gravels cap the London Clay on the higher ground in the south, and also on the lower ground in the district where the Lambeth

Group contained large quantities of sand: notable examples included Batchworth Heath, Bushey Heath, Letchmore Heath, Watford Heath, Barnet Heath, Cow Heath (Broxbourne) and Hertford Heath. Further north, heathland developed on glacial sands and gravels over Reading Beds at Bernard's Heath near St Albans and at Patmore Heath, Albury. Glacial sand and gravel overlying Chalk gave rise to Codicote Heath and to Pottersheath and Mardley Heath in Welwyn; deposits of Plateau Drift, decalcified and acidic, underlie the former heath at Breachwood Green in King's Walden; while Bower Heath and Marshall's Heath, to the west of Wheathampstead, occupy areas of pebbly clay and sand overlying Chalk at depth. In some places very thin acidic deposits, usually comprising coversand or brickearth, overlying Chalk give rise to the distinctive 'chalk heaths', in which grasses such as sheep's fescue are dominant and heather and gorse constitute subsidiary elements, as at Berkhamsted and Northchurch Commons and Caddington Heath in the Chilterns, and at Therfield Heath in the far north-east of the county. Such areas shade off, almost imperceptibly, into true chalk downland, with its great range of wild flowers and butterflies. Little of this habitat survives in Hertfordshire, in part because there was never very much – the northern boundary of the county only extends onto the escarpment of the Chilterns, where the main exposures of Chalk are found, in a few places – but also because much of the escarpment was ploughed up in the Middle Ages and, in some locations, much earlier.[81] Such areas of open grazing as existed, many probably chalk heaths rather than true downland, were generally converted to arable following enclosure in the early nineteenth century, and today downland can be seen at just a few sites, including Roughdown and Sheethanger Commons in Hemel Hempstead, in Tring Park and on the escarpment south and east of Hexton.

Large areas of heaths and other commons survived into the eighteenth century, but many were then subject to enclosure and 'improvement'. Those on the Southern Uplands were, as we have seen, largely destroyed in a wave of parliamentary enclosures in the early nineteenth century, while on the Chiltern Hills almost all have been lost to agriculture. Caddington Heath, for example, was enclosed in 1800 and shortly afterwards Arthur Young was able to record, with obvious approval, that 'Mr Pickford of Market-street [Markyate], hired on a long lease 509 acres of the heath, at 14s an acre; and he has improved it with admirable spirit indeed.' Within 18 months, all but 127 acres had been ploughed up.[82] Significant areas of common land have nevertheless survived in the Chilterns, at Berkhamsted, Aldbury and Hudnall; and numerous smaller areas of heathland still exist, scattered throughout the county, including Nomansland Common on the boundary between Sandridge and Wheathampstead. The great chalk heath at Therfield remains, and numerous small greens and commons can still be found on the boulder clays in the east of the county. Whatever their precise origins, all are semi-natural environments which owed their character and appearance to regular grazing and, in the case of heaths, to the regular cutting of bracken, gorse, broom and heather (above, pp. 102–3). As such traditional management practices declined in the course of the twentieth century the majority of

Hertfordshire commons were gradually colonised by birch, hawthorn and other scrub, and subsequently by oak: most have now returned, or are returning, to woodland. It is only through the strenuous efforts of organisations such as the Herts and Middlesex Wildlife Trust that a proportion are kept free of woodland and scrub, through the cutting back of vegetation and, where possible, by reintroducing grazing. The latter is often difficult because most commons are crossed by roads: the increasing number of cars in the first half of the twentieth century was one of the main reasons for the decline in the use of commons for grazing livestock (Figure 6.5).

Some areas of heathland or other kinds of common land on freely draining soils were enclosed by manorial lords in order to create rabbit warrens or 'coneygarths' – 'coney' being the usual term for the adult animal before the eighteenth century. Rabbit farming made good use of poor, acid soil, or of slopes too steep to plough, and warrens were particularly numerous on the gravels and sands in the Vale of St Albans and in the major valleys draining into this.[83] Other warrens were established within deer parks, or in some convenient place beside a manor house. The rabbit is not indigenous to England, in spite of its current ubiquity, but was introduced by the Norman aristocracy in the twelfth century as a source of meat and fur.[84] In the early fifteenth century the abbot of St Albans made a warren by enclosing 30 acres of Colney Heath next to his manor of Tyttenhanger: a document drawn up in 1542 described how the warrener 'was allowed to have firebote in the warren, and sufficient timber for hutches and traps for destroying all manner of vermin, and to cover, ditch, plassh, and lay the burrows'.[85] Warrens were unpopular with commoners, as we have already noted in the case of Northaw in 1544, because rabbits

Figure 6.5. Sheep grazing on Harpenden Common at the turn of the nineteenth and twentieth centuries – a practice which died out as the volume of motorised traffic increased. Reproduced by kind permission of the Harpenden & District Local History Society (vis 328).

Figure 6.6. Detail of a map of the manor of Digswell *c.*1599, showing the rabbit warren at Black Fan, complete with rabbits, the warrener, his house and his dogs and a mixture of standard and pollarded trees. Reproduced by kind permission of HALS (DE/P/P1).

competed with their livestock for the often meagre herbage and caused damage to arable crops in neighbouring fields. The lord of the manor of Tring was taken to court by his own tenants in the 1580s when he tried to establish a warren within the common woods of The Leith and Chickenswick. He had constructed 'burrows' – artificial breeding mounds – and the rabbits were said to be damaging the wood and causing 'great spoil and waste'.[86] Even when established on private land warrens caused problems for local people. One example, extending over more than 80 acres, was created on farmland by the lord of the manor of Wigginton in the late sixteenth century but the rabbits soon escaped, ruining the crops on neighbouring farms and the grazing on the nearby common called Mayne Wood.[87]

A map of the manor of Digswell, drawn up in 1599, shows the manorial warren in wonderful detail, complete with the warrener, dogs and rabbits (Figure 6.6).[88] The pollards depicted were probably used in part to provide supplementary food during the winter months, the branches being lopped in full leaf and then dried and stored as 'leafy hay' until needed. A lease for the warren at Knebworth dating from 1722 thus allowed the tenant 'once in every yeare' during the 21-year lease 'to Lopp sixty Pollards for Browse for the Rabbitts'.[89] Rabbits were kept for their meat – a luxury food into the seventeenth century – but also for their fur, and many warrens specialised in rabbits of a particular colour. A lease drawn up in 1634 for the warren at Hertingfordbury ordered that the warrener was to leave, at the end of the nine-year term, 'one hundred and twenty female Conyes and twenty male Conyes wheirof the better halfe to bee black'.[90] The warren at Knebworth, in contrast, was stocked with 'Silverhaired Rabbits' when leased in 1722. Warrening was still going strong at this time. Cesar de Saussure, a Swiss visitor to the county in 1728, described how:

> Most country houses have parks and rabbit warrens. ... Rabbit-warrens are likewise very lucrative to proprietors, if not too far from London. Some rabbits are of a grey colour, mingled with white and black; these skins make fine furs, and are sent to Dantzig and Hamburg, and from thence to Poland, where they are much sought after, for you know of course that in that cold country furs are greatly worn. The skin of this sort of rabbit is expensive, bringing in more money than its flesh does.[91]

But rabbit warrens were not just economic enterprises. Because the right to establish them was a manorial privilege they also served as important expressions of status, like deer parks; and, like parks, they might, on occasions, be used as elements within the designed landscapes laid out around country houses, as we shall see. They became less and less fashionable in the course of the eighteenth century, however: to agricultural improvers such as Arthur Young they were old-fashioned, wasteful forms of land use, and developments in farming technology allowed more profitable uses of the kinds of poor, light land they generally occupied. The large warren at Lockleys near Welwyn survived long enough to be recorded on the first edition Ordnance Survey map of 1884 but most warrens had ceased to exist by 1840, their sites commemorated in innumerable field names scattered across the county.

One other important type of traditional non-arable land use, now very rare in the county, was meadow land – permanent grassland from which livestock were excluded during the spring and early summer, so that the grass could grow long and be cut for hay, the main form of winter fodder. Few meadows are now managed intensively and many were ploughed up, or otherwise destroyed, in the course of the twentieth century. Meadows are of considerable importance for biodiversity and several of those which survive – including King's Meads between Hertford and Ware and Hunsdon and Eastwick Meads – are now in the care of the Herts and Middlesex Wildlife Trust (Figure 6.7). Because they were closed to livestock for much of the year plants intolerant of grazing were able to flourish, flower and set seed without disturbance, providing a display of summer colour. Most meadows were located in the valleys of rivers and major streams, where high water tables ensured rich grass growth even in summer drought, although even in medieval times some were found in upland locations, a tendency which increased in the post-medieval period. Meadows were usually divided into strips held by individual farmers, like arable open fields – as shown beside the river Lea on a map of the Bayfordbury estate in 1807[92] (Figure 6.8) – but they were grazed in common after the hay harvest. In some parts of England

Figure 6.7. Buttercups and ragged robin on Hunsdon Mead, a nature reserve managed by the Herts and Middlesex Wildlife Trust. Photograph by Clare Gray. Reproduced by kind permission of the Herts and Middlesex Wildlife Trust.

Figure 6.8. Detail from a map of the Bayfordbury estate, surveyed in 1807, showing meadows beside the river Lea which are still in part divided into strips. Reproduced by kind permission of HALS (DE/X629/P1).

meadows were 'floated', or irrigated, firstly in the winter to force an early growth of spring grass, which could be grazed for a month or so before the area was closed to livestock to allow the hay to grow; and again in the early summer, to counteract any deficiency in soil moisture and produce a bumper crop. Complex systems of channels and networks of raised ridges were often constructed to achieve these ends. In Hertfordshire floating was only sporadically practised, and in a distinctive form. Arthur Young noted in 1804 how the main concentration of such meadows was on the river Colne near Watford, and that these were only irrigated in the summer months, and never in the winter: the enhanced growth ensured that they could be mowed twice in an average year.[93]

Hedges and farmland trees

In the past hedges featured far more prominently in the Hertfordshire landscape – and in the lives of ordinary people – than they do today. They not only provided stock-proof barriers and a clear definition of ownership but were also a major source of fuel. Lying far from any coalfields and with very limited supplies of peat, the county was reliant upon local sources of wood for cooking and heating until the advent of the railways in the nineteenth century made coal more widely available. While coppiced woods and wood-pastures provided much fuelwood, hedges were also carefully managed for this purpose, and as late as 1804 Arthur Young described those in Hertfordshire as the 'collieries' of the county.[94]

Unlike the thin and often gappy rows of shrubs that we see today, hedges in earlier centuries were more rigorously managed features, and often more substantial, some resembling narrow strips of coppiced woodland rather than hedges as we would understand them. Sir John Parnell asserted in 1769 that he knew:

> No part of England more beautiful in its stile than Hertfordshire. Thru'out the oak and Elm hedgerows appear rather the work of Nature than Plantations generally Extending 30 or 40 feet Broad growing Irregularly in these stripes and giving the fields the air of being Reclaim'd from a general tract of woodland.[95]

Features like these were commonly referred to not as hedges but as 'hedgerows', or even as 'hedge groves', and they are described in documents in terms of the area that they covered. The 22 acres of woods and underwoods growing on St Lawrence Farm in Cheshunt in 1556 were largely comprised of hedgerows and were listed as follows:[96]

In A thynne hedgerow on the west side of Hiefilde cont	1a
In A hedgerowe on the south side of the same filde cont	1a
In A hedgerowe ioyning to Middle filde cont	2a
In A hedgerowe and A littill Grove ioyning to Middle filde	2½ a
In A hedgerowe ioyning to HeyReke cont	½ a
In A hedgrowe ioyning to Welcrofte Fild cont	3a
I[n] hedgerowes ioyning to Wormeleyfielde cont	4a
In A Grove ioyning to the Common Woodd cont	8a

A little earlier, in 1543–4, Henry VIII's woodward in Hertfordshire recorded the sale of eight acres of underwood 'from divers hedges in the parish of Flampsted called Hedgerowys' worth £10 13s 4d.[97] At Broxbourne in the mid-eighteenth century 'Whitestubbs Hedge Row', 'Pimbridge Hedge Row' and 'Palmer's Hedgerow' were all incorporated into the same 12-year coppicing rotation which applied to the woods and wood-pastures on the Broxbournebury estate.[98]

Several estate maps drawn in the eighteenth century clearly distinguish between narrow 'hedges' and broader 'hedge rows': a map of Earl Cowper's estate at Hertingfordbury in 1704 thus depicts 'hedges' with green lines and 'hedge rows' as broader green strips containing trees; a slightly later plan of the Gaddesden estate shows broad hedgerows bordered by trees extending around three fields, with narrower boundaries around the others; while a plan of Herons Manor Farm in Wheathampstead, drawn up in 1768, similarly distinguishes these different kinds of field boundary.[99] The majority of Hertfordshire field boundaries were by this time, and probably always had been, of the more normal, narrow kind. Some may also have been managed by coppicing but most, at least by post-medieval times, were laid or 'plashed': that is, cut and woven at intervals to create a strong stock-proof barrier. The Hertfordshire version of this ancient practice was described by Arthur Young in 1804.[100] A row of upright living stems – 3–4 feet apart and preferably of whitethorn – was left behind when the bulk of the hedge was cut and cleared away. These provided a sturdy framework: other stems, cut and trimmed to a uniform height, were kept as vertical hedge-stakes but the remainder were almost (but not quite) cut through near the base, then bent over to a near-horizontal position and woven between the hedge-stakes. Small twigs called 'edders' or 'ethers' were woven along the top of the completed hedge to bind the hedge-stakes together. Both upright and horizontal stems remained alive and quickly sprouted new growth, so that the hedge soon became a dense green wall impenetrable to stock.

Plashing is recorded at Essendon as early as the late fifteenth century and is described in numerous farm leases from the seventeenth and eighteenth centuries: hedge management was an important matter which was carefully regulated by landlords.[101] Such documents commonly state that ditches were to be scoured and pollards lopped at the same time as the associated hedges were plashed.[102] Young recorded that hedges were generally plashed every 12 years, but leases show that intervals of between 9 and 12 years were common and in the late eighteenth century copyholders of the manor of Barnet were expected to plash every 7 years.[103]

A seventeenth-century lease of farmland at King's Walden mentions the tenant's right to take the 'offal' – the material cut from the hedge when it was laid – for 'firebote',[104] and in general it is clear, as already noted, that hedges as much as hedgerows were a major source of fuel in the county. Arthur Young described in 1804 how the need for firewood had 'induced the farmers to fill the old hedges everywhere with oak, ash, sallow, and with all sorts of plants more generally calculated for fuel than fences'.[105] Pehr Kalm, who came to England in 1748, similarly noted how in the Hertfordshire Chilterns hedges were commonly planted with a mixture of hawthorn and sloe, but that in addition the farmers 'planted – either at regular intervals or more casually – small saplings of *Salicibus* [willow], beech, ash, rowan, lime, elm and other deciduous trees'.[106] Farmers and landowners planted a range of shrubs partly because it was often hard to source large quantities of hedging thorn but also because, while the latter made a good stock-proof barrier, it had few other practical uses and made indifferent firewood, awkward to handle. Only in the course of the eighteenth and nineteenth centuries did the planting of single-species hawthorn hedges become normal, especially for the new boundaries established by parliamentary enclosures.[107]

Those who believe that hedges can be accurately or even approximately dated by counting the number of species they contain in a standard 30-yard length, in the manner suggested in the 1960s by Max Hooper, should bear this pattern of development in mind.[108] Even if hedges did indeed acquire new 'colonists' at a relatively even rate, as Hooper

Figure 6.9. Typical species-rich east Hertfordshire hedge containing hawthorn, blackthorn, dog rose, hazel, field maple and wild clematis, or Traveller's Joy.

believed, counting the number of species that a particular hedge now contains can tell us little about its age if we do not know how many species it was originally planted with. Most of the county's surviving hedges are, to varying degrees, mixed but the kinds of species present, and probably the number, is at least as much a consequence of environmental as of historic factors. In the east, on the damp and fertile boulder clays, mixed and often species-rich hedges containing hazel, maple and dogwood, with significant quantities of ash, elm, hawthorn and blackthorn, and with scattered examples of hornbeam, wayfaring tree and spindle, are common (Figure 6.9). On the more calcareous soils of the district a particular feature are the great masses of Old Man's Beard (or Traveller's Joy), noted as early as 1548 by William Turner as growing 'plentiously between Ware and Barckway in the hedges, which in summer are in many places al white with the downe of thys vine'.[109] In the Chilterns, hedges containing a similar range of species are commonly found on the more calcareous ground. But on the acidic Plateau Drift dogwood and spindle are rare, hazel more common than maple, and the total number of species present generally less. In the south of the county the most species-rich hedges again occur on patches of more calcareous ground – where the Chalk is exposed in major valleys, or where patches of chalky boulder clay occur. On the heavy, acidic London Clays or on the spreads of Pebble Gravel even hedges of medieval date commonly boast no more than four or five species in a 30-yard length. On these poor soils blackthorn and, to a lesser extent, elm tend to out-compete other shrubs, suckering vigorously and gradually spreading sideways along the hedge. Many hedges are largely or even entirely composed of these two species. This is especially the case where they are no longer rigorously managed, and it may well be that, a century ago, many would have contained rather more species. Here, of course, is another aspect of hedge development not sufficiently acknowledged by believers in 'species dating' – hedges can lose species over time, as well as gain them.

In the past most hedges contained large numbers of trees: indeed, the majority of trees in the county, both standards and pollards, probably grew in hedges, rather than in woods and parks, or on commons. Some also grew in the 'hedge greens', unploughed strips around the margins of arable fields, while others were thinly scattered across pasture closes or – on occasion – densely filled them, forming diminutive private wood-pastures.[110] Timber trees were always excluded from leases and reserved for the landlord, as were the trunks of pollards. One example drawn up in the seventeenth century for a farm at King's Walden typically excluded 'timber and timber trees, bodies of all pollards and other trees', but allowed the tenant to take the lops and tops of pollards for firebote and his own use but not to sell.[111] The maintenance of future timber supplies was also a concern and landlords frequently sought to preserve (as they did in woods) promising young trees – variously termed spires, spars, storers, tellers or saplings – of oak, ash and elm.[112] The seventeenth-century tenant of Lannock Manor at Weston was even instructed to 'plant ten young trees every year (oak, ash or elm), cherish and preserve them'.[113] Penalties for failing to nurture future tree stocks could be harsh, and a late eighteenth-century tenant at

Barnet was expected to 'do every Thing in his Power for the Encouragement, and growth of the young Timber Shoots, under the Penalty of Twenty Shillings for every Shoot or Sapling which shall be wilfully hinder'd from growing'. He could also be fined 20s if he was to 'stub up, prune, or injure any Sapling', and five pounds for destroying or injuring any young timber tree.[114] Care was also taken to regulate the management of pollards, which were usually cut on a rotation of around 10 to 12 years, at the same time as the hedge was 'new made' and the accompanying ditch scoured out. A lease drawn up in 1657 for a farm in Ridge in south Hertfordshire thus allowed the tenant 'to lopp all the pollard hasells maples sallows willows hawthorns & hornebeame trees growing in the severall hedges, fields, dells & hedgrows' provided that he 'lopp and cutt but one tenth part of all the pollards hasells, maples sallowes willowes hawthorns & hornbeams every year for and during the last nyne yeres of the term'.[115] An agreement from 1693 concerning a farm in Aldenham stipulated that the tenant 'shall not lopp or cutt or cause to be lopped or cut any of the pollards growing upon the premises but when the hedges shall be new made and ditches scoured where the sayd pollards do grow'.[116]

Some farmland trees were self-sown, but most were not and the varying proportions of different tree species found in different parts of the county was largely the result of human selection and deliberate planting over many centuries, although influenced to some extent by environmental factors. Surveys and maps which describe the trees growing on a total of 3,000 acres of enclosed farmland, mostly from the last quarter of the eighteenth and the early nineteenth centuries, enable us to compare the regions of Hertfordshire discussed in earlier chapters, excepting the 'champion' north, where most land still lay in open fields at this time.[117] Although a relatively small sample, this is probably reliable in showing that oak was the predominant timber tree on Hertfordshire farms, accounting for a half or more of trees in the west, north-east and south of the county (Figure 6.10). The proportion of ash

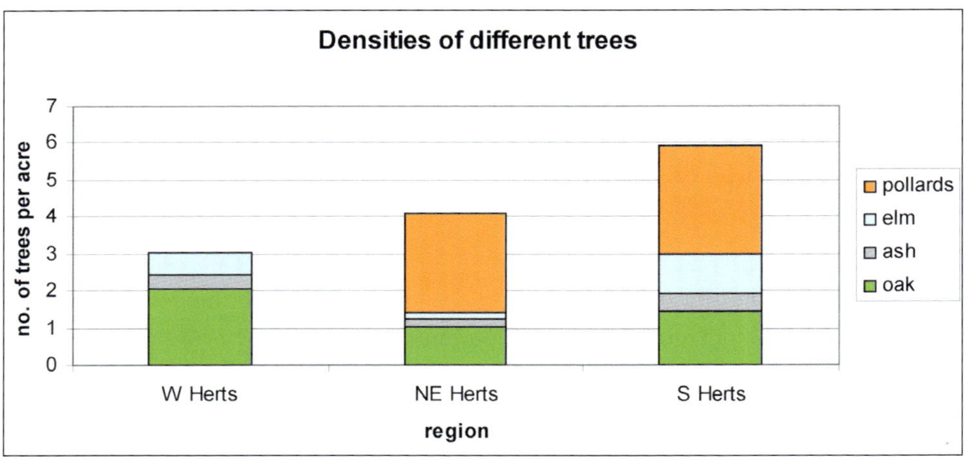

Figure 6.10. The varying densities of the three main timber trees and pollards in different areas of the county, as suggested by eighteenth- and nineteenth-century sources.

appears to have been similar throughout the county, comprising around 15 per cent of the total. Elm, in contrast, seems to have been more abundant in the south of the county than elsewhere – mirroring its frequency as a hedge shrub – where it comprised around a third of timber trees. Three farms covering 260 acres (105 ha) in the south-east of Aldenham parish in 1765 had 333 oak (53% of the total), 251 elm (40%) and 43 ash (7%) timber trees of all sizes but on 192 acres (78 ha) of farmland in other parts of the parish in 1803 elms predominated, with 757 elms, 684 timber oaks and 359 ash trees. If, however, the number of saplings of each species is included in the totals, oak and elm were equally abundant as hedgerow timber trees, each comprising about 40 per cent of the total, with ash comprising about 20 per cent.[118] Six farms or holdings on the manor of Barnet in 1786, covering 195 acres (79 ha), had 359 oak (51%), 269 elm (38%) and 71 ash (10%) timber trees of all sizes.[119] But of the 442 timber trees in the hedgerows of Sheep's Farm, London Colney, in 1852, 61 per cent were elm.[120]

Most farmland trees, however, were not standards but pollards. Surveyors seldom recorded their species, which was evidently considered unimportant: on the 175 acres of Olives Farm at Hunsdon in the mid-sixteenth century, for example, 493 unspecified pollards were recorded but the timber species, comprising 403 oaks, 23 ash trees and 11 elms, were carefully described.[121] In north-east Hertfordshire pollards comprised over 60 per cent of the trees, a figure which would doubtless have been higher in earlier centuries: landowners had become increasingly hostile to the practice of pollarding in the course of the eighteenth century.[122] There were fewer recorded in the west of the county, but significantly more in the south-east.[123] On seven farms on the Broxbournebury estate in 1784, for example, covering 373 acres (151 ha) in total, there were 3,012 pollards but only 299 timber trees. The prize for the property with the greatest density of pollards on the estate would have been won by West End Farm in Wormley where, spread fairly evenly through the 28 fields, which covered a mere 94 acres (38 ha), there were no fewer than 1,358 pollards, together with 131 oak, 14 ash and 3 elm trees, not to mention 569 'spars' (saplings), plus 18 apple and 2 walnut trees in the orchard.[124] But even this was beaten by an 89-acre (36-ha) farm at Hammond Street, Cheshunt which in 1650 contained '927 lopt Pollardes of oake Ash and Hornebeame'. Assuming the majority of these were growing on the 48 acres (19 ha) of enclosed fields on the farm, the density was about 19 pollards per acre.[125] The survey recording these pollards is remarkable: not only does it note the predominant species but the composition of those species is unusual. The few documents which do record the species of pollards suggest that these were most commonly elm, followed by ash and then oak. Although hornbeam was often the most frequent pollarded tree on commons, it was rarely recorded on farmland.

These are very broad patterns: in addition the sources reveal much local variation, a consequence of both environment and choice. On the enclosed meadows belonging to Broxbourne Mill in the eighteenth century willows were, perhaps unsurprisingly, by far the most abundant tree.[126] Pehr Kalm described 'many large beeches' in the hedgerows of the Chilterns around Little Gaddesden in the mid-eighteenth century.[127] But more puzzling is an early eighteenth-century map of a farm in Flaunden, covering 49 acres (20 ha) on the Chiltern dipslope.[128] This details 27 oaks, 18 elms and 9 ash growing in the hedges, together with 15 'asps' – probably aspen. But these are outnumbered by 59 apple trees and no less than 165 cherries, growing not only in the boundaries close to the farmhouse but also scattered more generally through the fields. It is likely that the commercial orchards which flourished in this part of the county by the nineteenth century had ancient roots.

If the total numbers of pollards and timber trees are combined, the enclosed farmland in south Hertfordshire had by far the highest density of farmland trees in the county, averaging about six per acre, compared with four per acre in the north-east and just three per acre in the west. It must be emphasised, however, that the sample is mainly drawn from relatively late sources and it is possible, even probable, that in earlier centuries both the total density of trees and the proportion managed by pollarding would have been greater. In comparison, the modern countryside contains remarkably few trees, and most of those recorded even in late eighteenth- and nineteenth-century documents have gone. The loss of pollards, which were regarded as ugly and old-fashioned even by the late eighteenth century, has been particularly dramatic.[129] There are now, for example, almost none to be found growing at Wormley West End, where over 1,300 were recorded only 230 years ago. We think of hedgerow trees as a timeless part of our landscape heritage but past generations regarded them in largely economic terms: their density in the past reflects the huge demand for wood and timber, and the reduction in their numbers in the course of the nineteenth and twentieth centuries was a consequence of the availability of other sources of timber and wood, and of alternative fuels. This said, even in the remote past trees were not regarded solely in economic terms. Large and old examples have always inspired affection and awe. In early medieval times prominent trees were chosen as meeting places for hundred courts, such as that after which the hundred of Edwinstree in east Hertfordshire took its name, or the ash tree at St Albans Abbey beneath which the courts of the hundred of Cashio were held until after the Dissolution.[130] The name of the village of Elstree is derived from *Tiðulfes treow* – Tithwulf's tree – which is recorded in a pre-Conquest description of the boundary of land granted to Westminster Abbey at Aldenham;[131] while the hamlet of Goff's Oak gets its name from a tree belonging to a member of the Goff family, who were living in Cheshunt by the mid-fourteenth century.[132] Trees were often used to mark boundaries between parishes, and were scored with the mark of the cross by perambulation parties. Remarkable trees were often portrayed on old maps, including, for example, the 'Cucumber Oak', 'Frame Oak' and 'Dog Kennel Oak (14 yards in Circ.)' on Northaw Common, depicted on the county map by Dury and Andrews of 1766; the 'Beautiful Oak' at Warner's End Farm, Hemel Hempstead, shown on a survey of 1809; and the Queen's Oak at Brocket Hall, neatly drawn on a map of 1798.[133] Several old and prominent trees, like the last, stood in the grounds of mansions, whose owners, more than the generality of men, could afford to value them for their aesthetic above their economic worth. At Panshanger, seven kilometres to the east, a visitor to the home of Spencer Cowper described in 1757 how 'The finest oak in all this country is in his woods – 5 yards & a half round, & not the least decayed – he has made a grand Walk thro the coppice to it.'[134] One of the largest maiden oaks in Britain, it still survives and is in good health. Its age is unknown, but it is unlikely to have been planted much after *c.*1400.

Notes

1. F. Vera, *Grazing ecology and forest history* (Wallingford, 2002); F. Vera, 'The dynamic European forest', *Arboricultural Journal*, 26 (2002), pp. 179–211.
2. Rackham, *Ancient woodland.*
3. J.C. Cox, 'Forestry', in W. Page (ed.), *VCH Hertfordshire*, vol. 4 (London, 1914), p. 276. At Tring in the second half of the sixteenth century, cattle were excluded for seven years after coppicing (TNA:PRO E134/43Eliz/East3).
4. TNA:PRO E315/458 accounts etc. of sales of woods, 1544–5.
5. HHA Court Rolls 11/4 bailiff's accounts for the manor of Little Hadham, 1322–1507, fos 83 and 69.
6. Hertfordshire lost 44% of this precious habitat during the twentieth century.
7. Countryside Management Service, 'Hertfordshire's woodland strategy', unpublished report *c.*1997, pp. 4–6, 18 <http://www.hertsdirect.org/docs/pdf/h/wstrat>, accessed 13 February 2012.
8. P. Austin, 'The leasing of Lord Burghley's Hoddesdon woods in 1595. An insight into woodmanship', *Hertfordshire's Past*, 41 (1996), p. 18 quoting J. Evelyn, *Silva*, II (York, 1801 edition), p. 170 and J. Fitzherbert, *Here begynneth a newe tracte or treatyse mooste profytable for all husbandmen* (London, 1523).
9. HALS DE/P/T262 Indenture between George Mym of Woodcott in the parish of Ebbisham, Surrey Esq and William Pryor of Hartingfordberie, warrener, 1637.
10. HALS DE/Lw/E33 account of timber sold from Marshalls and Butlers, 1539.
11. TNA:PRO SC6/873/21 and SC6/873/25 accounts for the manor of Weston, 1304/5 and 1396/7; HALS K100 accounts for the manor of Knebworth, 1370/1; 9607 accounts of the Capel estate at Hadham and Walkern, 1591–1640; B/86 account book for manor of Cheshunt, with Broxbourne (mainly timber sales), 1657–94; DE/R/E114/1–10 accounts for Wain Wood, 1739–67; HALS 27252 wood accounts for the Lamer estate, 1760; DE/Bb/E26 wood accounts for the Broxbournebury estate, 1719–95.
12. W. Branch Johnson (ed.), *Memorandoms for ... the diary between 1798 and 1810 of John Carrington* (London, 1973).
13. HALS DE/K/E15 Knebworth estate papers, 1840.
14. HALS 9607: A note of money received By Gyles Humberston for wood Sold by him out of the new Coppye in Walkerne Parke, 1612.
15. Branch Johnson, *Memorandoms*, pp. 46, 84 and 98.
16. Austin, 'Lord Burghley's Hoddesdon woodlands', pp. 11–21. Thorowgood and Keeling reckoned they could produce 12–25 loads of faggots and 4–8 loads of roundwood per acre.
17. HHA Court Rolls 11/4 fo. 147 manorial account for Little Hadham, 1327/8; King's College, Cambridge KCAR/6/2/034 Cheshunt manor accounts, 1435/6.
18. B. Adams (ed.), *Lifestyle and culture in Hertford ... Wills and inventories for the parishes of All Saints and St Andrew, 1660–1725*, Hertfordshire Record Publications 13 (Hitchin, 1997), pp. 77, 124 and 41.
19. HALS AH/2476 An Abstract of Faggotts Cut in the woods & on Other parts of the Ashridge Estate by William Pratt & Company From December 31st 1834 to (blank).
20. HALS 21600 Bushey manor, copy of court roll, 1738.
21. HALS *County Press* p. 3, 5 July 1834, *Hertfordshire Mercury* p. 3, 4 May 1861 and p. 5, 15 July 1871; *Hertfordshire Mercury* p. 3, 26 May 1888.
22. Branch Johnson, *Industrial archaeology*, p. 14; HHA Court Rolls 10/16 account for the manor of Bedwell, 1484/5.
23. HHA Hatfield Manor Papers I, 1396 Michaelmas, p. 265.
24. HHA Court Rolls 10/20 Hoddesdon lands, 1493.
25. HALS K116 manorial account for Knebworth, 1408/9; HHA Court Rolls 10/16; Court Rolls 11/4 fo. 22 Little Hadham manorial accounts, 1463/4.
26. P.C. Archer, *Historic Cheshunt* (Cheshunt, 1923), p. 150.
27. The Broket Archive, <http://www.brockett.info/hertfordshire/Hertsto1599.htm>, accessed 26 July 2012.
28. HALS 9607. The account for 1606 is headed 'A note of the Chardges for fellinge hewinge and Cuttinge of wood for Coles for yor worshipps house'.
29. *Ibid.*
30. HHA Court Rolls 11/4.
31. HALS DE/Lw/E33 account of timber sold from Marshalls and Butlers, 1539.
32. C. Saunders, 'A sixteenth-century tannery in St Albans', in *Hertfordshire's Past*, 3, 1977, pp. 9–12. The earliest known record of the Gape family of tanners is of John Gape, tanner, who died at St Albans in 1495 and left his tenement in Sopwell Lane to his son Henry (S. Flood (ed.), *St Albans Wills 1471–1500* Hertfordshire Record Society Publications 9 (Hitchin, 1993), p. 128).

33. Adams, *Lifestyle and culture in Hertford*, pp. 100–103.
34. P. Austin, 'Traditional woodland management. The woods of the Manor of Digswell 1776–1843', *Hertfordshire's Past*, 51 (2001), p. 9; J. Hill (ed.), *Welwyn*, Hertfordshire Family & Population History Society, Militia Series No. 64 (1996), p. 10 and *Layston*, Hertfordshire Family & Population History Society, Militia Series No. 16 (1990), p. 19.
35. HALS DE/R/E114/1–10.
36. Rackham, *Ancient woodland*, p. 164.
37. Cox, 'Forestry', p. 276; The law was strengthened in 1570 during Elizabeth's reign.
38. Page (ed.), *VCH Hertfordshire*, vol. 2, p. 330. East Barnet was said to contain just 10 acres of woodland (*Ibid.*, p. 337).
39. *Ibid.*, p. 341.
40. TNA:PRO E315/391.
41. HALS 54569–72 deeds of Monkenfrith Farm, 1583–4.
42. BL Ordnance Survey Drawing 151, Chipping Barnet, 1805.
43. Cox, 'Forestry', p. 280.
44. Countryside Management Service, 'Hertfordshire's woodland strategy', p. 6.
45. R. Liddiard, *The medieval park: new perspectives* (Macclesfield, 2007); S. Mileson, *Parks in medieval England* (Oxford, 2009); Rowe, *Medieval parks*.
46. BL Cotton MS Tiberius Bii, fos 140–41 survey of manors of the bishopric of Ely, 1222, which records tenants called Ralph le Parker and Walter le Parker.
47. Cambridge University Library EDR G/3/27 Old Coucher Book of Ely, fo. 80; HHA Hatfield Manor Papers I, 1392/3, p. 253 records that each tenant was to make 120 perches of fence.
48. Rowe, *Medieval parks*, p. 166.
49. HALS DE/B2067B/E24 includes 'a wodd called Gorambury parke' containing 20 acres.
50. Only 5 of the 30 medieval parks known to have contained 200 acres or more had failed before 1500 (Eywood, Albury, Scales, Cheshunt old park and Great Munden).
51. The parks containing 300 acres or more were: Hatfield great, Berkhamsted, King's Langley, Bedwell, The More, Hatfield Woodhall, Standon, Shingle Hall, Eywood, Benington, Walkern, Hunsdon old, Weston great, Hatfield Millwards, Knebworth great, Ware and Tyttenhanger. The only one of these which does not seem to have survived into the sixteenth century is Eywood park, which belonged to the abbey of St Albans. The acreage of this park is not known but it may have been as large as *c.*418 acres.
52. *Calendar of Close Rolls, Edward I, 1272–1279* (London, 1900), p. 378; TNA:PRO SC6/868/20 bailiff's account for the manor of Standon, 1328/9; M.C.B. Dawes, *Register of Edward the Black Prince, part 4, 1351–1365* (London, 1933), p. 265; HALS K117 and K119 bailiff's accounts for the manor of Knebworth, 1410/11 and 1412/13.
53. TNA:PRO E317/Herts/27.
54. TNA:PRO E317/Herts/19 parliamentary survey of Cheshunt Park, 1650.
55. HALS DE/Bb/E27 timber survey of farms and tenements on the Broxbournebury estate, 1784.
56. In 1291–2 John Tailboys of Titburst released to Walter abbot of Westminster all his right of common pasture in the woods of the abbot at Aldenham Frith (Page (ed.), *VCH Hertfordshire*, vol. 2, p. 155); Monkenfrith is mentioned in Page (ed.), *VCH Hertfordshire*, vol. 2, p. 341 citing Riley, *Gesta Abbatum*, III, p. 218 and I, p. 474; references to Boreham Frith Farms and frith names at Bricket Wood can be found in HALS DE/Cp/L1 Dispute between Lord Essex and Falcon re Brickett Wood and Garston manor including deeds to manors of Park and Meriden and various other papers relating to Frythe House, and manors, 1607–16.
57. Page (ed.), *VCH Hertfordshire*, vol. 2, p. 331, citing *Calendar of State Papers Domestic* 1689–90, p. 79.
58. Mead, *Pehr Kalm*, pp. 43–4.
59. HALS 10996 A/B Notes in case concerning cutting down of trees on Cheshunt common, n.d; this figure is based on the stated area of the common in 1785 of 1186.5 acres, which had changed little since 1669 (HALS DE/Cr/125/2 Plan of the Liberty and Manor of Cheshunt together with that part of Theobalds Park within the Parish of Northaw in the County of Hertford, and also that part of Theobalds Park within the Parish of Enfield in the County of Middlesex, 1785; DE/B1767/M1 survey and custumal of Cheshunt, 1669).
60. HALS B/86. The calculation of a precise average is not possible owing to a minor hiatus in the records on the death of the lord of the manor.
61. This figure assumes that the extent of Broxbourne common was the same as that shown on a map of the parish of Broxbourne of 1829 and on the tithe map of 1839 (HALS 71014; DSA4/25/2). Sir Richard Lucy was lord of the manors of both Cheshunt and Broxbourne until 1667, when they passed to Sir John Monson, bart., who was succeeded by his grandson, Henry Monson, in 1683 (Page (ed.), *VCH Hertfordshire*, vol. 3, p. 432).

62. Archer, *Historic Cheshunt*, p. 150; HALS DE/Bb/E26.
63. M. Grieve, *A modern herbal* (London, 1931), <http://www.botanical.com/botanical/mgmh/a/ash--073.html>, accessed 9 November 2012.
64. H.M. Colvin (ed.), *The history of the king's works, vol. III 1485–1660* (London, 1975), p. 203. The oak cost six shillings a load.
65. TNA:PRO E315/391.
66. HALS 80870–87 Bovingdon estate timber accounts, Lane Farm and Little Shantocks, 1800–10.
67. HALS K117 and K119; Page (ed.), *VCH Hertfordshire*, vol. 2, pp. 345, 405.
68. HALS K117 and 119; Rowe, *Medieval parks*, p. 34.
69. HHA Hatfield Manor Papers: Summaries I, p. 339; Page (ed.), *VCH Hertfordshire*, vol. 3, p. 99.
70. P. Austin, 'Hatfield Great Wood and its inclosure', *Hertfordshire's Past*, 38 (1995), pp. 6–7, citing HHA Sals. General 44/1 and Sals. Legal 66/7.
71. HALS 80226 court rolls manor of North Mymms, 1427; Page (ed.), *VCH Hertfordshire*, vol. 3, p. 430.
72. James, *Flora of Hertfordshire*, p. 112.
73. Rowe, *Medieval parks*, p. 34.
74. Page (ed.), *VCH Hertfordshire*, vol. 3, p. 104; Austin, 'Hatfield Great Wood', p. 2, citing TNA:PRO Ancient Deeds E326/4213.
75. TNA:PRO E315/391, fos 8 and 84.
76. J.E. Cussans, *History of Hertfordshire*, vol. 2 (London, 1876), p. 104; Austin, 'Hatfield Great Wood', p. 6.
77. HHA Court Rolls 9/25 bailiff's account for Little Hadham, 1435/6; see also Rackham, *Ancient woodland*, pp. 174, 223 and 224. Despite its recorded use as browsewood for deer, Rackham comments that 'they do not enjoy it'.
78. James, *Flora of Hertfordshire*, p. 117.
79. Countryside Management Service, 'Hertfordshire's woodland strategy', p. 6.
80. Monken Hadley was first recorded simply as Hadley *c.*1136 (R.B. Pugh (ed.), *VCH Middlesex*, vol. 5 (Oxford, 1976), p. 260).
81. Snail assemblages in the soil-wash accumulated in a dry valley near Pitstone indicate that the Chiltern scarp above was cleared of woodland around four thousand years ago, used as pasture for grazing livestock and subsequently ploughed by the Late Iron Age (Catt and Cheshire, 'Quaternary deposits', p. 184).
82. Young, *Agriculture*, pp. 150–51.
83. Rowe, 'Rabbit warrens', pp. 110–11.
84. J. Sheail, *Rabbits and their history* (Newton Abbot, 1971); T. Williamson, *Rabbits, warrens and archaeology* (Stroud, 2007).
85. Page (ed.), *VCH Hertfordshire*, vol. 2, p. 387, citing Jno. Amundesham, *Ann. Mon.* (Rolls Ser.), i, 254–60 and Convent. Leases, Herts. and Glouc. iv, 4, 41-year lease of the warren to John Bowman of Colney, 1532.
86. TNA:PRO E134/43Eliz/East3.
87. TNA:PRO E134/4Jas1/Mich19.
88. HALS DE/P/P1 map of the manor of Digswell, 1599.
89. HALS 46606 lease of Knebworth warren, 1722.
90. HALS DE/P/T264 lease of Hertingfordbury warren for nine years, 1634.
91. C. de Saussure, *A foreign view of England in the reigns of George I and George II* (London, 1902), pp. 305–8.
92. HALS DE/X629/P1 map of the Bayfordbury estate, 1807.
93. Young, *Agriculture*, pp. 178–9.
94. Young, *Agriculture*, p. 52.
95. London School of Economics Coll Misc 38/3 f.8 journal of Sir John Parnell.
96. TNA:PRO E315/391, fol. 144.
97. TNA:PRO E315/457 sales of wood in divers counties 35 Hen VIII, 1543–4.
98. HALS DE/Bb/E26.
99. HALS DE/P/P4; 15594 plan showing The Golden Parsonage, mid-eighteenth century; DE/V/P2.
100. Young, *Agriculture*, pp. 50–52 (with illustrations).
101. HHA Court Rolls 10/16.
102. A good example is HALS DE/Cd/E101 lease of Aspenden Hall Farm, 1745.
103. Young, *Agriculture*, p. 52; HALS DE/B513/E1 lease of the farm, Weston, 1665; DE/Cd/E101; DE/Lw/E7–8 Bowers tenement, Harpenden, 1706; DE/B983/E1 survey and valuation of copyhold estates of the manors of Chipping Barnet and East Barnet, 1787.

104. HALS DE/B513/E1 lease of 320 acres at King's Walden, 1662.
105. Young, *Agriculture*, p. 49.
106. Mead, *Pehr Kalm*, p. 116; W. Johnson, 'Hedges: a review of some early literature', *Local Historian*, 13 (1978), pp. 195–204.
107. G. Barnes and T. Williamson, *Hedgerow history: ecology, history and landscape character* (Macclesfield, 2006), especially pp. 129–34.
108. E. Pollard *et al.*, *Hedges* (London, 1974). For a useful critique of the 'Hooper hypothesis', in a Hertfordshire context see F. Dulley *et al.*, 'Hedges and local history', *Hertfordshire's Past*, 3 (1977), pp. 15–24.
109. J. Britten (ed.), *William Turner's The names of herbs* (London, 1802), p. 81.
110. See, for example, HALS A/2832–3 and DE/H/P16 plan of Stockers and Pearces Farms in Sawbridgeworth and Thorley, 1807 HALS DE/P/P4 – an enclosure in Broad-Mead beside the Mimram is depicted full of pollards with timber trees in the enclosing hedge; see also fields around Cole Green.
111. HALS DE/B513/E1 lease of Wilches and Gillies Farms in King's Walden, 1662.
112. See, for example, HALS DE/Am/E3 and DE/Am/E11 leases of Hartsbourne Manor Farm, Bushey, 1633–90.
113. HALS DE/B513/E1 lease of Lannock Manor, Weston, 1663.
114. HALS DE/B983/E1.
115. HALS DE/Cd/E14 lease of Ridge Hill Farm, 1657.
116. HALS DE/Am/E3.
117. This discussion is based on an analysis of the following timber surveys and estate maps held in HALS: DE/X905/P1; 27251 Tarrants Farm, Lamer estate, 1734; 11788 Three Houses Farm, St Stephens, St Albans, 1719; DE/B944/E2 Sheeps Farm, London Colney, 1852; DE/Cr/55 Beeson End Farm, Redbourn, Westwick Farm, Hemel Hempstead and St Michael's, Market Street Farm, Flamstead, and Folly Farm, Albury, 1838; DE/B1622/E9 The Holt Farm, St Stephens, 1817; DE/Cn/E1 Parnell estate, Meesden, 1762; A/2832–3 farms on the manor of Great Barwick, Standon, 1778; 43754; DE/B1768/P2 land at Colliers End, Standon, 1794; DE/Cn/P4 plan of an estate in Much Hadham, 1803; 81750 plan of land at Whempstead, Little Munden, 1808; DP/117/25/7 plan of three fields called Church lands, Watford, 1754; 54804; DE/Bb/E27; DE/B983/E2 farms and lands on the manor of East Barnet, 1786; QS/E3; TNA:PRO E315/391. There are no known comparable surveys of farmland trees for the area of the county north of the Chalk escarpment.
118. HALS 54804; DE/X216/B2 valuation of timber upon lands allotted and exchanged under the Aldenham Inclosure Act 1802–3.
119. HALS DE/B983/E1.
120. HALS DE/B944/E2.
121. TNA:PRO E315/391.
122. S. Petit and C. Watkins, 'Pollarding trees: changing attitudes to a traditional land management practice in Britain 1600–1900', *Rural History: economy, society, culture*, 14/2 (2003), pp. 157–76.
123. A contributory factor for the relatively low numbers in the west is the fact that most of the sources for that area are later than those for the east.
124. HALS DE/Bb/E27.
125. TNA:PRO E317/Herts/24 Parliamentary survey of Theobalds manor, 1650, p. 37.
126. HALS B/479 map of lands on the Broxbourne Mills Estate, late eighteenth century.
127. Mead, *Pehr Kalm*, p. 52.
128. HALS DE/X905/P1.
129. Petit and Watkins, 'Pollarding trees'.
130. Page (ed.), *VCH Hertfordshire*, vol. 2, p. 321.
131. Gower *et al.*, *Place-names of Hertfordshire*, p. 60.
132. In 1368 a charcoal burner called John Goffe was living in the parish of Cheshunt (Archer, *Historic Cheshunt*, p. 150).
133. A. Dury and J. Andrews, *A topographical map of Hartfordshire* (London, 1766); HALS DE/X195/P1–2 map of Warner's End Estate, 1809; DE/P/P15 plan of Brocket Hall Park, 1798.
134. Hampshire Record Office 9M73/958 diary of the Reverend George William Harris, Rector of Egglescliffe, Durham, 1757.

– 7 –

Traditional buildings

Parish churches

Buildings – whether cottages and farmhouses or ancient parish churches – make up a large part of our 'traditional' landscape. Built in local styles and with local materials, they provide much of the character of any particular district. Parish churches have an additional importance in historical terms. Not only do they often mark the oldest areas of settlement in any locality, providing vital clues about the way that the landscape developed in the course of the Middle Ages, but they were also important shared spaces, the principal focus for the communities which, over generations, worshipped and were buried there. In early times they were the largest and most durable building in the landscape. Even in the most built-up areas of the county they usually survive, testimony to hundreds of years of history.

The earliest churches were 'minsters', usually attached to royal estates. They were staffed by teams of priests and served the spiritual needs of extensive territories. The system of local parishes only developed gradually, from late Saxon times, as population rose and as wealthy families erected churches to serve their own estates. This process was well under way by the tenth century. The will of Æthelgifu, drawn up in the 980s, included bequests to minsters at St Albans, Hitchin, Ashwell, Welwyn and Braughing, but also to individual priests.[1] These included Edwin, who was 'to have the church for his lifetime on condition that he keep it in repair'. Church-building was a sign of status, one of the legal requirements for a man to claim the rank of thegn, or lord;[2] and the role of landowners in church foundation, both before and after the Conquest, is clear from the way that, even today, church and manor house often stand in close proximity, forming what some archaeologists call a 'hall–church complex', as at Barkway, Digswell, Hunsdon, Little Wymondley, Westmill or Widford. At Benington, Anstey, Therfield, Great Wymondley and Pirton churches lie beside or even within castle earthworks of eleventh- or twelfth-century date, although in the last of these cases, as we have seen, the church was probably moved to a new site within the outer bailey when the castle was erected.[3] But parish churches were erected not merely by lay landowners for reasons of status or convenience. St Albans Abbey built churches on its own estates, most strikingly the group established, in late Saxon times, within what was effectively the minster territory of the abbey. The *Gesta Abbatum* records how Abbot Wulsin built the three churches of St Michael's, St Peter's and St Stephen's in the mid-tenth century, although in reality they were probably erected early in the eleventh.[4] The proliferation of churches and the disintegration of large minster territories was to some extent the simple consequence of population growth and the expansion of settlement into new areas.

By the time of the Norman Conquest there were probably around 70 churches in Hertfordshire, perhaps a few more, a figure to be compared with a total of 141 surviving parish churches in the county.[5] Yet remarkably few of the latter, probably no more than 13, contain any clear architectural evidence of pre-Conquest origins: in large measure a reflection of the fact that in Hertfordshire, as elsewhere in England, most Saxon churches were built of wood, rather than stone, and were swept away in the subsequent centuries.[6] Some surviving examples of Anglo-Saxon architecture are striking and substantial: at St Mary Northchurch, for example, where the southern and western walls of the nave are both largely of Anglo-Saxon date; or at St Michael's in St Albans, which has both nave and chancel walls preserved, complete with the simple, single splayed arches typical of the period.[7] But in many other cases obvious evidence for pre-Conquest work is slight, comprising minor survivals from successive rebuilds. At Reed, for example, it is largely restricted to a doorway in the north wall and characteristic 'long-and-short work', comprising large rectangular blocks of stone extending alternately vertically and horizontally, at each corner of the nave (Figure 7.1); while the church at Braughing, in origin a Saxon minster, now looks almost entirely fifteenth century in date, and its early history is revealed only by fragments of late Saxon interlace sculpture built into the walls.[8] It is likely that much further evidence of pre-Conquest fabric remains to be discovered in our churches, and there is certainly scope here for further research by local historians and others. Yet at the same time it is noteworthy that those churches where probable examples of early remains are known mainly fall into two categories.[9] Most were either erected by St Albans Abbey on their estates in the west of the county (St Michael's, St Stephen's, Sandridge and Great Gaddesden), or else represent former minsters (Northchurch St Mary, Wheathampstead, Braughing and Walkern). This said, a handful of others, including Reed, Tewin, Westmill and Little Munden – all in the wealthier, more populous east of the county – can never have been anything other than local churches erected by lay owners. Evidently, a shift to constructing minor local churches in stone was under way in the county before the Norman Conquest, unless of course these examples – as has been suggested by some architectural historians – actually date from soon after the Conquest, when English masons were still building in the old style.[10] It is notoriously difficult to distinguish pre- from post-Conquest architectural details and long-and-short work, for example, was certainly used until the end of the eleventh century.

While more evidence of Anglo-Saxon fabric within our churches doubtless awaits discovery there is little doubt that – as in England as a whole – the period following the Norman Conquest saw a revolution in ecclesiastical architecture, as new lords rebuilt local

Figure 7.1. The nave of Reed church features Anglo-Saxon 'long-and-short work' at each corner (see inset). Decorated stonework around a small door on the north side of the church dates from around the time of the Norman Conquest.

churches in stone and established additional ones in places where none had been before.[11] There are several striking examples of Norman architecture in the county, including St Albans Abbey, St Mary Hemel Hempstead, and St Leonard Bengeo with its apsed nave (an unusual survival of something once common). Such churches take a variety of forms, with some (such as Bengeo) comprising only a chancel and nave while others – the largest, on the whole – having cruciform plans and a central crossing tower, as at Hemel Hempstead. Some churches of the latter type, as at Wheathampstead, represent wealthy former minsters. Numerous other parish churches display some clear evidence of Norman work – a total of perhaps 41 in all, spread fairly evenly across the county – although once again very many others probably contain twelfth-century fabric hidden from view or without clear diagnostic features.[12] Indeed, there are grounds for believing that *most* churches in the county contain some fabric predating the thirteenth century, for the overwhelming majority of them appear to have been constructed of stone by this time: even some of our smallest and poorest parishes, such as Bygrave, contain evidence of Norman architecture, while in broad terms the pattern of parishes seems to have fossilised the distribution of wealth and population as it was in the eleventh and twelfth centuries, with fewer and larger parishes in the west and south-west, on the whole, than in the east and north-east.

A number of larger parishes, however, gained small 'chapels-of-ease', erected in their more remote parts to serve the populations there. Most disappeared at the Reformation, such as those at Broad Colney, and Tiscot and Wilstone in Tring, but others – including Flaunden and Bovingdon in Hemel Hempstead, Long Marston in Tring, Harpenden in Wheathampstead and Nettleden in Pitstone (Buckinghamshire) – developed into parishes in their own right in the course of the post-medieval period.[13]

From the early thirteenth century the Norman style, with its round-arched Romanesque doors and windows, was replaced by the successive forms of 'gothic', with pointed openings, conventionally divided into 'early English', 'Decorated' and 'Perpendicular', each style with larger windows and a lighter, more upright feel than its predecessors. The first of these phases is generally held to have dominated in the early–mid thirteenth century, the second in the late thirteenth and early fourteenth and the last from the later fourteenth into the early sixteenth: but we need to be cautious. Both the terminology traditionally employed by students of ecclesiastical architecture and the neat chronology implied are misleading, in Hertfordshire as elsewhere. As one recent commentator has observed, in reality the different styles 'all merged imperceptibly with each other and overlapped considerably in time': 'Medieval masons did not recognise periods of architecture'.[14]

The thirteenth century was, as we have seen, a time of rising population and rising prosperity. Any remaining wooden churches were rebuilt in stone for the first time, while existing structures were altered and enlarged on some scale, often with the addition of aisles. These were sometimes erected on both sides of the nave but sometimes on only one – usually the north. Chancels were also almost universally rebuilt and extended in this period as a consequence of changes in the character of the liturgy. Curved apses, like those which have survived at Bengeo, Great Amwell and Great Wymondley, provided insufficient space for the new, more elaborate rituals and were demolished: in a large proportion of our churches, especially in the east of the county, a thirteenth-century chancel appears the earliest part of the building. After the Black Death, with the population now much lower, there was less need to increase the size of churches to house local congregations, but higher levels of disposable income, a rising class of lesser gentry and merchants and an obsessive interest in purgatory and intercession ensured other changes: the rebuilding or the addition for the first time of western towers, and the construction of porches, new roofs, clerestoreys and chapels.[15] In most cases Hertfordshire churches thus developed over the centuries by piecemeal addition and alteration, producing that fascinating mixture of features and fabrics which make them so rewarding to investigate. But in some places drastic rebuilding occurred, so that what we have today is largely of one period, although the work could extend over many decades. We thus have no idea of what the church at Ashwell may have been like before the fourteenth century, although it was presumably a large building, as there was a minster here by the tenth century and the place was a Domesday borough. What we see today is the result of a massive rebuilding campaign which extended through much of the fourteenth century, despite the terrible events of the middle of the century, including not only the arrival of the

Figure 7.2. St Mary's Ashwell is largely of fourteenth-century date and mainly constructed of the local Totternhoe Stone.

Black Death but also severe weather conditions, recorded in rare graffiti scratched on the tower wall (Figure 7.2).

The county's churches thus display a bewildering degree of variety: some are mainly of one period, most the product of several; some essentially simple in form, with no more than a nave and chancel, while others have complex plans, featuring aisles to north and south and perhaps subsidiary chapels, or retaining an early cruciform plan. The reasons for such variations in size and complexity are not straightforward. To some extent the structural development of any church is related to the demographic and economic fortunes of the parish that it served, but the links are complex and often indirect. Apart from the obvious fact that the parishioners paid for the nave and the incumbent (or impropriator) the chancel, building activity often reflected the fortunes of individual families, whose wealth might derive from lands or interests elsewhere, rather than entire communities. This said, to some extent the character of a church mirrors the size and wealth of the medieval population it served. Eileen Roberts has usefully divided Hertfordshire churches into five broad categories according to the complexity of their ground plans – 'primitive', 'simple', 'moderate', 'elaborate' and 'cruciform'.[16] These variations were the consequence both of the original form of the church and of later developments, as individuals and communities added in a piecemeal fashion towers, porches and the rest. As Table 1 shows, there is in general terms a correlation between the size of parish population and average wealth (as recorded in the lay subsidy returns of both 1307 and 1545), and the complexity of its church.

There is also a spatial dimension to this, in that churches with the most complex plans ('moderate', 'elaborate' and 'cruciform') tend to be associated with settlements in the principal valleys – market towns, the larger villages, old estate centres and minster settlements. Churches on the upland interfluves, in contrast, generally serving smaller and

Year	*Church plan*	*No. of parishes*	*Total no. of taxpayers*	*Total tax paid (shillings)*	*Average no. of taxpayers*	*Average tax paid per parish (shillings)*
1307	Simple	24	593	1213	24.7	51
	Moderate	34	1663	3112	48.9	92
	Elaborate	20	1469	2887	73.5	144
	Church plan	*No. of parishes*	*Total no. of taxpayers*	*Total value of goods or land assessed (£)*	*Average no. of taxpayers*	*Average value of goods or land assessed per parish (£)*
1545	Simple	24	634	£3,719	26.4	£155
	Moderate	34	1592	£10,598	46.8	£312
	Elaborate	20	1503	£9,272	75.2	£464

Table 1. Comparison of 78 parishes in the lay subsidy returns for Hertfordshire in 1307 and 1545 shows that churches with more complex plans tended to be in parishes with larger populations and higher levels of prosperity.

poorer parishes, tend to be of 'primitive' or 'simple' type. This pattern is particularly clear in the Chiltern dipslope region, as Nicholas Doggett has observed. Here 'the churches in the valleys have generally had the best opportunity to develop sophisticated plans as a result of serving long established, prosperous communities ... while upland churches have remained small to the present day'.[17] The pattern is not quite so marked in the east of the county, where the contrast between valley and upland, in terms of agricultural productivity and therefore wealth and population, was less. Across the county as a whole, moreover, the more elaborate church plans are fairly evenly distributed because a higher proportion of the churches in the west and south-west are of this kind, churches of all kinds being fewer in these districts – as we have seen – than in the north and east. This was because parishes were larger here, and generally remained so (albeit with the addition, as already noted, of chapels-of-ease in more remote places) even as the size of the gap between the two districts in terms of wealth and population narrowed significantly in the course of the Middle Ages.

Sometimes the connections between community history and architecture are particularly straightforward. We have already noted how some of the churches in the 'champion' areas of north Hertfordshire, located in diminutive nucleated villages such as Newnham, Letchworth, Radwell and Caldecote – places with small populations and dominated by minor local gentry – are small and simple in character, lacking aisles and, in the case of Radwell and Letchworth, with bell turrets rather than towers. At the other extreme, those found in the county's principal towns are usually large, complex and often substantially expanded in late medieval times, so that any early fabric has been removed or obscured. Hitchin, for example, may retain some fabric from the Saxon minster that unquestionably existed here, but while the tower has evidence of having been begun in the twelfth century the church as a whole appears fourteenth- or fifteenth-century in date: 'a building representing the commercial wealth of a late medieval town'.[18]

Variations in the appearance of the county's churches are also related to the materials with which they were constructed. Hertfordshire has little to offer in the way of good building stone and before the arrival of the canals in the eighteenth century and the railways in the nineteenth church builders had to use local materials for the main structure of the building. The most important of these is flint – the irregular nodules of silica which occur both in the Chalk which underlies most of Hertfordshire and in many of the superficial 'drift' deposits. Over 90 per cent of the 102 pre-Reformation parish churches examined by Percy and Enid Evans incorporated flint as a major building material.[19] The flint is often mixed with pebbles and lumps of stone and in a few churches the pebbles are laid in courses, as at the eleventh-century church at Ippollitts and around the apse of St Mary Great Wymondley. It is possible that some rubble walls were meant to be rendered, as at St Nicholas Norton and Holy Trinity Throcking, and traces of the old plaster can be seen butting up to the quoins on the tower at St Giles Wyddial.[20] In other cases in the late medieval period the flint was knapped (deliberately broken by a skilled craftsman) to reveal the glassy inner core which was used to make the attractive and weather-proof exterior face of the church wall, the inside of which was composed of unknapped material. Outstanding flint walls of this type include those at Aldbury, Great Hormead, South Mimms, Walkern and Watton.[21] Those at St Botolph Shenley have square-knapped flint and brick dressing, and some of the churches in the west of the county, including Puttenham and Long Marston (where only the tower survives), have flints knapped with particular skill and combined with Portland limestone to create a chequer-board pattern.[22]

Indeed, flint was never used alone in churches: some kind of good-quality masonry or brick was always required for the corners and for the surrounds to doors and windows. One stone commonly used for this purpose was Jurassic limestone, mainly dug from the quarries at Barnack in Northamptonshire. This was employed in the county in the Anglo-Saxon period, featuring in, for example, the long-and-short work in the corners of the nave at Reed and Westmill, in the quoins of the Saxon flint wall of St Mary Northchurch and in an arch above a blocked Saxon door in St Helen Wheathampstead.[23] Reused Roman brick and tile were also used to supplement flint before the Conquest. The Anglo-Saxon churches of St Michael and St Stephen at St Albans made extensive use of material plundered from the ruins of the Roman town.[24] Roman tile and brick, probably looted from 'villas' or Romanised farmhouses, also feature in the walls of St Mary Hitchin, St Mary Welwyn, St Andrew Much Hadham and St Mary Northchurch, all of which probably originated as minster churches, as well as in the Anglo-Saxon walls of St Mary Westmill. The recycling of Roman material was not exclusive to the Anglo-Saxons, however, and can also be seen in the fabric of such post-Conquest churches as St Mary Redbourn (*c.*1100), the twelfth-century St Mary Great Wymondley and St James Stanstead Abbots, while at St Margaret Bygrave Roman tile was used in the quoins at the east end of the nave.[25] Perhaps the most striking example is the central tower of the cathedral at St Albans, built as the church of the abbey from 1077, where Roman tiles and Barnack Stone, both salvaged from the ruins of Verulamium, were combined for decorative effect.

Also common was the use of Totternhoe Stone (or 'clunch'), a form of hard Chalk which was quarried to the north of Barkway and near Ashwell but primarily in the vicinity of Totternhoe, just over the county boundary in south Bedfordshire.[26] It was a valuable freestone which was easily carved for decorative purposes and was transported long distances even in Roman times. Some Anglo-Saxon churches incorporate features made from it, including the carved figure of Christ (or rood) on the wall of Walkern church; it forms a bonding course in the flint wall at Northchurch and a window jamb at St Stephen St Albans.[27] It was much used in St Albans Abbey from the end of the twelfth century, and can be seen in the remains of the cloister arcades on the south side of today's cathedral and in many of the churches erected on the abbey's manors: the abbey owned at least two of the quarries near Totternhoe from the mid-fourteenth century and Abbot John of Wheathampstead (1420–40) purchased two more.[28] Typically, Totternhoe Stone was used decoratively as ashlar (thin rectangular blocks) on walls of flint rubble, as at St Lawrence Abbot's Langley, Our Lady Watford and Holy Cross Sarratt; it can also be seen in the late fifteenth-century window tracery of the chancel at Bramfield.[29] Its use was not confined to churches belonging to St Albans Abbey, however, and from the end of the twelfth century it was almost universally employed for dressings and interior work in Hertfordshire's parish churches: for arches, parapets, door-frames, windows, capitals on columns, fonts and religious sculptures.[30] It also formed a major component of the exterior walls of a handful of churches, all in the north of the county and close to the quarries, as at St Mary the Virgin Ashwell (high-quality ashlar blocks) (Figure 7.2) and St Peter and St Paul Tring.[31] However, although it survives well inside buildings, Totternhoe Stone is much less well suited to external use as it absorbs moisture and weathers badly if not protected by an overhanging roof and regular applications of lime wash.[32] Much was thus badly decayed by the nineteenth century and was replaced with more durable stone during the extensive restoration programmes undertaken in many parishes during the Victorian period. At St Peter Great Berkhamsted, for example, all the Totternhoe Stone quoins, buttresses, window reveals and tracery were replaced with Oolitic limestone in the Butterfield restoration in the 1870s.[33]

A variety of other materials was employed in the construction of the county's churches. Hertfordshire Puddingstone is a hard conglomerate composed of flint pebbles set within a silicate matrix which is found in surface deposits in many parts of the county. It sometimes takes the form of large slabs which were used to provide strong and damp-proof foundations for medieval churches, as at Hemel Hempstead (beneath the tower), Great Munden and Sarratt.[34] Smaller lumps were incorporated in the flint and rubble walls of many churches, including at North Mymms, Abbot's Langley and Much Hadham (Figure 7.3).[35] Another

Figure 7.3. A lump of Hertfordshire Puddingstone in the lower wall of the tower of St Andrew's Much Hadham.

Figure 7.4. Fifteenth-century brickwork at St Dunstan's church, Hunsdon.

Figure 7.5. St Andrew's Much Hadham, a typical Hertfordshire village church: constructed of flint, mainly of thirteenth-, fourteenth- and fifteenth-century date, and with a tower surmounted by a 'spike'.

locally available conglomerate, formed from gravels and sands bound together in an iron-rich cement and identifiable by its rusty colour, has been found in small quantities in the rubble walls of 14 Hertfordshire churches: Aldenham, Bengeo (St Leonard's), Bishop's Stortford, Bramfield, Cheshunt, Great Munden, Hatfield, Little Hadham, Little Munden, Sandon, Sawbridgeworth, Stanstead St Margaret, St James Stanstead Abbots and Watton-at-Stone. John Potter, who has carried out a major study of this material, argues that its presence indicates churches of Anglo-Saxon construction,[36] but its absence from most of the churches in the county which are unquestionably of pre-Conquest date perhaps makes this unlikely.

From the early fifteenth century brick-making was revived in the county and in the succeeding centuries numerous small-scale and ephemeral brick-works sprang up to take advantage of widespread deposits of clay suitable for the making of bricks. Sources included the Gault, Reading and London Clay formations, glacial boulder clays and glacial lake deposits, as well as 'brickearths' – windblown deposits (loess) – found on the Chiltern dipslope and in the lower Lea valley.[37] The clay was generally mixed with sand and/or crushed chalk, both of which were also widely available, before firing. Brick was used to build the chapel and eastern part of the chancel of St Dunstan Hunsdon *c.*1450, at the same time that Sir William Oldhall was building Hunsdon House, also with brick (Figure 7.4).[38] At St Mary Redbourn ornamental brickwork in the parapet over the south aisle and in the side chapel is dated to about 1478.[39] By the sixteenth century brick was being more widely used, especially in porches and chapels: examples include the south porch added to the small twelfth-century church at Meesden about 1530 and the north chapel of 1532 at St Giles Wyddial.[40]

Although these and other kinds of material were commonly employed in the construction of medieval churches in the county, flint was, as we have noted, by far the most important, usually in the form of rubble. Providing the top is protected from the weather, flint walls are very durable, and their physical characteristics have largely determined the appearance of the typical Hertfordshire church. They have to be relatively thick (at least two feet) and are often strengthened with large and prominent buttresses; towers in particular are thick-walled and are often stepped back at each stage in order to reduce the weight carried by the lower walls.[41] The traditional 'Hertfordshire spike', a short, thin timber spire, usually covered with lead, seems to have developed out of a need to minimise the weight that had to be supported by the flint rubble walls of the church tower (Figure 7.5). In spite of its common name, the Hertfordshire spike is not evenly distributed across the county, and is also found in some numbers across the county boundary. It is essentially a feature of east Hertfordshire, north-west Essex and south Cambridgeshire (Figure 7.6). Although in part a reflection of the dominance here of towers built of flint rubble, this cannot be the entire explanation for such a pattern, as similar towers are more widely distributed: spikes appear to be a local fashion, copied by a number of neighbouring parishes in the course of the fourteenth and fifteenth

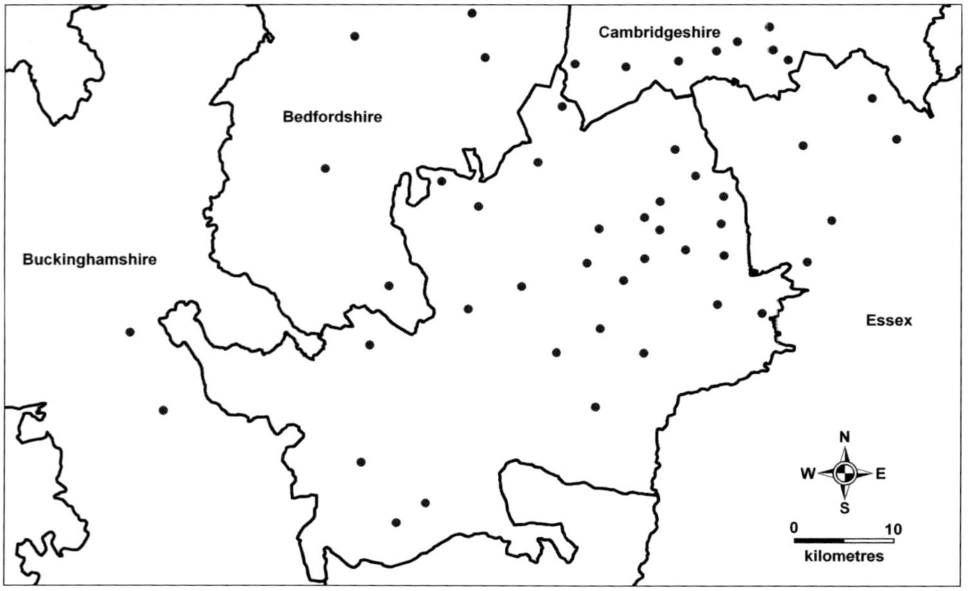

Figure 7.6. The distribution of churches with 'Hertfordshire spikes'. In spite of their common name, these diminutive spires are found in some numbers in the neighbouring counties of Cambridgeshire and Essex.

centuries. But other aspects of church design or decoration cannot be so explained as the outcome of using particular building materials; nor did they necessarily spread, like a contagion, from church to nearby church. By the high Middle Ages sculptors and masons were often highly mobile, and wealthy patrons might adopt ideas picked up at court or through other connections at a high social level, transcending local geographies. In a detailed study of window tracery, columns and mouldings created around arches and doorways, Eileen Roberts has detected a 'simple and unadventurous' style employed in the fifteenth century in churches located in the north and west of the county, which she attributed to medieval masons working at the Totternhoe quarries in south Bedfordshire, with examples at Tring, Little Gaddesden, Northchurch, Berkhamsted, Hemel Hempstead, King's Langley, Redbourn, Kimpton, St Paul's Walden, King's Walden, Ippollitts, Hinxworth, Wallington and Wyddial.[42] A different and more sophisticated set of mouldings was detected in another group of churches: in the chancels of Bygrave and Little Munden, the chancel chapels of Hitchin and Wallington, at Kelshall and in the west towers of Aston, Codicote, Knebworth, Willian and Great Wymondley.[43] The common factor linking these latter churches was not geography but social connections and patronage. These mouldings are very similar to some created a generation or more earlier in Canterbury cathedral under the direction of the king's masons, and Roberts has suggested that a young mason from Totternhoe worked for a time on the nave there (built 1379–1405) before returning to his home area. He subsequently reproduced this style in the church at Knebworth for the lord of the manor, John Hotoft, in the early fifteenth century.[44] Hotoft was a high-ranking civil servant in the household of Henry Prince of Wales, later Henry V. In 1423 Hotoft was appointed Keeper of the Wardrobe, a post which brought him into contact with the most skilled craftsmen of his day and perhaps explains how he came to employ a mason who had worked at Canterbury to carry out improvements on his church at Knebworth.[45] Within Hertfordshire Hotoft was part of an elite group of prominent gentry from whose ranks government officials, sheriffs, Justices of the Peace and members of parliament were regularly chosen, and who held manors in the parishes of the ten churches listed above.[46] Hotoft's patronage would naturally lead to further commissions from his social peers for the mason and his pupils, and perhaps their pupils, and it is this which explains the shared style of mouldings in these churches.

Churches continued to be built and altered throughout the post-medieval period. Interiors were transformed by the Reformation in a series of stages through the sixteenth and seventeenth centuries but there was little rebuilding of the main fabric, with notable exceptions, all of brick, being the north transept at Little Hadham, the north chapel at St James Stanstead Abbots, the entire church at Buntingford and the chapel at Oxhey in Watford.[47] In the eighteenth century the most dramatic changes were generally made to churches near major mansions, such as the fine chancel at St Paul's Walden, which was remodelled in 1727.[48] The chancel of St Mary Magdalene Offley was entirely remodelled in Portland Stone in 1777 by Sir Thomas Salusbury of Offley Place,[49] while at Ayot St Lawrence the medieval church, which lay in the village on the southern edge of Sir Lionel Lyde's park, was partially demolished in 1778 and replaced with a Neoclassical temple designed by the architect Nicholas Revett on the west side of the park in order to enhance the views from Lyde's house.[50] The nineteenth century, a period of renewed energy in the established church, saw restoration and rebuilding on a much wider scale, and the erection of many entirely new churches. Interiors were again transformed, with old box pews and other fittings accumulated since late medieval times usually being swept away and replaced with rows of neat pitch pine pews: only at a few places, such as Stanstead Abbots, do the earlier arrangements survive. 'Restoration', at its most drastic, fades imperceptibly into wholesale rebuilding, making definitions difficult: but in broad terms the total rebuilding of existing churches and the addition of new churches to serve newly created parishes was a particular feature of the poorly churched south and south-west of the county. Here relatively low medieval population levels had often ensured the existence of small, poorly maintained churches, many originating as chapels-of-ease, which were insufficient to meet the challenges posed by rapidly rising populations in these economically expanding and suburbanising areas (Figure 7.7). Churches at Totteridge, Elstree, Northaw, Rickmansworth, Bovingdon and Harpenden were entirely rebuilt (in the last case, keeping the medieval tower), while Flaunden was replaced on a new site and at Hoddesdon an eighteenth-century private chapel, upgraded to parochial status, was largely rebuilt in 1844.[51] Rebuilding also

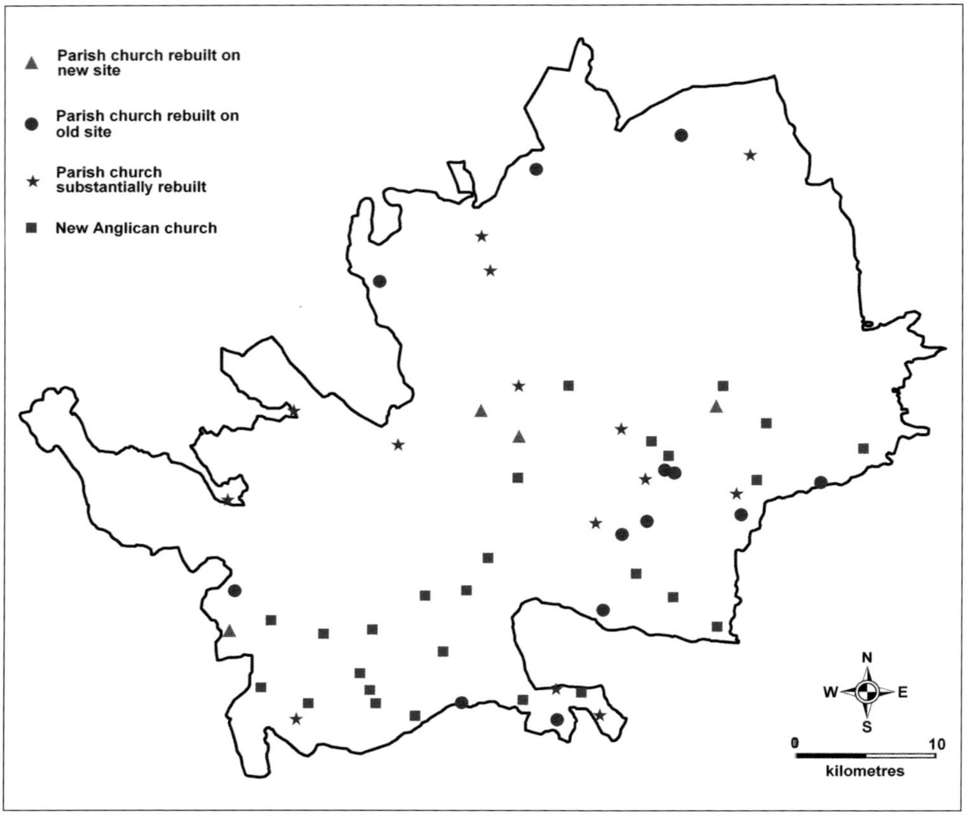

Figure 7.7. The distribution of new Anglican churches erected in the eighteenth and nineteenth centuries. Most new building occurred in the south of the county, where population growth was greatest. Source: Pevsner and Cherry, *Hertfordshire*.

occurred in the Hertford area: in the town itself, where both St Andrew's and All Saints' were drastically remodelled, and at Little Berkhamsted and Bayford.[52] In contrast, there was little rebuilding in the north and north-east, Therfield being one of the few exceptions, declining populations in these districts in the course of the nineteenth century ensuring little incentive for investment.[53] These new buildings – whether replacements of existing churches or new churches to supplement them – sometimes employed flint as their main building material, but often used materials brought from a distance, something now feasible given the improvements in the transport infrastructure. The church of St John the Evangelist Boxmoor was a chapel-of-ease to the parish church of Hemel Hempstead until 1844, when the parish of Boxmoor was formed, and the church was rebuilt in 1874; the new parish of Lemsford was created from part of the ancient parish of Hatfield in 1858, and a new church built the following year as a memorial to the sixth earl Cowper; Langleybury was a new ecclesiastical parish carved out of Abbot's Langley in 1878, 14 years after William

Jones Loyd, a London banker, had built a new church on his Langleybury estate.[54] All three of these were built using Oolitic limestone. The church at Waterford built for Robert Smith of Goldings in the 1870s, in contrast, was constructed of Kentish Ragstone, as were another 12 Hertfordshire churches built between 1846 and 1911.[55]

Our oldest and most architecturally complex religious buildings are parish churches, originally Catholic but eventually inherited by the Church of England. But we must not forget the importance of Dissenting congregations in the history of the county, and the traces which they have left in the landscape. Between 1692 and 1836 more than 500 places of dissenting worship were officially certified in a total of 76 Hertfordshire parishes; the Independents (later the Congregationalists), Baptists, Presbyterians (later Unitarians) and Society of Friends, or Quakers, were all represented. All these denominations seem to have declined in importance from the later seventeenth to the later eighteenth century but, as elsewhere in England, experienced a revival with the arrival of 'New Dissent', the Methodists, in the 1770s; by 1851 27,800 people were attending nonconformist chapels, compared with 40,400 worshipping in Anglican churches.[56] The physical marks left by nonconformity in Hertfordshire are, however, rather less than in some other English counties, such as Norfolk. Most surviving chapels are located in the main towns: in spite of some successes in establishing rural congregations in the first part of the nineteenth century, nonconformity seems always to have been strongly urban in character, in part, perhaps, because there was little support in Hertfordshire for the Primitive Methodists, builders *par excellence* of small rural chapels in the nineteenth century. Within towns there are some important examples of nonconformist architecture, with early Quaker meeting houses being particularly well represented. Those at Hitchin (late seventeenth century) and Hemel Hempstead (early eighteenth) no longer serve their original purpose but that at Hertford, erected in 1670, is one of the oldest meeting houses still in use in the country. These buildings are all fairly simple and 'vernacular' in character. Those of other denominations are all of later eighteenth- or nineteenth-century date, such as the Congregational chapel of 1819 in Buntingford or the Baptist of 1844 at Hitchin, by which time more sophisticated, classical styles were adopted. Only in the middle and later decades of the nineteenth century was the more familiar Victorian gothic style employed for numerous Baptist and Congregational chapels, especially in the urbanising south of the county – buildings often little different in appearance from contemporary Anglican churches.

Vernacular houses

'Vernacular' houses – farms and cottages built in local styles and with local materials – are a critical component in the landscape, but form a large subject which can only be dealt with briefly here: others have written at greater length, and much more knowledgeably, on the subject.[57] Such houses were, from early times, timber-framed: that is, they had a sturdy frame of posts, usually of oak, and the spaces between were filled with 'wattle and daub' formed from woven hazel covered with a mixture of clay, straw, lime and (often) animal dung. Only from the later seventeenth century were houses like this replaced, at vernacular level, by ones constructed entirely of brick. Whenever they were constructed, houses were constantly altered and modified by their owners – they have histories, not 'dates' – and in many cases this has ensured that the true antiquity of a dwelling is concealed behind a later façade.

There are no surviving vernacular buildings in the county dating to the period before *c.*1150, and only a handful remaining from before *c.*1250. Most of these are fragments, incorporated into later structures, and some are not, perhaps, strictly speaking, 'vernacular' at all, being manor houses such as Burston Manor, St Stephen's, near St Albans, erected in *c.*1200; Westwick Cottage at Leverstock Green, dated by dendrochronology to between 1184 and 1219; and Almshoe Bury, *c.*1245 (the term 'vernacular' and its precise boundaries with architecture of a 'polite' character are poorly defined and vague). These early buildings were of 'single pile' construction – that is, they were only one room deep – but to gain additional floor space they were aisled, like churches or barns (it is a striking indication of its antiquity that one of the capitals surmounting an aisle post at Burston has scalloped decoration of Norman form).[58] Such buildings also have complex roof structures featuring many long braces but no longitudinal timbers – 'purlins' – running the length of the roof.[59] Why builders should have been so concerned with lateral strength, yet were happy to neglect longitudinal stability in this way, remains unclear. To judge from these surviving fragments, thirteenth-century and earlier buildings also had earth-fast posts, like those of prehistoric houses: that is, the main structural members were sunk directly into the ground. With their main post thus highly susceptible to rot, and with roofs likely to 'rack' and collapse under strong winds, it is perhaps unsurprising that even fairly high-status buildings such as Burston Manor are rare survivors.

The increase in the numbers of surviving buildings from the late thirteenth century reflects both the adoption of fully framed forms of construction – in which frames were prefabricated and the principal posts rose from a cill-beam protected from the damp earth by a plinth of rubble – and the development of roof structures which featured varying forms of longitudinal stiffening; both in turn probably reflecting the greater wealth now available at peasant level. Farmhouses built in the fourteenth, fifteenth and sixteenth centuries were all single-pile buildings with a central 'hall' which formed the main living room and which generally contained the only fire, for warmth and cooking. Because there was no chimney this room was open to the roof, the smoke gradually drifting out through the thatch or through openings in the gables. Most surviving examples of such houses appear to have had a parlour at one end of the hall (a kind of bed-sitting room, the private space of the owner), with a chamber above, and, at the other end, two service rooms for food storage and preparation, again usually with a chamber above. These were separated from the hall by a cross passage running the full width of the house, with doors at each end. The end of the hall nearest the parlour, the 'upper end', was the more prestigious part of the room, where the owner and his family would take their meals.[60]

Figure 7.8. A picturesque range of vernacular houses, some with medieval origins, at Much Hadham in east Hertfordshire. Note the 'close studding' typical of the district.

The larger houses often had their service and/or their parlour ends in separately framed wings ranged at right angles to the hall range, and in this form they are often still easily recognisable in the hamlets and villages of east Hertfordshire especially. One or both of these ends is often 'jettied': that is, the upper floor projects slightly beyond the line of the ground floor, something which may have been done for structural reasons or to increase the amount of floor space upstairs, but which was probably in the main a stylistic device which also signalled to the outside observer that these parts of the house were floored. Hertfordshire can also boast around 15 surviving examples of so-called 'Wealden houses', such as Morris Cottage and the Crown House at Much Hadham, and Oak Beams at Walkern, in which a single hipped roof extends over both hall and jettied ends, giving the former a recessed appearance.[61]

Although there are some very small examples of late medieval buildings with this kind of three-part plan, as at Nobles Corner, Smith's End in Barley,[62] there is little doubt that surviving houses provide an unreliable guide to the size and sophistication of peasant homes. Only the largest or more substantial dwellings were worth keeping and adapting as fashions changed and, as life became more comfortable in the course of the sixteenth and seventeenth centuries, smaller and less substantial buildings were swept away. Of 11 peasant houses described in a survey of Weston in 1548, for example, 3 had only a hall and chamber, with one or more 'lofts' (roof spaces or storage), while a further 5 had only three rooms, variously described, some again with lofts.[63] None appears to have survived, added to and improved, to the present.

In England as a whole late medieval and sixteenth-century buildings displayed much regional variation, and Hertfordshire – typically – had no real style of its own, but lay instead at the junction of two broad architectural zones.[64] The medieval buildings in the east of the county display styles typical of those found in adjacent areas of Essex and, to a lesser extent, in East Anglia and Kent. They were universally of box frame construction: that is, the body of the building and the roof were structurally distinct, although both were raised together, as a series of trusses defining a number of bays, each truss comprising principal posts, principal rafters, ties and collars in a variety of combinations. Between were smaller members, or studs, which held the panels of wattle and daub which formed the wall itself. By the second half of the later fifteenth century 'close studding' was fashionable – flamboyant, wasteful displays of upright timber, especially on the principal public elevations, with so much wood that façades sometimes appear more 'black' than 'white' (Figure 7.8). Almost all medieval buildings in this district had so-called crown-post roofs in which a tie-beam ran the width of the building, preventing the walls from pushing apart under the weight of the roof, and from this rose a vertical crown post, often ornately carved, which supported a collar purlin running down the length of the roof. Each pair of rafters was tied together by a transverse collar, and each collar was supported at its centre by the collar purlin (Figure 7.9). It is also in the east of the county that the few surviving Wealden houses are found – an architectural form particularly associated with Kent, but also widely found in Essex and south Suffolk. It must be emphasised once again, however,

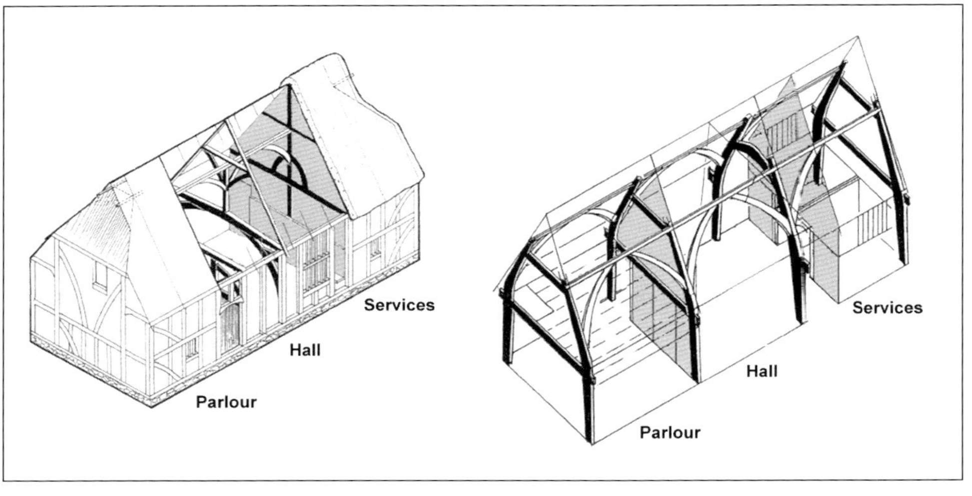

Figure 7.9. Reconstructions of the original forms of late medieval buildings in Hertfordshire (after the late Adrian Gibson). Right: Moor Cottage, Water End, Hemel Hempstead, showing cruck construction. Left: Nobles Corner, Smith's End, Barley, with a crown-post roof: this is a rare example of a very small late medieval house.

that the real wealth of early buildings in the county is obscured by later façades, although the overall shape of a house will often betray its origins, so that, for example, medieval houses with former open halls are indicated 'by their front doorways, set off-centre at the low end of the hall, and sometimes the later inserted chimneystack set off-centre to the roof apex to avoid an extant crown-post assembly'.[65]

Early houses in the west of the county displayed rather different styles, typical of the Midlands and areas further west. A small number, extending as far east as Offley, are of cruck construction.[66] That is, they have pairs of inclined timbers which sweep from floor to the apex of the roof, like a great 'A' frame, supporting both walls and roof, which were not structurally distinct (Figure 7.9). Good examples, albeit much altered, include Westwick Row Farm and Moor Cottage, Water End, to the east and north of Hemel Hempstead, respectively. Most, however, are again of box-frame construction, although close-studding is much less common than in the east, houses instead often having wide gaps between the posts and studs (Figure 7.10). The large panels of wattle-and-daub so defined are often crossed in places by short, straight braces which run from vertical posts upwards to the horizontal 'plates'. Bold, curving arched braces also rise to the wall plates. In addition, while crown posts are widespread in this district other forms of roof construction are also found. Some houses, for example, had (like crucks) purlins in the plane of the roof, rather than (as in crown-post roofs) running centrally through the roof void. Such roofs usually,

Figure 7.10. House in Aldbury, showing the wide square framing – infilled in part with brick – typical of the western side of the county.

Figure 7.11. The Old Rose and Crown, Braughing, a sixteenth-century house with the main chimney stack placed against the cross passage (indicated by the front door). The continuous jetty along the front signals that the house is fully floored throughout.

like those in the Midlands, have curving wind braces running between the principal posts and the side purlin to provide extra strength.[67]

In the middle and later decades of the sixteenth century there were a number of major changes in the design of houses in Hertfordshire, as elsewhere in England. They were still only one room deep, and still divided into three main sections (service, hall, parlour). But they were now provided with brick chimney stacks, usually massive in character and often contained within a separate 'bay' or section of the frame; and, instead of having an open hall, they were fully floored throughout. Existing houses were modified along similar lines.[68] The stack was sometimes located within the hall and placed against the cross passage, as at the Old Rose and Crown in Braughing – an arrangement which allowed traditional ways of using the hall to be maintained: the end nearest the parlour continued to be the more prestigious end, where the owner sat when eating, with cooking taking place at the 'lower' end, nearest the service rooms (Figure 7.11). Other houses, however, such as the Old Rectory in Great Hormead, have the stack placed between the hall and the parlour, providing fireplaces in both. In both cases, a continuous jetty running along the front elevation proclaims that the house had been built, from the start, fully floored throughout. With the hall now floored over there was little incentive to display elaborate crown posts, and these were anyway rendered impractical by the fact that the chimney stack got in the way of the central collar purlin. The roofs that replaced them were simpler structures featuring purlins which were clasped between the principal rafter and the collar.

Experts have long argued over why open halls went out of fashion. The old view was that as the price of bricks fell so chimney stacks became affordable to middle-class farmers and merchants and, once stacks were inserted, there was no longer any need for an open hall to disperse the smoke from the fire. But more complex explanations of a social character have also been advanced, not least because the earliest chimneys in the county, dating from the start of the sixteenth century, seem to have been timber-framed anyway.[69] A few examples still survive, as at 2 High Street, Barkway. Documentary sources suggest that such features were common in the sixteenth century but were subsequently replaced by brick stacks, which posed less of a fire risk.[70] The flooring over of the hall may well have resulted from the increasing popularity of chimneys but this perhaps stemmed not so much from a fall in the price of bricks as from an increasing desire to heat rooms other than the hall, either through radiant heat from the stack or from additional fireplaces. This provided the owner with more privacy from servants and farmhands at a time of growing social polarisation.[71]

These social processes – the division of the household along more hierarchical lines – soon culminated in the widespread adoption, from *c.*1600, of the standard 'lobby entry' plan, which was common across much of southern and eastern England. The cross passage was now dispensed with and the main door opened into a small lobby immediately in front of the stack – an arrangement which allowed direct access to the private parlour without trekking through the more public space of the hall. Many were created from the conversion and alteration of older structures, of medieval or sixteenth-century date: the stack was simply placed in the cross passage. In Hertfordshire, a high proportion of the houses newly built in this form had two rather than three rooms on the ground floor, with a central chimney stack providing fireplaces in a parlour and a combined hall/kitchen. Such an arrangement reflects the importance of a symmetrical façade as Renaissance ideas became popular in a fashionable county close to the capital; and, while it reduced the number of rooms at both first and ground floor level, additional space was often now provided in a usable, properly floored attic.

There were further important changes in the course of the seventeenth century. Clasped purlins declined in popularity, and purlins were now normally butted directly into the principal rafters. More importantly, in terms of the outward appearance of houses, from the late sixteenth century it became fashionable to cover over the studs and timbers with plaster, in part because increasing pressure on timber supplies ensured that ostentatious displays of close-set studs were no longer easily affordable. The plaster was

Figure 7.12. A variety of patterns is shown in the pargeting on a house at Furneux Pelham. This style of decoration is particularly common in the east of the county.

sometimes moulded to form the decorative patterns known as *pargeting* (Figure 7.12), which mainly takes quite simple forms, with rectangular panels defined by scratched lines or beaded borders; but more elaborate forms include repeated chevron patterns within panels (especially common in the area around Sawbridgeworth), basket-weave patterns, or interwoven fan designs. The most flamboyant displays, featuring deeply moulded floral decoration and human and animal figures (inspired by wider European Renaissance fashions), are largely a feature of town buildings, especially inns (as in 3–11 Fore Street in Hertford). Pargeting probably reached a peak of popularity in the mid–late seventeenth century, declining thereafter. It was quintessentially a style of eastern England, although sporadically practised elsewhere and perhaps more widely distributed in the past. It is thus found especially in Suffolk and Essex, to an extent in Cambridgeshire and Kent, and – like so many other features of Hertfordshire's built heritage – is principally a feature of the eastern side of the county, with fewer examples in the west, largely in urban contexts (as at Godmersham House in Fishpool Street, St Albans). It was mainly in the eastern part of the county, also, that house façades in this period were often covered in horizontal weatherboarding, a common practice in the adjacent areas of south-central Essex. In the west, in contrast, a distinctive development was the infilling of the large spaces between the beams with brickwork. In some cases such brick nogging may be original but in most

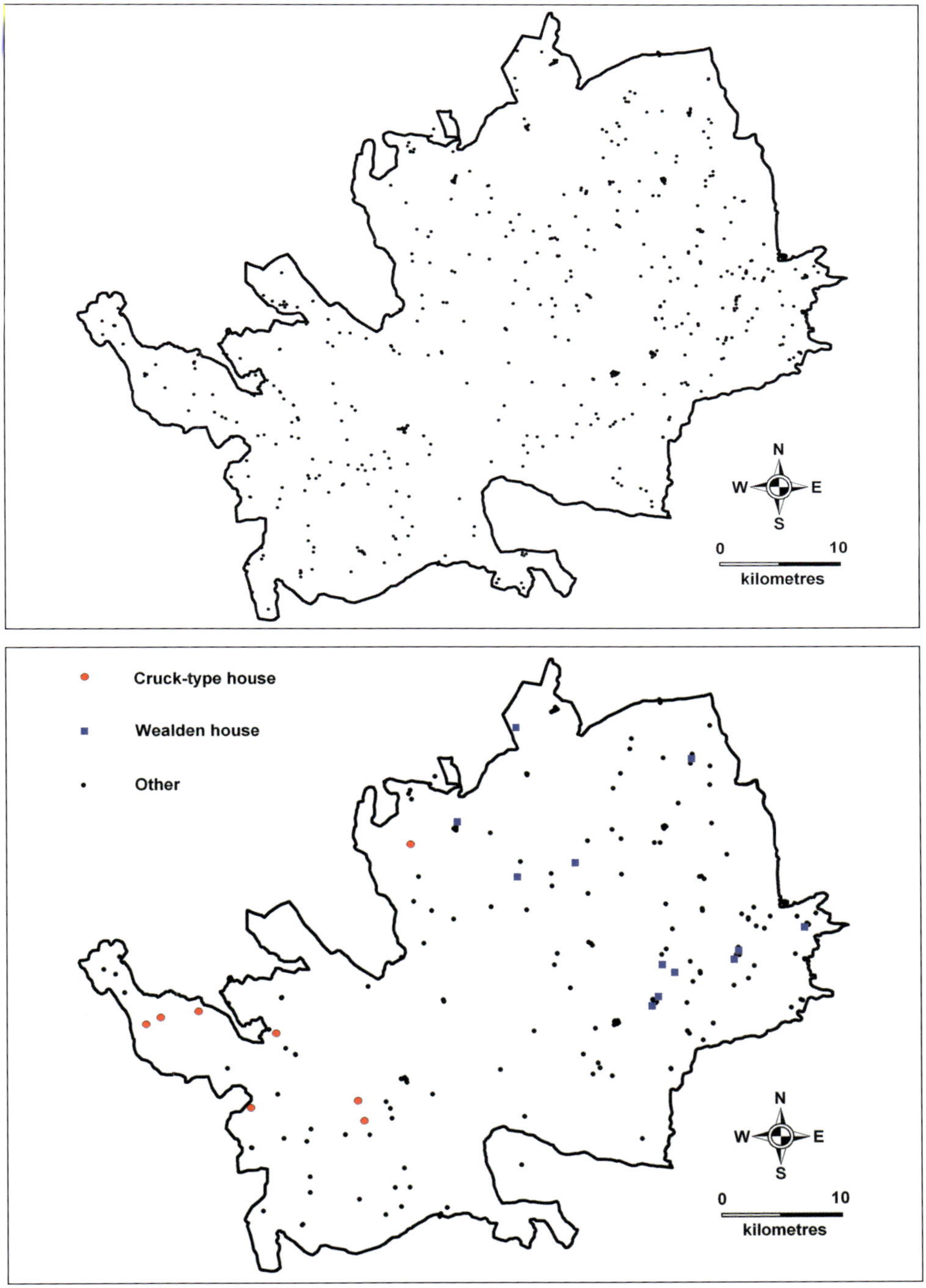

Figure 7.13. (a) The distribution of houses in Hertfordshire erected before the late seventeenth century (based on the information held in the Hertfordshire Historic Environment Record). (b) The distribution of medieval houses in Hertfordshire (i.e., erected before the mid-sixteenth century: based on the information held in the Hertfordshire Historic Environment Record).

it appears to have replaced wattle and daub. The contrast between the vernacular styles of the east and west of Hertfordshire were thus deep and enduring, although more work is needed to ascertain whether these stylistic zones were stable or changed over time – another useful area of research for the county's local and landscape historians.

Some vernacular houses have thatched roofs – mainly of straw. But tiles were already being used at vernacular level in parts of the county by late medieval times, and they were widespread by the seventeenth century. Hertfordshire uses plain tiles – simple straight tiles, pegged into roof battens – rather than the curved pantiles found in East Anglia, only a few examples of which can be found covering outbuildings in the far north-east of the county. Tile production, like later brickfields, was widely distributed through the county, exploiting all the main clay sources, and tile kilns are often referred to in minor placenames, as well as in a variety of early documents. Some exploited the London Clay in the south of the county: there are medieval records of tilers and potters at Tylers Causeway in Hatfield; in 1538 William Marston of St Peter's was leased the manor of Napsbury by the abbot of St Albans 'except one house called le Tylehous and the land where they dig clay for tiles and bricks'; and in 1547 there are references to a 'Tyle Kilne' in the nearby parish of Ridge.[72] But the boulder clay deposits were also used, and in 1422 William Colt of Standon was granted the use 'of a house called Tylhous' on the manor of Milkley (now Mentley) 'with free entrance and exit to and from the said house by three ways, viz. towards Pokerych [Puckeridge], Eldehallegrene [Old Hall Green], and Schakelockyslane, for carrying tiles'. In return for pasture for his horse, two cows and three hogs, Colt was to furnish the lord of the manor with 'a thousand tiles yearly for roofing the house there'.[73] The pasture field south of Mentley Farmhouse was called Tile Croft in the nineteenth century and within and around it are earthworks of probable extraction pits, a building platform and hollow ways which appear to be the remains of the medieval tile works.

Houses erected before the late seventeenth century are found throughout the county: Figure 7.13(a), based on information in the Hertfordshire Historic Environment Record, is neither complete nor entirely accurate, but it is reliable in showing the overall distribution of early houses and indicates that, while there are slightly more in the east of the county than in the west, the difference is not very great. The distribution of known medieval buildings (Figure 7.13(b)), again based on HHER information, is broadly similar. Both maps are slightly surprising, given that the lay subsidy returns of 1307 and 1543–5 both suggest that the east was both more densely populated and, on average, wealthier than the west (above, pp. 20–2, 26–7). The current distribution of old houses is a consequence of population densities and middle-class prosperity at the time the houses were erected, but also of later patterns of attrition, and it is possible that old houses were maintained and updated (especially with new façades) to a greater extent in the west in the course of the eighteenth and nineteenth centuries because, as we shall see, the influence of London encouraged an expansion of wealth and population in this part of the county at that time: the north-east and east, in contrast, suffered relative economic decline and absolute population decline in the nineteenth and probably in the eighteenth century, perhaps ensuring that a higher proportion of early houses could not be adequately maintained and were consequently lost. But this is not an entirely satisfactory explanation and at present it must be admitted that the rather even distribution of early houses across the county is hard to understand. Whatever the explanation, the county's heritage of timber-framed houses is a surprisingly rich one, and note should also be made of such gems as Piccotts End, just to the north of Hemel Hempstead, with its magnificent late medieval wall paintings – no longer, unfortunately, open to the public.

It is not only timber-framed houses which survive in some numbers in Hertfordshire from the period before the later seventeenth century. Timber-framed farm buildings, especially threshing barns, can also be found. These are, once again, scattered fairly evenly across the county, although perhaps with more medieval structures – and certainly some of the most impressive examples – being found in the west. These include the huge barns erected on the St Albans Abbey manors at Abbot's Langley, Langleybury, Croxley Green, Kingsbury, Parkbury and St Julians in St Stephen's parish, all with crown-post roofs and of fourteenth-century date, and all probably built at the same time, part of a 'grand refurbishment' carried out by Abbot John Moote. At Great Gaddesden is the county's only surviving cruck barn, of probable fifteenth-century date, although radically rebuilt following its collapse in the late twentieth century. We should note that while surviving buildings of sixteenth-century or earlier date are all timber-framed, there is some documentary evidence for the early use of Totternhoe Stone in St Albans in the form of a mysterious reference in 1661 to 'a very large barn of 12 great bays of great height enclosed with strong stone walls covered with tiles'.[74]

By the time of the Restoration brick and tile were becoming the standard building materials in the county, used to construct new houses and to mask, behind a fashionable façade, the real character of others. There were many small brickyards in operation, such as that near Baldock owned by Thomas Goodwyn who, dying in 1665, left 'all loose bricks & tiles in my yard' to his son John.[75] Tile widely replaced thatch on many existing buildings, so much so that by the early eighteenth century a building in Hoddesdon could be identified by name as 'The Thatched House', presumably because it was by then unusual. Pehr Kalm, travelling in 1748 across the county from Waltham Cross to Little Gaddesden via Cheshunt, Bell Bar, London Colney, St Albans and Hemel Hempstead, noted:

> The houses that we saw on our journey were for the most part of brick. … In many instances … their walls consisted of a cross-work of timber with bricks in between. Houses frequently had two or three storeys … the uppermost floor … consists … of rooms and chambers which are usually occupied by the domestic servants. … In the upper storeys, the walls commonly consist of thin boards and lathe plastered over on both the inside and outside of the house with clay and lime so that they look like stone.

He observed that roofs were mostly of tile but many outhouses as well as some of the brick-built cottages had thatched straw roofs, as did most farmsteads on the Chiltern dipslope west of St Albans.[76] The use of brick and tile among the wealthier farmers, the adoption of coal and the construction of smaller hearths and chimneys allowed double-pile houses to be erected, two rooms deep – rather than single pile, as all earlier vernacular buildings had been. The adoption of brick was not total, however: light timber framing continued sporadically to be used for the construction of cottages into the nineteenth century, often now employing elm or even softwoods rather than oak. Farm buildings also often continued to be timber-framed, and with black weatherboarding, right through to the nineteenth century, although the later examples often have slate roofs.

The availability of pattern books was, by the start of the eighteenth century, beginning to ensure that design was becoming increasingly standardised throughout England and building materials (such as slate from Wales) also began to be sourced more widely as the transport network improved and developed during the eighteenth and nineteenth centuries. Regional differences did not disappear entirely, but they became less pronounced with the passing decades, and by the time that large swathes of the county became suburbanised in the course of the twentieth century all trace of local architectural distinctiveness had vanished.

Notes

1. Crick, *Charters of St Albans*, no. 7.
2. R. Morris, *Churches in the landscape* (London, 1989), p. 253.
3. D. Renn, *Medieval castles in Hertfordshire* (Chichester, 1971), pp. 13, 14, 21–2, 25.
4. Riley, *Gesta Abbatum*, p. 22; Niblett and Thompson, *Alban's buried towns*, pp. 188–9; Hunn, *Reconstruction and measurement of landscape change*, p. 70.
5. See Williamson, *Origins of Hertfordshire*, pp. 207–9.
6. N. Doggett, 'Medieval parish churches', in Short (ed.), *Historical atlas*, pp. 126–7.
7. T.P. Smith, *The Anglo-Saxon churches of Hertfordshire* (Chichester, 1973), pp. 17–19; J. Sherwood, 'Influences on the growth and development of medieval and early modern Berkhamsted', in Slater and Goose (eds), *County of small towns*, pp. 228–9.
8. HHER 4332.
9. We are relying here on the information provided by Doggett, 'Medieval parish churches'.
10. Smith, *Anglo-Saxon churches*, pp. 17, 19, 34; H.M. Taylor and J. Taylor, *Anglo-Saxon architecture* (Cambridge, 1965), pp. 509–10, 645–6.
11. Morris, *Churches in the landscape*, pp. 160–67.
12. Doggett, 'Medieval parish churches', pp. 126–7.
13. Doggett, 'Medieval chapels', pp. 113–17.
14. N. Pevsner and B. Wilson, *Norfolk* (London, 2002), p. 49.
15. Doggett, 'Medieval parish churches'.
16. E. Roberts, 'Parish church studies: an analysis of Hertfordshire plans', *Hertfordshire's Past*, 4 (1977), pp. 22–7.
17. N. Doggett, 'The parish churches of Dacorum: how they reflect social history', *Hertfordshire's Past*, 11 (1981), p. 28.
18. N. Pevsner and B. Cherry, *The buildings of England. Hertfordshire*, 2nd edn (Harmondsworth, 1977), p. 197.
19. J. Catt, C. Edmonds, H. Bailey and C. Green, 'Mineral resources and church building stones', in Catt (ed.), *Hertfordshire geology and landscape*, p. 308.
20. English Heritage List Entry Summary for churches of St Nicholas (1174104), Holy Trinity (1175183) and St Giles (1101811).
21. E. Roberts, 'Totternhoe Stone and flint in Hertfordshire churches', *Medieval Archaeology*, 18 (1974), pp. 86–7.
22. Catt *et al.*, 'Mineral resources and church building stones', p. 317.
23. Smith, *Anglo-Saxon churches*, pp. 6–8.
24. Niblett and Thompson, *Alban's buried towns*, p. 188.
25. HHER Number 4299 St Margaret Bygrave.
26. Catt *et al.*, 'Mineral resources and church building stones', p. 304. The quarry was less than two miles from the boundary of the parish of Kensworth which, until 1897, was part of Hertfordshire.
27. Smith, *Anglo-Saxon churches*, p. 10.
28. E. Roberts, *The hill of the martyr: an architectural history of St Albans Abbey* (Dunstable, 1993), pp. 10–11.
29. Page (ed.), *VCH Hertfordshire*, vol. 2, pp. 326, 464, 441.
30. Roberts, 'Totternhoe Stone and flint', p. 71.
31. English Heritage List Entry Summary for churches of St Mary Ashwell (1102715) and St Peter and St Paul Tring (1084356).
32. Roberts, *Hill of the martyr*, p. 7.
33. Catt *et al.*, 'Mineral resources and church building stones', p. 319.
34. Roberts, 'Totternhoe Stone and flint', p. 87; Page (ed.), *VCH Hertfordshire*, vol. 2, p. 441; Catt *et al.*, 'Mineral resources and church building stones', p. 314.
35. Page (ed.), *VCH Hertfordshire*, vol. 2, p. 257; Catt *et al.*, 'Mineral resources and church building stones', p. 312.
36. J.F. Potter, 'The London Basin's gravel churches: indications of geology, medieval history and geographical distribution', *Landscape History*, 23 (2001), p. 7.
37. Catt *et al.*, 'Mineral resources and church building stones', p. 305.
38. HHER Number 4364.
39. Page (ed.), *VCH Hertfordshire*, vol. 2, p. 369.

40. HHER Number 4346; English Heritage List Entry Summary for church of St Giles (1101811).
41. Roberts, 'Totternhoe Stone and flint', p. 87.
42. *Ibid.*, pp. 74–6.
43. E. Roberts, *A school of masons in 15th century north Hertfordshire* (Hertford, 1979), p. 1.
44. *Ibid.*, pp. 15–16.
45. *Ibid.*, pp. 2–7.
46. *Ibid.*, p. 1. The gentry families include the de Benstedes, Pastons, Prisots, Pulters and Thornburys.
47. English Heritage List Entry Summary for churches of St Cecilia Little Hadham (1290608), St James Stanstead Abbots (1078730), St Peter Buntingford (1348000) and Oxhey chapel (1174300).
48. English Heritage List Entry Summary for church of All Saints St Paul's Walden (1347042).
49. Catt *et al.*, 'Mineral resources and church building stones', p. 317.
50. Pevsner and Cherry, *Hertfordshire*, pp. 79–80; Munby, *Hertfordshire landscape*, pp. 161–2.
51. Pevsner and Cherry, *Hertfordshire*.
52. The medieval church of All Saints, Hertford was destroyed by fire in 1891.
53. Pevsner and Cherry, *Hertfordshire*.
54. Page (ed.), *VCH Hertfordshire*, vol. 2, pp. 228, 323; vol. 3, p. 110.
55. Catt *et al.*, 'Mineral resources and church building stones', p. 319.
56. A. Ruston, 'Protestant nonconformity', in Short (ed.), *Historical atlas*, pp. 134–5.
57. See in particular the works of J.T. Smith and A. Gibson: J.T. Smith, *English houses 1200–1800: the Hertfordshire evidence* (London, 1992); J.T. Smith, *Hertfordshire houses. Selective inventory* (London, 1993); A. Gibson, 'Timber-framed building in Hertfordshire', in D.F. Stenning and D.D. Andrews, *Regional variation in timber-framed buildings in England and Wales down to 1550* (Chelmsford, 1998), pp. 22–30.
58. E. Mercer, *English vernacular houses* (London, 1975), pp. 40–50; Gibson, 'Timber-framed building', p. 23–4.
59. C.A. Hewett, 'Aisled timber halls and related buildings, chiefly in Essex', *Transactions of the Ancient Monuments Society*, 16 (1968), pp. 45–99.
60. Smith, *Hertfordshire houses*.
61. *Ibid.*, pp. 32–3, 148–9.
62. Gibson, 'Timber-framed building', p. 26.
63. HALS 49464.
64. An observation first made in G. Bailey and B. Hutton, *Crown post roofs in Hertfordshire* (Chichester, 1966). For an up-to-date summary, see Gibson, 'Timber-framed building'.
65. Gibson, 'Timber-framed building', p. 25.
66. A. Gibson, 'A base cruck hall at Offley', *Hertfordshire Archaeology*, 4 (1976), pp. 153–7.
67. Gibson, 'Timber-framed building', pp. 22–3.
68. Mercer, *English vernacular houses*, pp. 50–60.
69. Smith, *English houses*, pp. 185–6; A. Gibson, 'A timber-framed smoke hood at Barkway', *Hertfordshire Archaeology*, 2 (1970), pp. 112–13.
70. For example, at Wellbury in Great Offley in 1552, where in the hall of the manor house there was 'a Chimney of Timber tonnell'd with Bricke (and) One Kitchin with a Tonnell of Wood' (HALS 72351 survey of the manor of Wellbury, Offley, 1553).
71. Mercer, *Vernacular houses*, pp. 58–9; M. Johnson, *Housing culture* (London, 1993).
72. HHA Hatfield Manor Papers I, 1396 Michaelmas, p. 265; HALS 44737 lease of the manor of Napsbury, 1538; 79423 lease of the manor of Blackhide alias Corsers, 1547.
73. TNA:PRO E 210/407 Ancient Deeds, Series D.
74. HALS DE/Pm/72439 Chancery bill in title deeds and estate papers of the Pym Family of Norton Bury, Norton, 1661, relating to the derelict state of the chancel of St Peters church and the barn.
75. HALS DE/Bw/28766 probate, Thomas Goodwyn of Baldock, 1665.
76. Mead, *Pehr Kalm*, pp. 36, 140.

Great houses and designed landscapes

Hertfordshire, as we saw in Chapter 1, is not for the most part characterised by very large landed estates. Ashridge, Cashiobury, Hatfield and Gorhambury were sizeable properties, but much more common were small gentry estates and 'villas' without extensive holdings. This circumstance was to some extent related to the character of local soils and land use, but was mainly a consequence of proximity to London. Men who made their money in the capital often sought a rural retreat within easy reach, where they could escape the noise, smells and diseases of the city. Competition for land forced up property prices, especially in the south, and militated against the acquisition of very large accumulations of land. Hertfordshire was attractive to such men not only by virtue of its accessibility: its countryside provided opportunities for establishing parks and for otherwise enjoying the thrill of the chase, and it possessed many attractive places in which to build country seats, as well as the raw materials from which designed landscapes could be created – there was an abundance of woodland and trees which could be incorporated within new parks and gardens, and the topography provided extensive vistas which drew admiring comments from travellers. One, crossing Bushey Heath in the early eighteenth century, described how:

> They had there on the right hand, the town of St. Albans in their view; and all the spaces between, and further beyond it, look'd indeed like a garden. The inclos'd corn-fields made one grand parterre, the thick planted hedge rows, like a wilderness or labyrinth, divided in espaliers; the villages interspers'd, look'd like so many several noble seats of gentlemen at a distance. In a word, it was all nature, and yet look'd all like art.[1]

True, other counties in the vicinity of London had their attractions. But Hertfordshire – lying as it does astride some of the main routes connecting the capital with the Midlands and North, and with Scotland and north Wales – made a particularly convenient staging point for men with interests in the capital but estates elsewhere.

Before the Restoration

There were major residences in the county in the early Middle Ages, the homes of local gentry or nobility or properties of the Crown, but relatively little is known of them. The royal castles at Berkhamsted and Hertford were imposing residences as much as fortresses, if only sporadically visited, and by the thirteenth century both were equipped with deer parks where the king could hunt. Hertford was substantially rebuilt as late as the fifteenth century – the imposing brick gatehouse was erected in the 1460s – and visited several times by Queen Elizabeth in the sixteenth century, parliament assembling here while plague raged in London.[2] A third important royal estate was at King's Langley, which Edward I's queen, Eleanor of Castile, developed into an elaborate royal residence in the 1270s. The park she established was subsequently expanded by both Edward I and Edward III at the expense of arable land, making a substantial impact on the local landscape.[3] Details of the gardens around the palace are scant but there was a vineyard in the early fourteenth century. The palace was in ruins by the end of the sixteenth century, although parts of the associated friary are incorporated within a private school, and the park was broken up in the seventeenth century.

Major ecclesiastical landowners also possessed imposing residences. St Albans Abbey owned a number, of which Tyttenhanger, the favourite resort of a succession of abbots, was probably the most important. The house was rebuilt on the south bank of the river Colne at the end of the fourteenth century and a deer park was created in the adjacent meadows and pastures by Abbot John of Wheathampstead (1420–40).[4] Nothing remains of this mansion,

Figure 8.1. The surviving wing of the former Bishop's Palace at Hatfield.

but at Hatfield one wing of the great palace erected by Bishop Morton of Ely in the 1480s, together with the gatehouse range, still stands beside the later Hatfield House (Figure 8.1). Like other similar residences of the period it is built of brick, but with windows still of late gothic 'Perpendicular' form. It is a striking reminder of the sophistication of late medieval domestic architecture. Like most such residences, the palace was ranged around a courtyard that was entered through an imposing gatehouse.[5] Such an arrangement, which often developed gradually over time as houses were expanded, was partly a consequence of the fact that many large houses were constructed within moats, but it was mainly because they were single-pile dwellings, like the vernacular farmhouses discussed in Chapter 7. In addition to the familiar 'core' of parlour, hall, cross passage and service rooms, great landowners also needed accommodation for extensive households, storerooms and chapels and, rather than creating a long, attenuated line of rooms, it made sense to form them into a square, an arrangement more convenient in terms of both access and security. Gatehouses were key features of such residences, designed to convey an impression of military strength, although few such buildings were seriously defensible. The best example in Hertfordshire is Rye House in Hoddesdon, where the brick-built gatehouse of an otherwise demolished mansion, erected in the 1440s, now stands alone within a moated enclosure.[6]

Most large late medieval houses in the county have disappeared without trace, replaced by more fashionable structures. The More in Rickmansworth, for example, was a substantial crenellated brick house built (or perhaps completed) by Ralph Boteler, first Baron Sudeley, about 1450 and later sold to George Neville, Chancellor of England, who apparently altered the building, furnishing it luxuriously. Neville became Archbishop of York in 1465 but his estates passed to the Crown when he was charged with treason.[7] During Henry VIII's reign The More became one of Cardinal Wolsey's residences and he, from 1521, enlarged and beautified both house and gardens, constructing a gallery which projected north from the house across two arms of the moat surrounding the adjacent garden. When complete The More was as sophisticated as Hampton Court, Wolsey's better-known building project, but it was demolished in the seventeenth century.[8] The moats – one surrounding the house, the other part of the gardens – survived until the start of the twentieth century, but they too have since been destroyed.[9]

Although there were clearly a number of imposing mansions in the county in the later Middle Ages, it was the establishment of a centralised Tudor state in the sixteenth century and the growing importance of a court based permanently in and around London that saw a marked increase in their numbers. The Tudors took a keen interest in the county. From early in his reign Henry VIII held the deer parks at Cheshunt, Hertingfordbury, Berkhamsted, King's Langley, Standon, Hunsdon and Weston.[10] He also made regular use of the Bishop's Palace at Hatfield from 1514, when his farrier, Hannibal Zenzano, became the lessee of the manor and keeper of its three parks.[11] The king did not become legal owner until 1538 but it was the home of Princess Mary until 1534, and Princess Elizabeth and Prince Edward spent much time there as children.[12] Henry also, from 1525, spent thousands of pounds improving the fifteenth-century house at Hunsdon, where he also laid out two new parks in addition to the one already there. In 1529 he acquired The More and its park when Wolsey fell from favour: it made a particularly convenient bolt hole within easy reach of London.[13] Henry's obsession with hunting ensured that by 1540 he owned around half the parks in the county, but within a decade of his death in 1547 the number had been reduced from 18 to 5 (2 at The More and 3 at Hunsdon), with a further 6 held by Princess Elizabeth at Ashridge, Berkhamsted, Hatfield (3 in all) and Hertingfordbury. Elizabeth was so fond of Hatfield that the estate was conveyed to her in 1550 and it continued to serve as a royal palace after she became queen in 1558.[14] Hunsdon, however, she granted to her cousin Henry Carey, while The More went in 1576 to Francis, Earl of Bedford.[15] Unlike The More, Hunsdon has not entirely disappeared, and the gatehouse and fragments of the north wing, built into a later house, are enough to suggest the size and sophistication of this great residence.

Perhaps more striking than the direct effects of the Tudor and early Stuart monarchs were the building campaigns of their officials and courtiers, of wealthy merchants and of lawyers grown rich in service to government or notable figures, in a litigious age when contests in court now regulated the affairs of great men. Many of their houses have, again, disappeared without trace. Nyn or North Hall, in Northaw, was built in the 1540s by William Cavendish, an auditor of the Court of Augmentations, and extensively rebuilt and extended by Ambrose Dudley, Earl of Warwick in the 1570s and 1580s.[16] The earl was a prominent courtier and his wife a gentlewoman of Queen Elizabeth's privy chamber. According to Chauncy, the earl 'raised here a stately house from the ground, and contrived it in very beautiful order, gracing it with delightful gardens and walks, and sundry other pleasant and necessary devices',[17] but his ambition to create a park on part of the adjacent Northaw Common was thwarted by fierce resistance from the commoners.[18] Nyn Hall was demolished in 1774 but subsequently rebuilt.[19] The house erected by the Freman brothers, Ralph and William, London cloth merchants, at Aspenden at the beginning of the seventeenth century was, by 1700, embellished with a loggia, a variety of fashionable walled gardens and a deer park, but in 1856 was, likewise, completely demolished.[20] Berkhamsted Place, constructed by Sir Edward Carey, keeper of the Queen's jewels, within the medieval park which Elizabeth had granted to him along with the derelict castle, survived (much altered) until it was demolished as late as 1965.[21]

In a number of other cases, however, a large sixteenth-century house survives, but embedded in later fabric or extensively altered. Stanstead Bury near Stanstead Abbots, for example, built in the 1560s by Edward Baeshe, a naval administrator, is a complex building, changed and added to over the subsequent centuries, but retaining much original fabric.[22] Hamels was built by John Brograve, Attorney-General to the Duchy of Lancaster from 1580, at the centre of an estate he had accumulated west of Braughing. A 'fair brick house … situated upon a dry hill where is a pleasant prospect to the east',[23] it remained the seat of his descendants for over a century and, although apparently 'rebuilt' in the latest fashionable architectural style on three subsequent occasions, much of the original house remains hidden within.[24] In a few cases a great house has completely disappeared but remains of its associated designed landscape survive, as at Benington Park. In 1614 Sir Julius Caesar, one of the leading civil lawyers of his generation, became Master of the Rolls and bought the estate, probably for his son Charles, also a lawyer. The house, which overlooked a park, was enlarged and by 1628 magnificent gardens had been laid out beside it. The house has gone but the earthworks of the terraced gardens – one of the finest examples of garden archaeology in the county – still exist to the south-east (Figure 8.2).[25] These remains are a

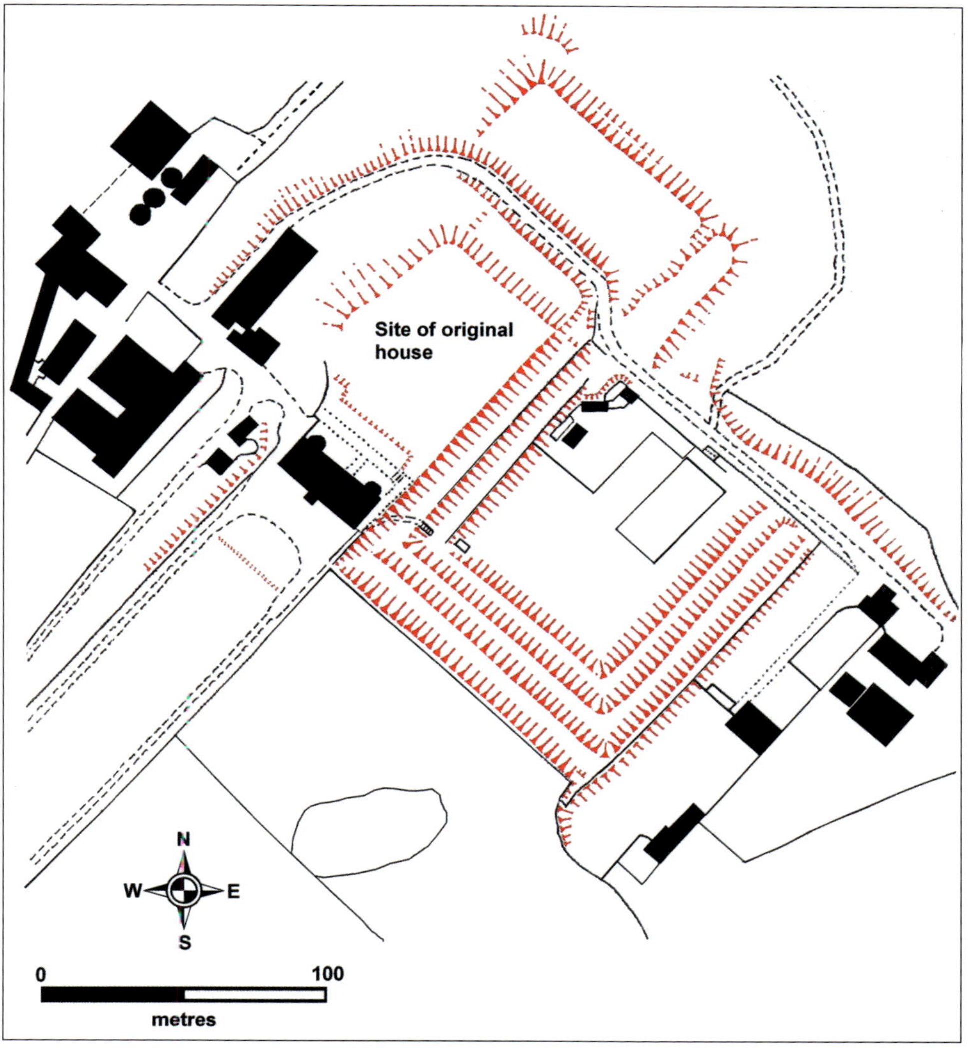

Figure 8.2. The earthworks at Benington Park, one of the most impressive examples of garden archaeology in the county. To the south-east of the site of the original house lie the remains of the huge terraced garden, created by modifying the natural contours. Further terraced areas lie to the north-east of the house site. The gardens are shown on estate maps of 1628 and 1743, and most of the surviving remains were probably created shortly before the former date.

rare survival. We know that Tudor and early Stuart mansions were provided with elaborate gardens – set within walled enclosures, often terraced, and featuring topiary and geometric displays of planting in knots and parterres – but only rarely do any traces remain. Another example is at Stanstead Bury, where parts of the garden walls survive to the east of the hall and extensive earthworks – probably of an elaborately landscaped formal approach – remain to the west.

Where more substantial traces of sixteenth- and early seventeenth-century houses survive it is possible to obtain a clearer idea of their spatial organisation. Standon Lordship is a large brick-built house which was erected in the 1540s by Sir Ralph Sadleir, Henry VIII's principal secretary, beside the river Rib in Standon, within a large medieval deer park. The building was still ranged, like a grand medieval house, around a central courtyard, and had embattled turrets and an imposing gatehouse. It was in ruins by the early nineteenth century and little remained upstanding, but the gatehouse range and part of the south wing were rebuilt by the second Duke of Wellington around 1872, and traces of the Tudor house can be seen as masonry projecting above the surrounding lawns.[26] Here, too, there are earthworks of elaborate gardens, partly adapted from earlier strip lynchets. A little later in date, but broadly similar in appearance, is Hadham Hall at Little Hadham, built by the Capel family in the 1570s on the site of an earlier house. All that remains is the western range – with the entrance gateway, flanked by castellated turrets – and part of the south range of a large courtyard mansion; the rest was demolished in the late seventeenth century.[27] Sections of the original garden walls also survive.

Well into the sixteenth century houses which were essentially of late medieval form, embattled and inward-looking, thus continued to be erected. But at the highest social levels, among leading courtiers, new Renaissance principals of design began to be adopted. Changed political and social circumstances ensured that great houses no longer affected an appearance of defensibility, instead displaying a familiarity with Renaissance culture. Battlements fell from favour, 'gothic' arches were gradually eschewed and greater emphasis was placed on symmetry – initially in terms of external elevation, later with regard to the layout of internal space. Houses began to incorporate forms of decoration derived ultimately from classical Italy and they began to be more outward-looking, no longer facing onto a courtyard, although in Hertfordshire this was a gradual development.[28] The surviving remains suggest that Hunsdon may have adopted some of these principals as early as the 1520s, but more obviously in the new mode was Gorhambury, erected by Sir Nicholas Bacon, Lord Keeper of the Great Seal, from the 1560s.[29] Only parts of the hall range and its imposing porch survive, a picturesque ruin in the grounds of the eighteenth-century mansion. The hall appears still in the gothic tradition, but the porch has been described as 'one of the first truly Renaissance structures of its type in England' and is replete with classical columns (Figure 8.3).[30] Larger and more completely Renaissance in character was Theobalds at Waltham Cross, built in the 1560s by Elizabeth's Secretary of State, William Cecil, Lord Burghley.[31] It was surrounded by fabulous gardens which were,

Figure 8.3. An engraving of the 'Remains of Gorhambury House' by J. Greig, from a painting by G. Arnald, 1808. Reproduced by kind permission of HALS (DE/X55/Z2/39).

in turn, set within a newly created park containing the household moat of the manor of Cullings, one of several properties purchased to create this fine seat. This was converted into a moated garden, an isolated pleasure ground of the kind beloved of late medieval and Tudor magnates. A map of 1611 shows how it was connected to the main gardens around the mansion by a series of formal tree-lined walks.[32] Robert Cecil expanded the park after his father died and created an intriguing artificial river with a series of formal pools alongside it.[33] But Theobalds did not remain his property for long. When King James VI of Scotland stayed here in 1603 on the way to his coronation as James I of England he was enchanted by the place and by the hunting opportunities it offered. Four years later he persuaded Cecil to accept Hatfield in exchange. James made further changes, purchasing farmland and enclosing parts of the commons in Cheshunt, Northaw and Enfield to extend the park, which by the 1620s covered over 2,508 acres, making it the largest ever created in the county.

Cecil in turn, now earl of Salisbury, set about transforming the old palace at Hatfield, demolishing most of the medieval building and erecting, with the help of the architect Robert Lyminge, the remarkable house which stands today – 'the most important mansion in the county and one of the four or five most important Jacobean mansions in England' (Figure 8.4).[34] It epitomised the new ideas of domestic design: it was ranged around three sides of a square, rather than around a closed courtyard; was in part two rooms deep; lacked

Figure 8.4. A view of the south front of Hatfield house.

a gatehouse, crenellations or any other obvious pretensions to defensibility; and was largely symmetrical in internal layout as well as external appearance, although it still retained an open hall with a cross passage. This was entered through a porch ornamented, like that at Gorhambury, with classical columns and niches.[35] The house was provided with gardens which contained innovative water features designed by the leading hydraulic engineers of the day. It remains one of the most impressive pieces of architecture in the county.

Great courtiers, eagerly embracing the new Renaissance ideas, vied with each other to create magnificent mansions and grounds: all different yet all variations, to some extent, on shared cultural themes. In 1601 Gorhambury was inherited by Sir Francis Bacon, Sir Nicholas' youngest son. He created an isolated water garden beside the river Ver by converting a series of fish ponds, which his father had constructed, in the form of a chain of moats. Bacon made pictures with coloured pebbles in the channels and erected a banqueting house on the island in the central pond. Nearby he built Verulam House, an elaborate lodge with a balustraded terrace on the roof, from which views across the gardens could be enjoyed.[36] This pleasure ground was, as at Theobalds, connected to the main mansion – a mile to the west – by tree-lined walks running through the park.[37]

Hatfield has survived to this day; Gorhambury was replaced in the eighteenth century by a new house on a different site, and only a small part remains standing; Theobalds was demolished during the Commonwealth but fragments can still be found in Cedars Park. Indeed, the English Civil War effectively marked the end of direct royal influence on Hertfordshire, as all the Crown estates in the county were sold off by the revolutionary government. High land values seem to have made Crown assets closer to London particularly vulnerable to liquidation.

As we have noted on a number of occasions, the most important mansions in the county in the sixteenth and seventeenth centuries were still provided with deer parks. About 18 parks – mostly those belonging to the Crown and royalist landowners – were lost during the seventeenth century, but the losses were soon to be counter-balanced by the creation of 14 new parks after the Restoration in 1660, and by the expansion of some older parks. Even when deer hunting became less fashionable in the course of the seventeenth century venison was still eagerly consumed by the wealthy, and the status conferred by well-stocked parks ensured that they formed the main backdrop of views from the mansion and gardens. Perhaps surprisingly to modern eyes, rabbit warrens were also often employed as elements in designed landscapes. A map of Theobalds drawn in *c.*1575, for example, shows the house and gardens surrounded to the west and south by two enclosures, covering 34 acres in all, both of which were labelled 'conygarth': one of the labels appears to have been added to the map by Cecil himself, signalling his intention to expand the warren.[38] An earlier warren had been established before 1544 just seven kilometres further west by William Cavendish, beside his newly built Nyn Hall.[39] The practice continued into the seventeenth century: an undated map, drawn up early in the century, shows that the main view from Sopwell House near St Albans was across a large rabbit warren.[40] The 'park' established at Upp Hall, Braughing, by London merchant Robert Dicer between 1656 and 1676 appears to have contained rabbits rather than deer, while at Lockleys near Welwyn Chauncy recorded that the owner 'made a fair warren to this Seat, stocked it with a choice breed of rabbits, all silver haired, and planted it with great store of excellent walnut trees'.[41]

The late seventeenth and early eighteenth centuries

After the Civil War few great mansions were built in the old way, around courtyards. Compact, symmetrical designs two or three rooms deep, variously influenced by French and Dutch models, were now the norm. Although elements of the old medieval plan, with hall, cross passage and the rest, lingered for a while, new and more symmetrical arrangements of rooms soon became universal. In fact, Hertfordshire was ahead of the game here, for it can boast one of the most innovative seventeenth-century houses in the country. In about 1640 Sir John Harrison, 'one of the Farmers of the Custom-house', built a house south of Hertford called Balls, now Balls Park, which already displayed many of these features. Indeed, it might easily be mistaken for an early eighteenth-century mansion, but is illustrated, much as it is today, in Chauncy's *History* of 1700, and described as 'a very fair stately Fabrick of Brick in the Middle of a Warren'.[42] After the Restoration such compact, symmetrical, double- or triple-pile brick houses were built in a range of simple, elegant classical styles, ranging from 'artisan mannerist' compositions such as Tyttenhanger (Ridge), of 1650–60, or Mackerye End (Wheathampstead) of 1665, through simple and elegant 'Queen Anne' buildings such as Brent Pelham Hall or Chipperfield Hall, to more elaborate houses displaying elements of the Baroque style, such as The Golden Parsonage in Gaddesden, originally built in 1705 – with stylish angle pilasters – as an addition to a house now demolished.[43] Gardens were still walled and formal, but they tended to be more closely integrated with the architecture of the house and ornamental landscaping increasingly spilled out across the surrounding countryside or parkland, the latter becoming more carefully designed in character. Wildernesses – extensive areas of ornamental woodland dissected by straight walks – were widely established and avenues, which had been planted sporadically since the sixteenth century, now became *de rigueur*. Perhaps the grandest Hertfordshire garden in this period was at Cashiobury near Watford, home of the earl of Essex, which is illustrated in Kip and Knyff's *Britannia Illustrata* of 1707 (Figure 8.5). There were walled gardens with parterres and topiary but also, to the east, a large wilderness or 'forest garden' crossed by tree-lined walks. This was laid out by the earl's gardener, Moses Cook, who recorded how 'In November 1672, I had the trees at Hadham Hall nursery taken up as carefully as I could, with good help, and carried them to Cashiobury, the place of their new abode – four rows of trees, 296, and of these I lost not one.'[44]

Avenues are often described as a French fashion and to some extent they were, although they also expressed power and ownership, for only absolute control over land allowed an owner to plant across it. Linear ponds or 'canals', in contrast, were a prime feature of

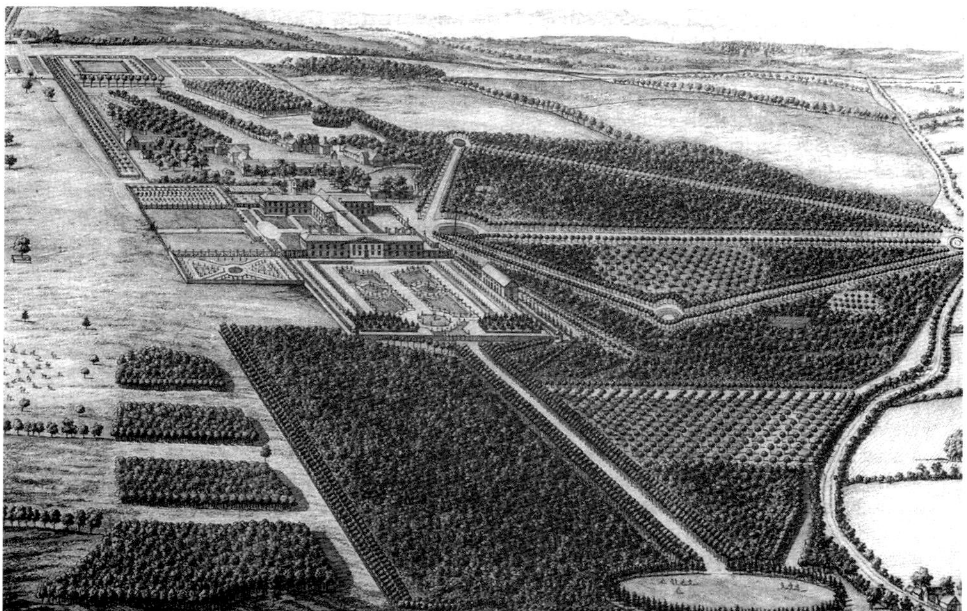

Figure 8.5. An engraving of Cashiobury by J. Kip and L. Knyff published in *Britannia Illustrata*, 1707. Reproduced by kind permission of HALS (WATF/58).

Dutch gardens. They generally doubled as fish ponds and were particularly common in Hertfordshire, especially where mansions were located on low-lying sites. At Holywell House, the home of the duke and duchess of Marlborough in St Albans, the river Ver was diverted southwards some time before 1722 and the former river bed remodelled to form an ornamental canal.[45] At Blakesware, near Widford, an example 100 metres long was dug to embellish the fine house built in the 1660s by Sir Thomas Leventhorpe:[46] the canal itself was perhaps the work of the next owner, John Plumer, who bought Blakesware in 1683 and was sheriff in 1689, or possibly even of his son William, who inherited in 1719.[47] At Tewin Water James Fleet, son of a lord mayor of London, made two canals on the north bank of the river Mimram some time before 1718, the larger of which ran parallel to an avenue to the east of the house and had a dovecote at its eastern end. The grounds at Hamels created by Ralph Freman MP also had two. One, constructed in 1716, extended for about 75 metres along a valley at the end of a formal garden below the south front; the other, the 'upper' canal, was dug in 1717–19 on the plateau above at right angles to the first, and on the same axis as the house – it was then deepened in 1720. Canals were thus not confined to riverside settings and about half the examples known in the county were established on higher ground, where topography and soils permitted: most were in areas of boulder clay (six examples) or London Clay (five examples). The first of these formations was not always completely impermeable, necessitating the provision of a clay lining. George Smith, the vicar of Braughing, described how at Hamels in 1716 'The canal sides & bott[o]m [were] part rammed with clay'; there was more 'claying' in September 1718 and the following month 'The mid[d]le [was] pitch[e]d'.[48]

Many of the seats created or remodelled in these new styles, as in the period before the Restoration, belonged to wealthy individuals based in London. Some were courtiers or government officials – men like Henry Guy, Groom of the Bedchamber to Charles II, who between 1687 and 1690 built a fine house at Tring designed by Sir Christopher Wren and embellished with extensive gardens.[49] Others were leading lawyers, such as Sir Walter Walker, advocate to Catherine of Braganza, who made his home at Bushey Hall. His son, George, inherited in 1674 and, according to Chauncy, 'erected a fair House upon the River Colne', around which he laid out a geometric landscape incorporating a walled garden with corner pavilions, avenues of trees and a circular bowling green on the north front of the house.[50] By 1685 the house had been acquired by another lawyer, Sir Robert Marsham, who had inherited his father's post as one of the Clerks in Chancery.[51] He made further alterations and by 1700 had laid out impressive water gardens here, partly by adapting an earlier moat, which could be admired from a roof-top terrace. Lawyers were active elsewhere. Spencer Cowper, a judge in the court of Chancery and attorney-general to the Duchy of Lancaster, bought Hertingfordbury Park in the 1720s where he, or his son William, laid out extensive formal gardens before 1732.[52] His brother William, Earl Cowper, Lord Chancellor, had established his family seat nearby at Cole Green at the beginning of the eighteenth century and had also acquired the neighbouring Panshanger estate in 1719.

As the century progressed, in a society increasingly organised around trade and finance, great houses were as likely to be built by successful merchants, financiers or bankers as by lawyers or politicians. Sir William Gore, who was lord mayor of London in 1702 and 1705, purchased Guy's house at Tring and his son William, who inherited in 1708 (and who was also a merchant and director of the Bank of England), laid out the surrounding park and employed Charles Bridgeman, the leading designer of his day, to create the grand landscape which is depicted in the bird's-eye view prepared by Badeslade.[53] In *c.*1715 Edward Rolt used some of the wealth inherited from his father, an East India Company man, to commission James Gibbs to design a new mansion at Sacombe and Bridgeman, once again, to create an innovative garden.[54] Jeremy Sambrooke of Gubbins, or Gobions, in North Mymms similarly employed Gibbs to alter his house and Bridgeman to redesign his gardens, again using wealth derived from the East India Company. Men involved in the South Sea Company also built mansions: Robert Chester, goldsmith of London and director of the company, acquired Briggens near Stanstead Abbots and Benjamin Styles used the vast profits made from trading in the company's shares to buy Moor Park near Rickmansworth.[55] Both Chester and Styles employed the architect James Thornhill to design their new houses and Bridgeman, once again, to design the grounds. This interesting group of entrepreneurs and speculators formed a close business and social network: William Gore's brother, John, was married to Jeremy Sambrooke's sister.[56] The fabulous wealth they accrued enabled them to buy their way into landed society and ensured a particular desire to make an impressive visual statement of their wealth, fashionable taste and aspirations.

The houses built or improved by such men and their contemporaries were more thoroughly classical in form than those erected in the decades either side of 1700, their design drawing on a range of imaginative and dramatic approaches to classical architecture loosely (and perhaps simplistically) labelled 'Baroque'. Some were built and faced, at least in part, in stone. Once again, the rapid turnover of estates and the rise and fall of families whose wealth was based on politics, trade and finance rather than landed incomes has ensured that much has been lost from the landscape. Tring and Briggens were much altered in the nineteenth century and Gobions was demolished.[57] But Moor Park, built in the 1720s by Thornhill for Styles, still survives, with its impressive free-standing classical portico and giant Corinthian columns. The gardens associated with such houses, designed in many cases by Bridgeman, were still for the most part formal and geometrical in style, but they were simpler and less cluttered than earlier designs, with less emphasis on formal parterres and more on grass lawns, neat gravel walks and extensive areas of ornamental shrubbery or wilderness, which often took up large areas of the design. Walls fell from favour and were concealed in a variety of ways. Turf ramps replaced brick terraces, and the sculpting of earth to make 'amphitheatres', mounts and other features was common. Typical of the new style was Sacombe, Charles Bridgeman's first commission in Hertfordshire, which featured extensive wildernesses crossed by straight paths ranged either side of a raised axial walk which led south-east from the house towards an ornamental canal. This was some 400 metres in length and terminated at an octagonal basin.[58] According to the historian Nathaniel Salmon, the impressive water feature was 'to be filled either from a Stream a Mile off, or a Spring which has been lately discovered about the Level of the Canal'.[59] Extensive earthworks of these great gardens still survive.

The stylistic novelty of these early eighteenth-century gardens should not be exaggerated, however. Avenues continued to be planted into the 1730s and 1740s, and the canal at Sacombe was one of many constructed in the county at this time: such features now became, if anything, even more popular than before. At Hitchin Priory Ralph Radcliffe, a merchant engaged in the Levant trade, widened and straightened the river Hiz to create one in his new park some time between 1727 and 1739.[60] Other examples include the 150-metre-long canal which still survives at Balls Park (Hertford), created by Edward Harrison, a director of the East India Company, between 1717 and 1732; that at Wormleybury, with a circular terminal basin, probably the work of Alexander Hume, another director of the East India Company; and one 140 metres long at Hertingfordbury Park, illustrated on an estate map of 1732, which was fed by a small stream and aligned at right angles to the river Mimram.[61] Its apsidal north end – emphasised by sculpted grass

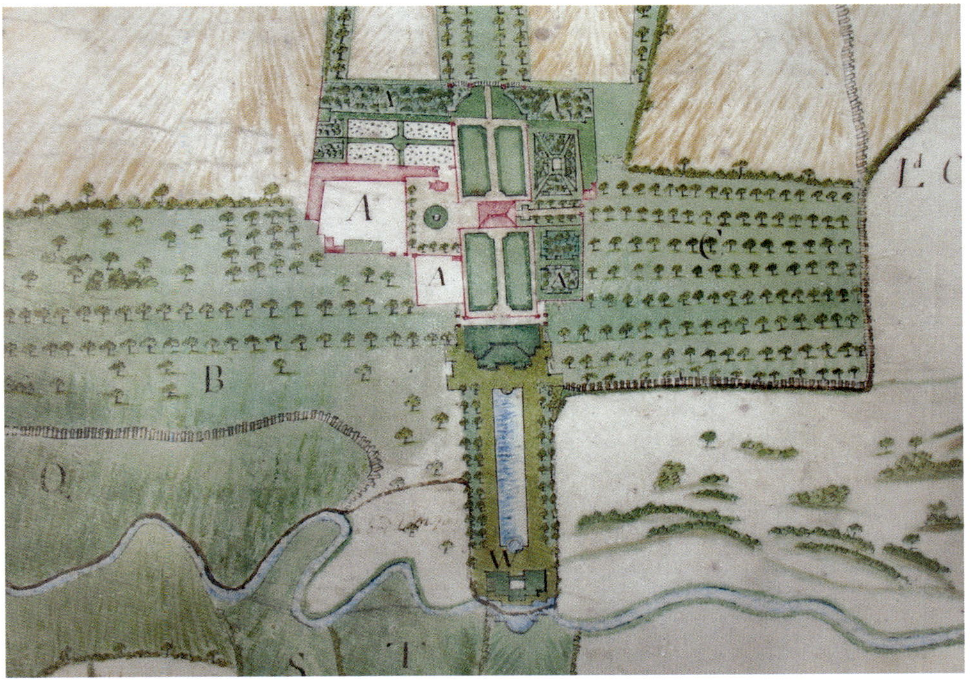

Figure 8.6. Detail of a map of 'Hertingfordbury Farm' surveyed in 1732, showing the formal style of gardens typical of the early eighteenth century laid out around the mansion in Hertingfordbury Park. The most notable feature is the ornamental canal laid out on axis with the house and apparently with terracing at each end. The semi-circular northern end of the canal (north is at the bottom) is carefully replicated in the opposite bank of the river Mimram. Reproduced by kind permission of HALS (DE/P/P7).

terracing – was echoed on the opposite bank of the river (Figure 8.6). This was the work of either Spencer Cowper, who purchased the estate in the 1720s, or of his son William, for whom the map was made. Bridgeman, as well as designing the canal at Sacombe, also provided examples at Briggens (275 metres long), Moor Park (300 metres) and Tring Park (335 metres), while the gardens he designed at Gobions were provided with two, both of which survive within the woodland which now covers the site.

Also popular in the late seventeenth century, but elaborated and extended in the first half of the eighteenth, were woodland gardens, and the more general practice of incorporating areas of semi-natural woodland within designed landscapes. Well-wooded Hertfordshire offered many opportunities for this kind of thing. A series of rides was cut through Whippendell Wood, in the north-west corner of Cashiobury park, in the mid-eighteenth century: they radiated from two *rond-points*, perhaps another of Bridgeman's schemes. The wood was reached along an avenue planted with lime trees, which still survives in the middle of the golf course. At Tring in the 1720s, this time certainly under Bridgeman's direction, the woodland growing on the steep Chiltern escarpment above the park became the setting for a series of formal *allées* radiating from a central obelisk and laid along terraces carved out of the slope itself. Both are examples of woods lying at some distance from the house: more often it was woodland in the immediate vicinity that was transformed in this way, although usually the paths have long since been abandoned and the wood, reverting to original state, gives little impression of this phase of its history. An undated mid-eighteenth century map of Kendals in Aldenham, for example, shows how most of the grounds comprised an extensive woodland garden, but although the canal shown survives as an elongated pond there is no trace among the hornbeam coppice of the open vista once focused on the house, nor of the complex network of winding paths, and without the map we would have little idea that they had ever existed.[62] In contrast, at St Paul's Walden Bury the basic structure of the woodland garden laid out by its owner, Edward Gilbert, to complement the house built in 1740 still survives, albeit restored and replanted over the centuries. The formal framework of radiating straight *allées*, more usually laid out on flat terrain, has here been successfully established across a small valley. Each vista is focused on a statue or a building, including a temple and the octagonal Organ House within the gardens and the parish church outside them. Probably the most important woodland garden in early eighteenth-century Hertfordshire was at Gobions, designed once again by the great Charles Bridgeman for Jeremy Sambrooke. It was ranged either side of a stream running in a steep-sided deep valley – typical of many in the London Clay uplands – and contained a mixture of straight and serpentine *allées* and clearings containing buildings, statues, canals and a bowling green. Extensive earthwork traces of many of these features can still be found. Avenues extended out through the surrounding fields, one leading to the 'Folly Arch', which still survives on a ridge to the south (Figure 8.7); a romantic ruin a little further along the same skyline was never built.[63] The wooded gardens were connected to the house, which was on higher ground to the north, by an avenue, but were quite separate

Figure 8.7. The 'Folly arch' at Gobions, probably erected around 1740, is the most impressive surviving feature of the elaborate gardens designed by Charles Bridgeman.

from it, not even sharing the same central axis. They were a distinct experience, replete with symbolism, and included – with the arch, a cascade and grotto – elements of drama and romance. All this ensured that they were very different from other gardens of their time, which were generally more predictable in character and more closely associated with the mansion, something which presumably explains why Gobions was one of the most famous gardens in England in the middle decades of the century.[64]

The Palladian Age

By this time architects were turning for inspiration not to French or Dutch interpretations of Renaissance ideas but directly to the works of the Italian Renaissance architect Palladio, and to Inigo Jones, his seventeenth-century English interpreter. The larger houses erected in Hertfordshire in the period between 1750 and 1780 – such as Brocket Hall (by James Paine), Gaddesden Place (by James Wyatt) or Gorhambury (Robert Taylor) – all had designs loosely based on Italian villas, with central porticoes, a rusticated basement and their principal rooms raised as a *piano nobile* at first-floor level.[65] Smaller country houses were built in stripped-down versions of this Palladian style. Internally, all trace of medieval arrangements had now gone. Two main public reception rooms, ranged one behind the other – hall and saloon – occupied the central axis of the building, with suites of rooms and further reception rooms, leading off from them, initially as a series of linear sequences but later as an informal circuit. At the same time fashions in landscape design were changing. The irregular, asymmetrical and romantic tendencies seen at Gobions were now taken further: the 'naturalistic' landscape style of Lancelot 'Capability' Brown had arrived. Geometric gardens were swept away and fashionable residences stood, 'free of walls', within open parkland scattered with trees, ornamented with carefully placed clumps and often surrounded in part by a perimeter belt. Avenues were, in theory at least, felled and the house was now approached along a curving – and often meandering – drive. Where the terrain was amenable, a lake would be provided in the middle distance by damming a stream or river within the park. This often involved complex engineering work: a notable example is that at Brocket Park, retained by a massive dam which was designed by James Paine in the form of a classical bridge with a cascade below it (Figure 8.8). Gardens, it must be emphasised, did not entirely disappear. There was usually an area of 'pleasure ground' comprising winding gravel paths, shrubs and specimen trees to one side of the house. But across the main façade there was an open lawn, separated from the park by a sunken fence, or ha ha, to ensure that the park appeared to be the mansion's main setting. Landscape parks were mainly grazed by sheep and cattle, but deer continued to be kept in around twenty parks in the county into the nineteenth century and, at a few places, into the twentieth.

Parks as ornamental landscapes were not entirely new: seventeenth-century deer parks had been designed as much for their appearance as for hunting, and many landscape parks

Figure 8.8. The elaborate bridge and cascade, serving in part as a dam for the lake, designed by James Paine for the park at Brocket Hall and built in 1774.

developed directly from these. Many others, however, were new creations, made at the expense of agricultural land. Gardens, barns and outbuildings were removed from around the house and the surrounding fields were thrown together, although a proportion of the hedgerow trees was retained in order to form an instant sylvan scene, while small copses and woods were often reused as clumps or as part of the perimeter belt. Some of these ancient trees usually still survive, often associated with the earthworks of old field boundaries or of hollow ways marking the line of former roads. Some parks also contain earthwork traces of lost settlements, as at Knights Hill, Westmill (above, p. 79), but in most cases these had been abandoned long before the park was laid out. Clearance of villages to make way for parks was rare in Hertfordshire, largely because, in a landscape characterised by early enclosure and scattered settlement, manors and mansions often stood isolated from village nucleations or could be rebuilt in places that were. More usual was the closure of roads and footpaths by Chancery writ or, from 1773, by a Quarter Sessions Road Closure Order. The ubiquity of road and footpath closures is one signal that eighteenth-century parks, although usually studied primarily in aesthetic or horticultural terms, also had a profound social significance. They were landscapes of seclusion, their belts excluding all close views of the local population or their homes, and providing their owners with privacy. The removal of walled gardens, moreover, was also accompanied by the destruction of orchards, dovecotes, fish ponds and farmyards – all the productive clutter with which the gentry had once been happy to associate themselves. This proclaimed that the fashionable owners of such places were distanced from the humdrum practicalities of domestic production and belonged to a class quite distinct from the farming and labouring population; while an emphasis on the *park*, rather than elaborate gardens, as the main setting for the mansion emphasised the superiority of the landed elite over the expanding middle classes. The creation of a park demanded the one resource which only they, and those wealthy enough to join them, possessed – an abundance of land.

Capability Brown himself worked at perhaps a dozen places in the county, including Moor Park, near Rickmansworth, where he softened Bridgeman's design; Cole Green Park, near Hertingfordbury; Beechwood Park, near Markyate; The Hoo, near Kimpton; Youngsbury, to the north of Ware; Pishobury, to the south of Sawbridgeworth; and Digswell.[66] At Ashridge he removed much of the existing geometrical design and cut back the surrounding woodland to form a number of large clumps and – most importantly – to create the 'Golden Valley' to the north of the house, by removing trees in the base of a dry valley but leaving woodland with carefully scalloped margins on the valley sides. One of the finest pieces of landscape design in the county, it survives intact (Figure 8.9). Other contemporary designers with national landscaping practices also worked in the county, including Nathaniel Richmond, who updated the grounds at Sacombe, created the park at Lamer near Wheathampstead and laid out Hitchin Priory Park; and Richard Woods, who made improvements to Newsells, in the far north-east of the county, and to the grounds of Brocket Park north of Hatfield.[67] Brown died in 1783 and from the end of the decade

Figure 8.9. The 'Golden Valley', Ashridge, the finest surviving example of Capability Brown's work in Hertfordshire.

Humphry Repton became the most celebrated landscape designer in England. He, too, worked on a large number of Hertfordshire commissions, including Panshanger, Tewin Water and Wall Hall in Aldenham, in each case preparing one of his trademark 'red books' containing design proposals complete with 'before' and 'after' views of mooted changes.[68]

The new style involved the creation of supposedly 'naturalistic' landscapes, but all were, of necessity, contrived to varying extents. Yet, at the same time, natural landforms determined what was possible. Lakes were invariably artificial creations – there are no large natural water bodies in the county – but they could be formed on a convincing scale only where appropriate valleys and watercourses existed. At only four of the landscapes designed by Brown could rivers be modified to provide lakes: the Mimram at The Hoo and at Digswell; the Stort at Pishobury; and the Rib at Youngsbury.[69] At the first of these topography dictated that the lake was about half a kilometre away from the house, which stood 35 metres higher on the rim of the valley and with restricted views of the water. At Youngsbury, in contrast, while the house occupied a similar topographical site the valley slopes were steeper and more concave, allowing views from the house to the lake, prompting Brown himself to comment that 'Nature had done so much, little was wanting, but enlarging the river.'[70] At Pishobury the topography was different: the house (built originally for Walter Mildmay in 1585) occupied low ground within a large meander of, and about 400 metres from, the river Stort. Brown diverted water from the river to fill a crescent-shaped lake closer to the house. Some owners were determined to create a lake despite the lack of a suitable valley and watercourse and this was successfully achieved at several sites, but only after a great deal of earth-moving. At Gobions (Gubbins) a large depression was dug into the gently sloping London Clay close to the house, the excavated earth forming a massive dam around its lower, western and southern sides, probably during the ownership of John Hunter, deputy director of the East India Company, from 1777.[71] There is no obvious source of water and the lake, which still survives although the mansion is long gone, is presumably fed by springs. Neighbouring Brookmans Park similarly acquired a pair of conjoined lakes before the end of the eighteenth century, likewise on London Clay and spring-fed. Similar high-level lakes were dug at Stagenhoe, near St Paul's Walden, and at Sacombe Park, where George Caswall remodelled the landscape around his new mansion in the early nineteenth century: both were excavated into the boulder clay plateau. The early nineteenth-century owner of Theobalds Park, Sir George Beeston Prescott, used a more novel source of water. A diminutive serpentine lake was here created by widening the course of the New River, the man-made watercourse constructed in the early seventeenth century to carry fresh water to London from springs at Ware.[72]

The later eighteenth and nineteenth centuries

The decades around 1800 saw a major change in architectural taste: instead of being modelled on Renaissance villas or classical temples, country houses began to be designed in 'gothic' style, drawing on architectural features characteristic of medieval castles and manor houses. In Hertfordshire the new fashion was pioneered at Cashiobury, which was remodelled by James Wyatt around 1801 (with Repton contributing to the design of the grounds), and soon followed by Ashridge, set high in the Chilterns, which was rebuilt as a rambling medieval fantasy by Wyatt and his nephew Sir Jeffry Wyatville from 1808 (Figure 8.10). Panshanger *could* have been the first such house in the county: in 1799 Repton was asked by Earl Cowper to advise on the design of a new house, as well as on the disposition of its grounds, and proposed two alternative architectural styles for a new building, one of which was 'gothic'. But the earl spent years deliberating before work finally started in March 1806 to remodel and extend the existing house in the latter style.[73] Panshanger and Cashiobury were demolished in the twentieth century but Ashridge survives, with its great soaring central keep, turrets and spire, and the longest façade of any house in England. Other smaller residences were also built or remodelled in a broadly gothic style in the first half of the nineteenth century, including Wyatville's Hillfield House in Aldenham (*c.*1805); nearby Wall Hall (renamed 'Aldenham Abbey'); Gilston Park; and Knebworth House, which was expanded as a castellated house in two phases in the course of the nineteenth century. The enthusiasm for gothic continued into the high Victorian period, although mansions were now more likely to take the form of Elizabethan manor houses than castles or abbeys.

Figure 8.10. Ashridge House, the great gothic fantasy designed by James Wyatt and Sir Jeffry Wyatville in the early nineteenth century. The terrace garden, typical of the nineteenth century, was restored in the twentieth century. Photograph by Chris Beddall.

As styles of country house architecture changed, so too did the design of their grounds. In 1813 Repton planned a series of gardens to accompany Wyatt's castellated mansion at Ashridge, many in an appropriately archaic, formal style. This enthusiasm for geometric gardens, with parterres and knots, and for the provision of a range of 'themed' gardens, intensified through the first half of the century and, by the 1850s, designers such as William Andrews Nesfield were providing formal terraces across the main façades of the house, complete with yew topiary and parterres (Figure 8.10). The latter were often based on seventeenth-century designs but generally planted with colourful displays of bedding plants, especially pelargoniums. Moreover, rather than being set within walled enclosures they occupied level terraces separated from the park by low balustrades, so that the owner and his guests could enjoy wide views from the one to the other.[74] Nesfield himself worked at The Grove, to the north of Watford, and at Moor Park, where he created a series of gardens, including an Italian Garden and a Fernery, a diversity typical of the period.[75] But while attention was focused once more on gardens, parks continued to proliferate, especially in the south of the county, while existing ones were altered and, on occasion, extended.[76] Their planting became more diverse. Although conifers such as cedar, larch and Scots pine had been planted alongside indigenous trees like oak, elm and beech in the parks of the eighteenth century, species such as horse chestnut and lime were now more widely used (sometimes in formal avenues, which also made a return to fashion), in addition to a wider range of foreign species, including the Sequoia, introduced in 1853. Fine examples of this tree, now so huge it is hard to believe they are little more than a century old, tower over the surrounding landscape at places such as Gorhambury, Langleybury and Panshanger.

As in earlier periods, many mansions in nineteenth-century Hertfordshire were owned not by long-established landed families but by businessmen, lawyers, retired government officials and – increasingly – industrialists. Charles Longman was John Dickinson's partner in the leading papermaking business in the county, which operated a number of mills scattered along the Gade valley (below, pp. 263–4). He built a new house at Shendish near Hemel Hempstead in 1853, designed by the architect John Griffith in a broadly 'Elizabethan' style.[77] It occupied an elevated site above the Gade valley within a new park which covered around 60 hectares, and had elaborate grounds designed by the noted landscape gardener Edward Kemp. His popular book *How to lay out a garden*, published in 1864, discusses how a footpath was diverted away from the house and the main drive carried over it on a low bridge, and how clumps were planted in the pleasure ground with Scotch Firs, Austrian pines, Spruce and 'deciduous trees ranging from five to twenty-five feet in height, which Mr Longman has been most successful in transplanting from a property in the same district'.[78] It also describes the creation of formal flower beds, cut into the lawn below the house and alongside the wall of the kitchen garden. A nearby chalk pit was made a feature of the grounds, 'the medium of displaying rugged masses of natural vegetation' of which 'wild Clematis (common here) will be a conspicuous feature'.[79] The house survives, now a hotel, together with some of the trees planted by Kemp and a fine octagonal summer house in 'gothic' style. The chalk pit also remains, with traces of ornamental rockwork. The London–Birmingham railway, opened in 1838, was included within the eastern boundary of the park, and the Apsley Paper Mill was in full view: no attempt was made to screen it. Men such as Longman – or John Dickinson himself, who erected a house with similar views on the far, eastern side of the Gade valley – adopted existing styles of landscape design, but were happy to include in the prospect these signs of the modern world, the source of their wealth.[80]

The character of Hertfordshire mansions and parks

The above account has discussed the development of country houses and their grounds as a series of distinct stages, but in reality most mansions and their landscapes are complex palimpsests which incorporate many phases, of both addition and contraction, with the latest in the sequence often obscuring the earlier.[81] As a consequence, the majority of Hertfordshire country houses appear, stylistically, to have been erected between 1700 and 1850, are set in landscape parks dominated by nineteenth-century planting and have gardens which are predominantly Victorian in their structure. Variations in the scale of their impact on local landscapes is principally a function of their distribution and location, which are in turn the outcome of the interplay of social, tenurial and environmental factors and, in particular, of changing responses of owners and designers to the county's natural topography.

Many of the greatest medieval residences were located in low-lying positions in major valleys, in part because they were associated with the core areas of early medieval settlement, but also because this allowed the construction of the moats, fish ponds and water gardens beloved of contemporaries. The More, Tyttenhanger, Bushey Hall, Rye House, Standon Lordship and the castles at Berkhamsted and Hertford all fall into this category. The owners of the greatest houses usually had a deer park lying close to the house and consequently occupying low ground, although those at The More, Standon and Berkhamsted extended onto the hills above. Most parks of this period – created by lords of lower rank – were, however, quite detached from the owner's house. They were usually located some distance away on higher ground, where reserves of wood-pasture and waste had survived into the twelfth or thirteenth centuries, although often containing an isolated 'lodge' which provided accommodation for the park keeper and a base for hunting parties. This tendency for great houses to be located in major valleys was particularly marked in the south and west of the county. The extensive uplands here were largely devoid of high-status residences, except for a scatter of monastic houses in the Chilterns. In the east of the county, in contrast, there was less distinction in this respect between river valleys and interfluves. This was in part because of the early settlement history of the district – many early manorial sites were established on the margins of the upland plateaux – but also, perhaps, because the water-retentive soils allowed the construction of moats and ponds which were, indeed, ubiquitous in the east.

Starting in the fifteenth century, there were a number of important changes in the location of great houses. Isolated deer parks became less common and more and more great landowners possessed one in the immediate vicinity of their homes. Parks, moreover, gradually became not simply an adjunct to a residence but its setting, and were often made to surround the mansion on all sides. In the 'champion' north and far west of the county, where manor houses and farms tended to be clustered together in nucleated settlements surrounded by open fields, it was difficult to achieve such an arrangement without major disruptions to the existing landscape. Sir Robert Whittingham, a successful draper who rose to become an MP in 1416, Sheriff of London in 1419, Receiver-General for the Duke of Bedford in the 1420s, Treasurer of Calais in 1436 and Mayor of the Calais staple in 1439, made Pendley, near Tring, his main residence in the 1440s. He built a new mansion and enclosed 200 acres of arable and pasture to make a park, sweeping away the settlement in the process. But this is the only real example of an 'emparked' village in the county. In most districts, as already noted, a more scattered pattern of settlement and the frequent isolation of the 'Bury' from the homes of tenants allowed new parks to be created, or existing parks to be expanded, with little or no movement of meaner dwellings.

Several major houses were built on new sites within old parks during the sixteenth century. Berkhamsted Place, as we have seen, was constructed in the 1560s within the park originally attached to Berkhamsted castle, and around the same time Thomas Fanshawe's new mansion of Ware Park was erected, as its name suggests, within the ancient park at Ware. In addition, the Dissolution also freed up a number of former monastic sites in upland locations, set among woods and commons which provided excellent opportunities for emparking. Beechwood House originated in the twelfth century as a small nunnery, St Giles in the Wood, in a remote part of the parish of Flamstead on the wooded Chiltern dipslope. At the Dissolution it was granted to Sir Richard Page, a courtier, and in the late seventeenth century it passed by marriage to Sir Edward Sebright. It became the Sebright family seat for the next two centuries: a new house was built in the reign of Queen Anne and a formal park landscape (making full use of the local woodland) was laid out in the early eighteenth century, and subsequently remodelled by Capability Brown in the 1750s.[82] Nearby Markyate Cell similarly incorporates fragments of the mansion built in 1539–40 by Humphry Bourchier from the remains of the dissolved nunnery, while Gorhambury, although not developing from an earlier monastic site, nevertheless owed its location – on high ground overlooking the Ver – to the availability of property released when St Albans Abbey was dissolved and may well occupy one of the Abbey's former parks.

The origins of Ashridge lie in the thirteenth century, when Edmund Earl of Cornwall constructed a substantial lodge within a new park some four kilometres from Berkhamsted castle, his main residence, a building sufficiently grand to accommodate Edward I and his entire court for three weeks at Christmas 1290. Edmund also established a religious foundation – a monastery of the Bonhommes, a French order – beside it, which survived, unlike the lodge, into the sixteenth century. At the Dissolution it became a royal residence, passing to the Egerton family in 1604.[83] The original choice of site for house, park and monastery was determined by the availability of extensive tracts of 'waste', but this remote landscape provided the raw materials from which a succession of designers, culminating in Capability Brown and Humphry Repton, was later to carve a design of great picturesque beauty.

This gradual shift from low-lying sites to higher ground was a consequence not only of a desire for a parkland setting and social isolation, or of the opportunities presented by dissolved monastic sites. Sixteenth- and seventeenth-century landowners appear to have relished extensive prospects in a way that their medieval predecessors had not, and contemporaries often commented on the excellence of a view. Balls Park, according to Sir Henry Chauncy at the end of the seventeenth century, stood 'towring upon an Hill, from whence is seen a most pleasant and delicious Prospect'.[84] Nevertheless, the seventeenth- and early eighteenth-century enthusiasm for canals and water gardens also ensured for a time the continuing popularity of low-lying locations, and it was only gradually that the balance moved decisively in favour of upland sites, as damp positions began to be regarded as unhealthy. By the middle of the eighteenth century lakes below the house, in the middle distance, were the height of fashion: moats and ponds in its immediate vicinity appeared archaic and uncouth. The old site of The More in Rickmansworth, beside the river Colne, was thus abandoned in the seventeenth century and replaced by a mansion located on a more elevated site within the deer park. A little later, in the 1720s, Benjamin Styles 'removed a hill' to the north-east in order to open up a prospect along the Colne valley towards Watford (Figure 8.11). By this time even mansions already occupying dry

Figure 8.11. A drawing entitled 'More Park on the river Colne' by J.M.W. Turner, 1824, showing the elevated site of the eighteenth-century mansion with its views along the valley. Reproduced by kind permission of HALS (RICK/5).

'upland' locations might be relocated to improve the prospect. The Halsey family lived at The Golden Parsonage at the eastern end of their estate in Great Gaddesden from the sixteenth century until the return of Thomas Halsey, a merchant in the family business in Hamburg, in 1763. He became MP for Hertfordshire in 1768 and commissioned James Wyatt to design a grand new house, called Gaddesden Place, about a mile to the south-west, to which the family moved in 1774.[85] This stands on the eastern rim of the Gade valley with parkland laid out on the slopes below. It was not possible to create a lake, but the beautiful views across the valley more than made up for this deficiency. At Woodhall Park, near Watton-at-Stone, the magnificent house built by the Boteler family in stages from the sixteenth century was damaged by fire in 1771 and abandoned. The property was sold to Sir Thomas Rumbold, an employee of the East India Company, who in 1777 commissioned a new house from the architect Thomas Leverton which, unlike its predecessor 400 metres to the south, enjoyed fine views to the west across the Beane valley. The river was dammed to form a lake and the park landscaped, probably by William Malcolm and Son, in 1782–3.[86]

In some cases families owning more than one property chose that in the most suitable topographic location to develop as their principal seat. The Cowper family's mansion at Cole Green was demolished in 1801 when the fifth earl, who had inherited unexpectedly on the death of his brother in 1799, decided he would prefer to live at Panshanger, which the first earl had acquired in 1719.[87] The house, which had been occupied by various members of the Cowper family during the eighteenth century, stood on the northern edge of the Mimram valley with extensive views to the south. Following advice given by Humphry Repton in his red book of 1799, Cowper set men to work to create a lake.[88] Thousands of trees were planted on the valley sides according to Repton's directions to complete what was, in its heyday, one of the finest parks in Hertfordshire. Prospects across the landscape were thus of crucial importance by the eighteenth century, and where houses occupying unfavourable sites could not be rebuilt in new locations owners and designers often laid out walks or rides leading to some elevated vantage point, as at Tring, where, in the 1720s, Charles Bridgeman laid out a broad tree-lined terrace 800 metres long just below the crest of the Chiltern escarpment, overlooking the house and park and with circular viewing platforms at each end, providing superb views across the landscape – surely the most dramatic garden terrace in the county.

The distribution of mansions and parks in Hertfordshire thus reflected, in part, the emerging aesthetic preferences of the post-medieval period: throughout the county many major residences are positioned on the margins of valleys, with parkland stretching away below them. This pattern is already discernible on the county map surveyed by Dury and Andrews in the 1760s, although the precise nature of the designed landscapes and parks shown, their extent and even presence, is not always clear (Figure 8.12a). By the 1820s, however, when Bryant's map of the county was produced, the relationship of parks to major valleys is plain (Figure 8.12b). But it was not simply the consequence of choices made by wealthy people in the seventeenth, eighteenth and nineteenth centuries. It was also a consequence of more ancient structures, going back to the early Middle Ages, when manor houses, and the farms from which 'villas' often developed in the course of the eighteenth century, had often been established on the upper margins of tractable valley soils. It was, moreover, a pattern modified in detail by a range of other economic, environmental and tenurial factors. Mansions and their designed landscapes were thus not evenly distributed along Hertfordshire's major valleys and those of the Lea and Mimram, to the west of Hertford, for example, appear to have accommodated particularly dense concentrations. From the beautifully landscaped Broadwater in Brocket Park, north of Hatfield, the Lea flowed past the grounds of Woodhall (Hatfield) and Bush Hall before becoming another lake, the Broadwater in Hatfield Park; it then continued eastwards through Woolmers Park, and on past the grounds of Bayfordbury, before reaching Hertford. The rather smaller valley of the river Mimram, just to the north, was even more densely crowded, providing – by the early nineteenth century – the setting for no less than nine large houses and their grounds, two of which were landscaped by Brown and two by Repton: The Hoo at Kimpton, Lockleys near Welwyn, Digswell Water, Tewin Water, Tewin House, Marden Hill, Panshanger, Epcombs and Hertingfordbury Park. Repton himself commented on the near-continuous parkland in the valley and considered that 'united woods and lawns will by extending thro' the whole valley enrich the general face of the country'.[89] With the

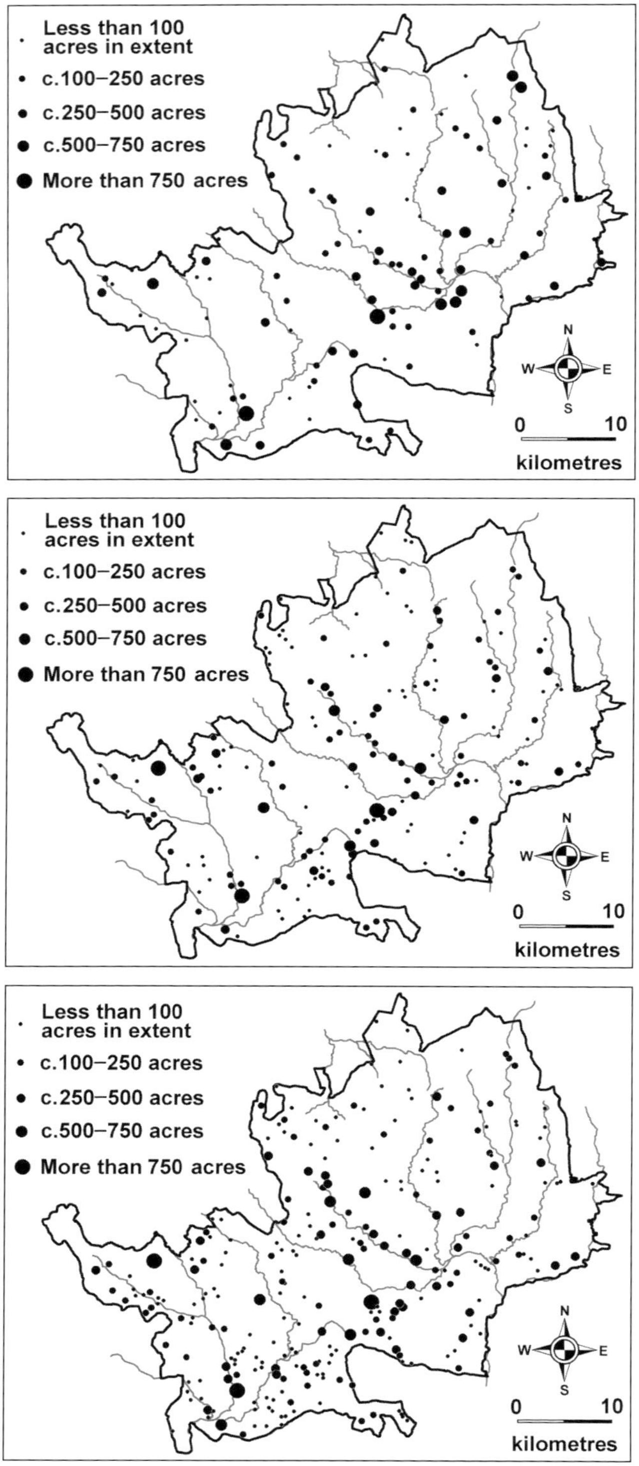

Figure 8.12. (a) The distribution of parks and large pleasure grounds in Hertfordshire, as shown on Dury and Andrews' map of 1766 (sites extending over less than *c*.5 acres excluded). (b) The distribution of landscape parks as shown on Bryant's county map of 1821–2. (c) The distribution of parks in Hertfordshire *c*.1880, as shown on the first edition 6-inch Ordnance Survey maps.

exception of The Hoo, all these houses were within easy travelling distance of Hertford, with all the amenities and opportunities for social gatherings it provided. Yet the valleys of the lower Beane and the Rib, and of the Lea downstream of Hertford, cannot boast this number of mansions and parks in spite of the fact that the first two are similar to that of the Mimram, providing intimate, varied and attractive rural views and with good communications to the county town. The difference may lie in patterns of land ownership. The parishes of Kimpton and Digswell in the Mimram valley were mainly in the hands of single owners from an early date and were largely enclosed by the middle of the eighteenth century. Only at Hertingfordbury was there still open arable within the valley: significantly, Panshanger Park (which lies in that parish) was the last of the valley landscapes to be laid out, after the fields were enclosed by parliamentary act in 1801.[90] Parishes in the Beane and Rib valleys, in contrast, had more complex patterns of ownership and many retained areas of intermixed arable until the nineteenth century.

In the middle of the eighteenth century there was thus a noticeable tendency for parks to cluster in the vicinity of Hertford, which functioned as an important social centre as well as serving as the location for the county assizes and other meetings or events which demanded the attention of the local elite. By the late nineteenth century, in contrast, the main factor structuring the overall distribution of parks in the county was proximity to London, which explains the marked concentration, of smaller parks especially, towards the south shown on the first edition Ordnance Survey 6-inch map (Figure 8.12c). Many were associated not with true landed estates but with 'villas' constructed by wealthy Londoners, accompanied by their immediate grounds but with little other property. In many cases, such residences occupied the sites of earlier farmhouses 'upgraded' in the course of the eighteenth and nineteenth centuries. The places selected for such treatment were, perhaps needless to say, often on the edges of upland plateaux, overlooking the valleys of the Gade, Colne or Ver, or the Dollis and Pymmes Brooks in Totteridge and Barnet. By the late nineteenth century, however, on the Southern Uplands especially, small parks were scattered in all locations and in some areas – such as East Barnet – more parks existed than true agricultural land. This district was already, in a sense, becoming suburbanised.

This density and number of small parks and associated 'villas' is a key characteristic of the county's designed landscapes.[91] But the influx of London money of which it was a consequence has shaped our heritage of country houses, as we have seen, in many other ways. In particular, it ensured a rapid turnover of owners, and thus more frequent and extensive rebuilding of country seats, than in many other parts of England, so that houses of obviously sixteenth- and seventeenth-century form are much rarer than in a county like Northamptonshire, and remains of walled gardens from this period are much thinner on the ground than in East Anglia. In addition, in the eighteenth century especially, many incomers had vast wealth yet found it difficult to convert this into broad acres: the result was that fortunes could be thrown at small properties, moving earth, levelling terraces and making lakes and canals. Subsequent amalgamations of estates or relocations of mansions to new sites in this restless world ensured that such grandiose grounds survive – in a remarkable number of places – as striking earthworks. Popes in Hatfield is a particularly good example of the landscape impact of transient wealth. In a valley a little to the north of the house, similar in many respects to that at Gobions a few kilometres away, an elaborate water garden was created at some time in the eighteenth century, possibly by David Mitchell (*né* Cook), who inherited the estate in 1722 and sold it in 1744, but perhaps more probably by Sir Benjamin Truman, one of many wealthy London brewers who moved out to Hertfordshire: he rented the estate from the 1750s, rebuilding the house which had been destroyed by a fire in 1746.[92] The gardens – shown on an estate map of 1785 – comprised a string of ponds and lakes, the largest of which had a boathouse on its northern shore.[93] Truman is reported to have been the builder of a Chinese boat which is shown sailing on the Broadwater in a drawing of Brocket Hall, seven kilometres from Popes, made in 1787, seven years after his death. Popes was absorbed into the Hatfield House estate in 1817 and the hall demolished: the first edition Ordnance Survey 6-inch map of 1883 shows woodland in the valley floor, echoing the outlines of the lakes that had been there a century before. The entire layout of the gardens still survives in earthwork form, covered in trees and undergrowth, including the massive retaining dams of the two lower lakes, an island and even the channel beneath the boat house where the 'junk' was presumably kept: a stunning and strangely evocative testament to a magnificent, but short-lived, designed landscape.

By the late nineteenth century the great days of the country house and its landscape were over. The major depression in agriculture which began in the late 1880s eroded the agricultural rents which had, at least in part, sustained landed estates. Death duties were introduced in 1894, while changes in the organisation of local government – the establishment of Hertfordshire County Council (with other county councils) in 1888, and of parish councils in 1894 – reduced the political role of landowners in the affairs of shire and parish. Great mansions with elaborate grounds became less important as markers of status and the attentions of leading landscape designers turned, to a significant extent, towards smaller properties, designed in Arts and Crafts mode – houses in the country, suitable for wealthy businessmen, rather than country houses in the traditional sense. Hertfordshire, especially the south and west of the county, has many important examples, including Amersfort at Potten End, where Gertrude Jekyll produced planting plans to complement the hard landscaping by Ernest Wilmott; and several gardens designed by Thomas Hayton Mawson, including Kilfillan, Staghurst and The Kraal in Berkhamsted, Thatched Rest at Welwyn and Roebuck and Bushey Rose Garden in Bushey. The last of these, originally part of the grounds of the now demolished Lululaund, home of the artist Hubert von Herkomer, is the only surviving example and has recently been restored for the enjoyment of the public.[94]

Some country houses were demolished, as we shall see, in the course of the twentieth century; many others were converted, often to institutional use. Parklands were often ploughed or given new roles, especially as golf courses, but many survive and make a major impact on the county's landscape. They have more than aesthetic or historical significance. The pasture in some old parks has never been ploughed or artificially 'improved' with chemical fertilisers and, in consequence, constitutes a rare and immensely important habitat. Although parks with medieval origins are the most ecologically valuable – such as Millwards Park at Hatfield and the surviving parts of Ware Park – later examples established at the expense of farmland are also significant in this respect, not least because, as we have seen, many when first created incorporated large quantities of existing hedgerow timber. Ancient oak pollards form miniature nature reserves which support rich communities of fungi and lichens as well as invertebrates, birds and mammals. Parks are also valuable archaeological assets. As oases of pasture in a largely arable county they often contain upstanding earthworks the likes of which have elsewhere been levelled. Ridge and furrow (the 'narrow' type, so typical of much of Hertfordshire) thus survives in the seventeenth-century Hamels Park near Braughing and in the nineteenth-century Weston Park, traces of old trackways and ancient field boundaries can be found in the seventeenth-century park at King's Walden, and a sunken road and evidence of former settlement exist in the eighteenth-century Coles Park at Knights Hilll near Westmill. Great houses, parks and gardens are often considered as external impositions on the countryside, but in reality they were an integral part of it, structured and shaped by local topography and by the development of the landscape over many centuries.

Notes

1. D. Defoe, *A tour thro' the whole island of Great Britain, divided into circuits or journies*, 2 Letter VI (1724–7), <http://www.visionofbritain.org.uk>, accessed 19 September 2010.
2. Renn, *Medieval castles*, pp. 17–18.
3. Rowe, *Medieval parks*, pp. 148–51.
4. Rowe, *Medieval parks*, p. 184.
5. Smith, *English houses*, p. 52.
6. Renn, *Medieval castles*, pp. 18–20; Smith, *English houses*, p. 2.
7. M. Hicks, 'Neville, George (1432–1476)', *Oxford Dictionary of National Biography* (Oxford, 2004); online edn, 2008 <http://www.oxforddnb.com/view/article/19934>, accessed 13 October 2011; Rowe, *Medieval parks*, p. 180.
8. H. Falvey, 'The More revisited', *The Ricardian*, 18 (2008), pp. 92–9.
9. A. Rowe, 'Hertfordshire's lost water gardens 1500–1750', in Spring (ed.), *Hertfordshire garden history II*, pp. 33–8.
10. The manors of Cheshunt and Hunsdon reverted to the Crown on the death of Henry's grandmother, Lady Margaret Beaufort, Countess of Richmond and Derby, shortly after his coronation.
11. Page (ed.), *VCH Hertfordshire*, vol. 3, p. 92. In 1538 the bishop of Ely accepted other properties in exchange for Hatfield.
12. *Ibid.*
13. G. Richardson, 'Henry VIII and travel', p. 2, <www.hrp.org.uk/Resources/Glen%20RichardsonFINAL.pdf>, accessed 30 December 2009.
14. Page (ed.), *VCH Hertfordshire*, vol. 3, p. 92. Hertford Castle was also maintained as an occasional royal residence (Colvin, *History of the king's works*, p. 179).
15. Page (ed.), *VCH Hertfordshire*, vol. 2, p. 377.
16. O. Harris, 'Nyn Hall, Northaw: an alternative history', *Hertfordshire Archaeology*, 13 (2004), p. 99. Cavendish went on to start the building of Chatsworth House in Derbyshire and was father to the first earl of Devonshire.
17. Chauncy, *Hertfordshire*, p. 501 – but apparently copying John Norden's *Description of Hertfordshire* (Ware, 1598). The earl's brother, Robert Dudley, Earl of Leicester, visited in 1585 and made payments to his brother's gardener and to '3 poore wemen at North hall that ware a wedeng in the garden'. (S. Adams (ed.), *Household accounts and disbursement books of Robert Dudley, Earl of Leicester, 1558–1561, 1584–1586* (Cambridge, 1995), pp. 286–7).
18. Jones, '"Commotion time"', p. 334.
19. Smith, *English houses*, p. 16.
20. D. Spring, 'The London connection', in Spring (ed.), *Hertfordshire garden history II*, pp. 23–5; Smith, *Hertfordshire houses*, p. 15.
21. Smith, *Hertfordshire houses*, p. 30.
22. D. Loades, 'Baeshe, Edward (*c.*1507–1587)', *Oxford Dictionary of National Biography* (Oxford, 2004); online edn, 2008 <http://www.oxforddnb.com/view/article/71698>, accessed 9 October 2011.
23. Chauncy, *Hertfordshire*, p. 226.
24. A. Rowe, 'Country house chameleon. The story of Hamels Mansion', *Hertfordshire's Past*, 43/44 (1998), pp. 44–54.
25. Rowe *et al.*, 'The earthworks at Benington Park, Hertfordshire: an exercise in dating an archaeological garden', *Landscape History*, 32 (2011), pp. 37–56.
26. Page (ed.), *VCH Hertfordshire*, vol. 3, pp. 354–5.
27. HALS DE/X10/Z1 W. Minet, *Hadham Hall and the Manor of Bawdes alias Hadham Parva* (Colchester, 1914); Smith, *Hertfordshire houses*, pp. 120–21; J. Milledge, 'Hadham Hall and the Capel family', in Spring (ed.), *Hertfordshire garden history II*, pp. 60–91.
28. Smith, *English houses*, p. 46.
29. R. Tittler, 'Bacon, Sir Nicholas (1510–1579)', *Oxford Dictionary of National Biography* (Oxford, 2004); online edn, 2011 <http://www.oxforddnb.com/view/article/1002>, accessed 9 October 2011.
30. Tittler, 'Bacon, Sir Nicholas (1510–1579)'; See also Spring, 'The London connection', pp. 7–9.
31. J. Summerson, 'The building of Theobalds 1564–85', *Archaeologia*, 97 (1959), pp. 107–26.
32. British Library, Cotton MS Aug.I.i.75, map by J. Thorpe, 1611. The catalogue describes this as a map of Cheshunt Park but it is Theobalds park (in the parish of Cheshunt).
33. Rowe, 'Hertfordshire's lost water gardens', pp. 39–41.

34. Pevsner and Cherry, *Hertfordshire*, p. 165.
35. Smith, *English houses*, pp. 60–61.
36. P. Henderson, 'Sir Francis Bacon's water gardens at Gorhambury', *Garden History*, 20 (1992), pp. 116–31; Spring, 'The London connection', pp. 13–14.
37. HALS DE/V/P1 Gorhambury estate map, 1634.
38. Bodleian Library MS Gough Drawings a.3. fol. 27. I am grateful to Robin Harcourt Williams, Archivist at Hatfield House, for his confirmation of Lord Burghley's handwriting.
39. Harris, 'Nyn Hall, Northaw', p. 99.
40. HALS XIII/30 map of the manor of Sopwell.
41. Chauncy, *Hertfordshire*, p. 319.
42. Chauncy, *Hertfordshire*, p. 265.
43. Smith, *Hertfordshire houses*, pp. 67–85.
44. M. Cook, *The manner of raising, ordering and improving forest trees* (London, 1676), p. 21.
45. K. Moody, 'The history of the gardens at Holywell House, St Albans', unpublished report for the Hertfordshire Gardens Trust (2003). The canal is shown on the plan of St Albans drawn by J. Andrews and M. Wren to accompany A. Dury and J. Andrews, *A Topographical Map of Hartford-shire*, 1766, republished by Hertfordshire Publications (Stevenage, 1980, reprinted 1986).
46. English Heritage Listed building description for Blakesware Manor (Building ID: 356043).
47. HALS DE/Of/6/380 Blakesware entry in volume of topographic paintings and commentary by H.G. Oldfield, vol. 6, p. 380.
48. A. Rowe (ed.), *Garden making and the Freman family: a memoir of Hamels 1713–1733*, Hertfordshire Record Publications 17 (Hertford, 2001), pp. xxxix, 11, 14, 19, 20, 23, 24.
49. Henry Guy was of local stock: his father was Henry Guy of Tring and his mother came from Ashlyns near Berkhamsted (A.A. Hanham, 'Guy, Henry (*bap.* 1631, *d.* 1711)', *Oxford Dictionary of National Biography* (Oxford, 2004); online edn, 2008 <http://www.oxforddnb.com/view/article/11798>, accessed 17 October 2011).
50. Chauncy, *Hertfordshire*, p. 541; HALS DE/Hx/P2 plan of the Bushey Hall estate, 1685.
51. Chauncy, *Hertfordshire*, p. 541; illustrations by J. Drapentier between pp. 540–41, 542–3. See also Rowe, 'Hertfordshire's lost water gardens', pp. 45–8.
52. HALS DE/P/P7.
53. P. Willis, *Charles Bridgeman and the English landscape garden*, with supplementary plates and catalogue (Newcastle upon Tyne, 2002), pp. 59–60; A. Fletcher, 'Charles Bridgeman at Tring Park: a reassessment', in Rowe (ed.), *Hertfordshire garden history*, pp. 41–7; J. Carswell, *The South Sea Bubble* (London, 1960), p. 278; C. Campbell, *Vitruvius Britannicus* IV (London, 1739).
54. Willis, *Charles Bridgeman*, pp. 59–60.
55. A.C. Skelton, 'The development of the Briggens estate, Hunsdon, since 1720', *Hertfordshire Archaeology*, 12 (1994–96), p. 114; Hertfordshire Gardens Trust and T. Williamson, *The parks and gardens of west Hertfordshire* (Hertford, 2000), pp. 26–8.
56. J. Burke and J.B. Burke, *A genealogical and heraldic history of the extinct and dormant baronetcies of England, Scotland and Ireland*, 2nd edn (London, 1841), p. 541; R. Clutterbuck, *The history and antiquities of the county of Hertford*, vol. 1 (London, 1815), p. 502.
57. Sacombe was not actually built.
58. K. Harwood, 'Some Hertfordshire nabobs', in Rowe (ed.), *Hertfordshire garden history*, p. 53; Bodleian Library MSGD a.4, fol. 64, design for gardens at Sacombe attributed to Charles Bridgeman, *c.*1715–17.
59. N. Salmon, *Hertfordshire* (London, 1728), p. 225.
60. Howlett, *Hitchin Priory Park. The history of a landscape park and gardens* (Hitchin, 2004), pp. 11–15.
61. Harwood, 'Some Hertfordshire nabobs', p. 62; K. Harwood, 'A Hertfordshire garden in the eighteenth century', *Herts Past and Present*, 3rd ser., 2 (2003), pp. 12–16; HALS DE/P/P7.
62. HALS DE/B1622/P3 plan of Kendals Hall, *c.*1740.
63. Shown on an undated plan, probably Charles Bridgeman's design for the site, in the Bodleian Library, Oxford (Bodleian Library MS Maps Herts. a.1 plan of Gubbins (Gobions), North Mymms, n.d.).
64. A. Rowe and T. Williamson, 'New light on Gobions', *Garden History*, 40 (2012), pp. 82–97.
65. Wrotham Park (by Isaac Ware) near Monken Hadley could also be added to this list, although not strictly within Hertfordshire at the time.
66. H. Leiper, 'Mr Lancelot Brown and his Hertfordshire clients', in Spring (ed.), *Hertfordshire garden history II*, pp. 92–120. He also worked at Littlegrove in Barnet, Porters Park at Shenley, Dyrham or Durham Park near Ridge and Wrotham Park near Monken Hadley. The latter two lay mostly and wholly in the parish

of South Mimms, respectively, and were transferred to Hertfordshire only in 1965.

67. Howlett, *Hitchin Priory Park*, pp. 25–44; F. Cowell, *Richard Woods (1715–1793): master of the pleasure ground* (Woodbridge, 2009), pp. 182–4, 221–2; E. Gatland, 'Richard Woods in Hertfordshire', in Rowe (ed.), *Hertfordshire garden history*, pp. 106–33.
68. S. Daniels, *Humphry Repton: landscape gardening and the geography of Georgian England* (New Haven and London, 1999), pp. 260–61; See also T. Way, 'Introduction', in H. Repton, *Humphry Repton's Red Books of Panshanger and Tewin Water, Hertfordshire, 1799–1800*, Hertfordshire Record Publications 27 (Hitchin, 2011), p. xx. Fourteen Repton sites are listed for the county.
69. A probable addition to this list is Durham, later Dyrham, Park, near Ridge, which was not in Hertfordshire in the eighteenth century.
70. HALS A/2845 Brown's plan for improving Youngsbury, n.d.
71. BL Ordnance Survey Drawing 149, Hatfield, 1805; HALS 34137 conveyance of Gobions estate, 1815 contains a plan showing the 'Great Pond' in the park; Harwood 'Some Hertfordshire nabobs', p. 55.
72. Page (ed.), *VCH Hertfordshire*, vol. 3, p. 450; not shown on Ordnance Survey Drawing 138 pt. 1, Enfield, 1799 but shown in a painting by Buckler in 1834 (HALS DE/Bg/4).
73. HALS DE/X3/9 the diary of John Carrington, 1806.
74. Elliott, *Victorian gardens* (London, 1986); C. Ridgeway, 'William Andrews Nesfield: between Uvedale Price and Isambard Kingdom Brunel', *Journal of Garden History*, 13 (1993), pp. 69–89.
75. Hertfordshire Gardens Trust and Williamson, *West Hertfordshire*, pp. 78–80. For Nesfield's involvement at The Grove see Bodleian Library, Oxford, MS Eng 2167 f.37 and f.95; for Moor Park, see *Journal of Horticulture*, 21 (1872), pp. 330–42.
76. Prince, *Parks in Hertfordshire*, pp. 159, 173, 175.
77. Apsley Paper Trail Documents/444.
78. Kemp, *How to lay out a garden* (London, 1864), pp. 367–70.
79. *Ibid.*, p. 237.
80. T. Williamson, 'Gardens and industry: the landscape of the Gade Valley in the nineteenth century', in Spring (ed.), *Hertfordshire garden history II*, pp. 121–47.
81. Smith, *English houses*, *passim*.
82. Hertfordshire Gardens Trust and Williamson, *West Hertfordshire*, pp. 29–30, 40–41; Leiper, 'Mr Lancelot Brown', p. 98.
83. Rowe, *Medieval parks*, pp. 162–5.
84. Chauncy, *Hertfordshire*, p. 265.
85. Pevsner and Cherry, *Hertfordshire*, p. 151.
86. Harwood, 'Some Hertfordshire nabobs', p. 59; English Heritage Register of parks and gardens of historic interest.
87. Leiper, 'Mr Lancelot Brown', p. 103; HALS DE/P/T3313 Release of the manor of Panshanger to Thomas Woodford in trust for the Right Hon. William Earl Cowper, 1719.
88. HALS DE/P/P21 Repton's Red Book for Panshanger; see also H. Repton, *Humphry Repton's Red Books of Panshanger and Tewin Water, Hertfordshire, 1799–1800*, with an introduction by T. Way, Hertfordshire Record Publications 27 (Hitchin, 2011).
89. HALS D/Z42/Z1 Repton's Red Book for Tewin Water, 1799.
90. H. Prince, 'The changing landscape of Panshanger', *Transactions of the East Herts Archaeological Society*, 14/1 (1955–57), p. 47.
91. S. Spooner, 'The diversity of designed landscapes: a regional approach *c.*1660–1830', PhD thesis (University of East Anglia, 2010); S. Spooner, '"A prospect two fields' distance": rural landscapes and urban mentalities in the eighteenth century', *Landscapes* 10/1 (2009), pp. 101–12.
92. Rowe, 'Hertfordshire's lost water gardens', pp. 48–53; H.C.N. Daniell, 'Popes Manor, Essendon', *Transactions of the East Herts Archaeological Society*, 7/2 (1924), p. 155.
93. HHA CPM Supp 69 'Plan of the manor of Holbatches alias Popes ... belonging to Mrs Rebecca Asheton surveyed by Taylor and Chilcott in the year 1785'.
94. Hertfordshire Gardens Trust and Williamson, *West Hertfordshire*, pp. 91–6; K. Harwood, 'Some Arts and Crafts gardens in Hertfordshire', in Spring (ed.), *Hertfordshire garden history II*, pp. 148–76.

– 9 –

Urban and industrial landscapes

Introduction

In the chapters so far we have been mainly concerned with Hertfordshire's countryside. Indeed, most volumes on landscape history are concerned primarily with rural areas, rather than with towns. But Hertfordshire, more than most counties, is extensively urbanised and suburbanised, and to appreciate its character we must now discuss these superficially less attractive built-up landscapes. For these are the environments in which most Hertfordshire people live and, even if they are in many ways less picturesque than the surviving areas of the county's countryside, their story is of no less interest. We will begin at the beginning, however, with the development of towns in the period up to *c.*1870, and thus with what are, perhaps, the more visually appealing aspects of our urban heritage.

The origins of towns

Urban life disappeared at the end of the Roman period and while it is clear that many of the towns established then – or earlier, in the later Iron Age – continued to be occupied right through the Saxon period, this was as elite residential centres rather than as towns *per se*. Such continuity could take many forms. At St Albans a middle Saxon defended centre – Kingsbury – developed within the walls of the old Roman city, but the late Saxon and medieval town grew up on the higher ground overlooking this – on the east rather than the west of the Ver, at the gates of a great Benedictine abbey.[1] This was dedicated to the second-century martyr Alban, who almost certainly lived, and was killed, at Verulamium. His grave was probably visited by Bishop Germanus of Auxerre in the 420s, during his mission to sub-Roman Britain to fight heresy, and it was still being venerated when the monk Gildas was writing around a century later, for he describes how the shrine of '*sanctum Albanum Verolamiensem*' could no longer be visited by Britons, presumably living in the west of England, because of the presence of barbarian settlers occupying the intervening area.[2] The abbey and the late Saxon and medieval town thus stand outside the Roman city in a manner reminiscent of the cathedrals at places such as Tours, Arles, Cologne and Lyon in France, or Bonn in Germany. It has long been known that these occupy the sites of late Roman cemeteries (placed by law beyond town walls) because they originally developed over the burial places of martyrs. Martin Biddle's excavations on the site of the abbey chapter house in the 1970s revealed a series of ditches of late Roman or early post-Roman date which ran 'precisely at right angles to the axis of the Norman abbey church, and so demonstrate that the arrangement of the abbey hill-top has preserved this alignment since at least as early as the 8th century and possibly since late Roman or immediate post-Roman times'.[3] Subsequent excavations revealed evidence for a late Roman cemetery directly underneath the south-western side of the cloisters, a cemetery which might well have included the grave (or supposed grave) of the martyr.[4]

Bede tells us that some time after Alban's death 'A church of wonderful workmanship was built, a worthy memorial to his martyrdom. To this day sick people are helped in this place and the working of frequent miracles continues to bring it renown.'[5] The 'F' version of the *Anglo-Saxon Chronicle*, probably produced at Canterbury in the eleventh century, records that this was replaced in 793 by a new church, erected by Offa, king of Mercia, to which the saint's relics were translated.[6] A third-century martyr's tomb in an extra-mural cemetery may thus have developed, in time, into a major Saxon church and eventually into a monastery associated with a major royal estate centre, a pattern of development familiar enough in Gaul, but unusual in England.

Yet it is also striking how many other places of ecclesiastical and administrative importance in middle and later Saxon Hertfordshire were located on, or beside, the sites of large Roman settlements. We have already discussed the development of Braughing, the second largest Roman town in the county, into a Saxon royal estate centre and minster site. The church at Northchurch, marking the original centre of an early estate which also embraced Berkhamsted (the two parishes are mixed in a complex way, and were apparently divided when the castle was built here in the immediate aftermath of the Norman Conquest), was also evidently a minster, and it stands only a few hundred metres from the Iron Age and Roman settlement centred on the Cow Roast Inn.[7] The Roman settlement at Ware continued to be occupied into the early and middle Saxon period, to judge from the finds made in the 1980s during excavations on the High Street and on the site of the library.[8] The church at Welwyn was described as a minster in the tenth century and middle Saxon burials have been discovered in its vicinity: it stands within the small Roman town which underlies much of the village.[9] Finally, Bishop's Stortford is associated with extensive areas of Roman occupation, and a measure of 'continuity' is suggested – as at Braughing – by the occurrence, to the west of the town, of a place bearing another example of a *wic-ham* name.[10] Evidently, many minor Roman towns maintained a significance as 'central places' right through the Anglo-Saxon period.

Nevertheless, the extent of continuity should not be exaggerated. Baldock, for example, was the third most important Roman town in the county and also the site of a major medieval town, but there was no direct connection between the two. The former was almost certainly abandoned in the immediate post-Roman period, while the latter developed on an empty site only in the twelfth century. And although Hitchin was, like St Albans, a defended *tun* of the Mercian kings which developed into a medieval town, it had no Roman predecessor – unless of course this is represented by Baldock, only seven kilometres to the north-east. Even where there was 'continuity', moreover, it was emphatically *not* continuity of urban life. There was no currency and, as far as we know, there were no markets in early or even middle Saxon times. Places like St Albans did not continue as towns, in which manufacture and trading took place, but rather as important meeting places and administrative centres, as centres of power: and it was this which often ensured, at a later date, their development once more as urban centres.

The revival of town life was gradual. In middle Saxon times tribal kings began to establish coastal trading centres called *wics*, where prestige goods could be imported or manufactured under close royal control:[11] no such sites exist in land-locked Hertfordshire, although it is possible that Ware, on the navigable Lea, fulfilled some similar role within the East Saxon kingdom.[12] But *wics* were not towns as we would understand them, and were few in number; true towns emerged only as a market economy, albeit one still to a significant extent under close royal control, developed in the later Saxon period. An important aspect of this was the creation, following the reconquest of England from the Danes and the development of a unified English kingdom under Edward the Elder, of a number of *burhs*, or royal fortresses, some, but not all, of which also functioned as market centres. This is how Hertford originated, established in 912 as a double fortress, astride the river Lea: it was a stronghold, supplied and garrisoned by the new shire that bore its name, but it also served, probably from the start, as a market centre.[13] Other *burhs*, a term now used in the sense of undefended settlements with market places and a resident non-farming population, came into existence in the course of the tenth and eleventh centuries. Domesday describes St Albans, Ashwell and Hertford as *burhs*, while 52 burgesses – merchants with special privileges which distinguished them from the broad mass of the rural population – are listed among the inhabitants of Berkhamsted. There were also 12 recorded at Stanstead Abbots; while further down the Lea, at Cheshunt, Domesday lists 10 *mercatores*, merchants. The founding of towns continued – indeed, intensified – in the two centuries following the Norman Conquest as lords granted special privileges to particular communities, giving them charters which made them boroughs, while the Crown granted places the right to hold a market as a mark of favour, or in return for services or cash. Once established, markets often generated recognisably urban communities around them, with a diverse occupational base.[14] Whether established as a town or developing into one, urban settlements in the Middle Ages had two main features: a market place, which has usually been much reduced in size (or in some cases completely obliterated) by gradual infilling with shops; and narrow plots running back from it, the frontage of which was occupied by workshops and trading premises.

Historians often assume that great landowners took the lead in medieval urban development. They were keen to create trading places, it is said, because they could make a

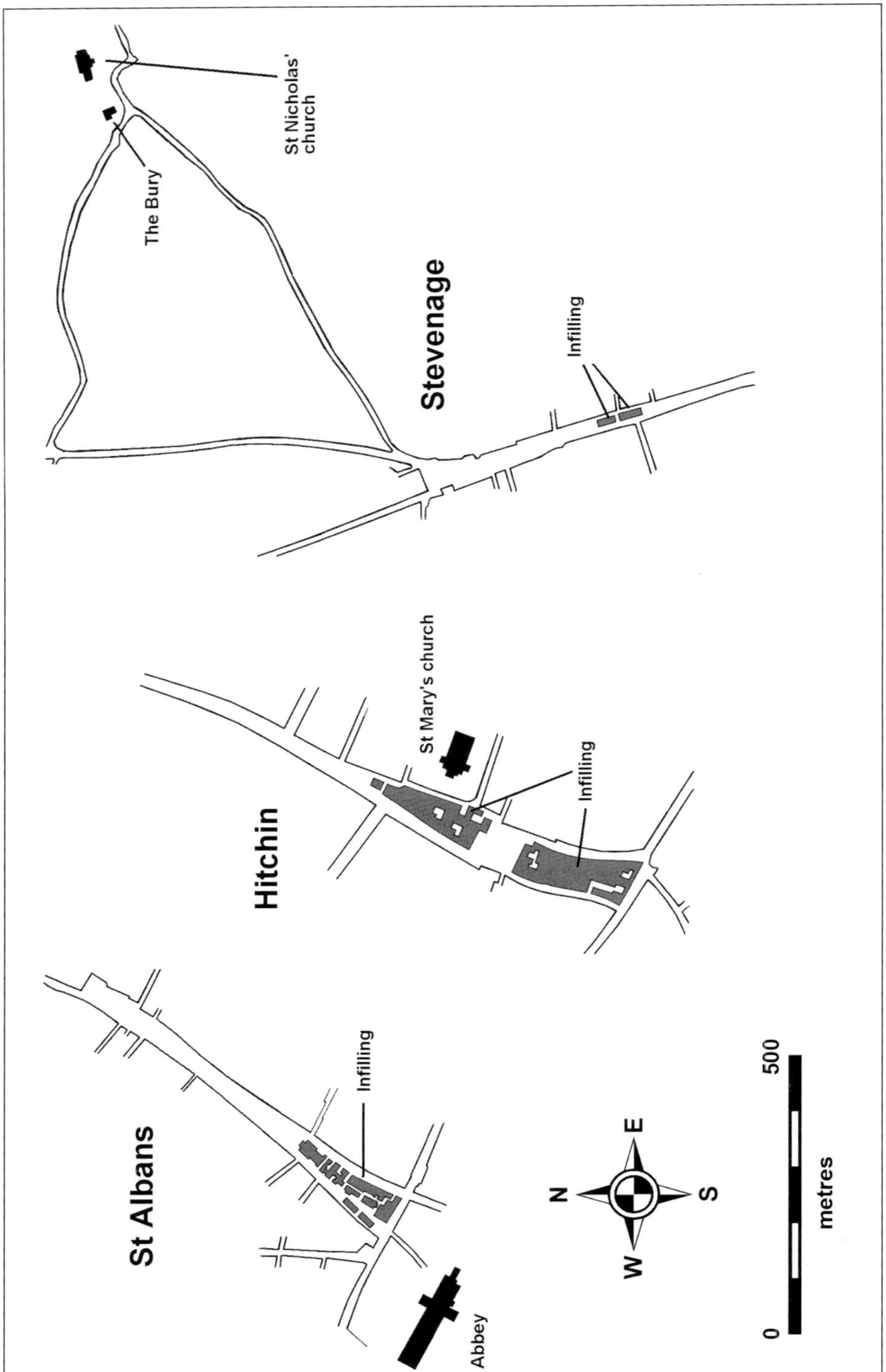

Figure 9.1. The market places at St Albans, Hitchin and Stevenage.

profit from renting out burgage plots and from the tolls charged on traders using markets. But it is possible that their role in all this has been exaggerated, and that, both before and after the Conquest, large landowners simply cashed in on spontaneous economic developments. The regular influx of people attending courts, religious meetings or other assemblies at places like St Albans would have provided a ready market for traders and craftsmen, and in most cases it may only have been after informal markets had developed that lords moved in to exploit them. It is also frequently assumed that the information provided by documentary sources – by Domesday and the various charters and market grants recorded in official documents – provides a good guide to the development of towns. But, if markets developed spontaneously, then official recognition may often have come some time, perhaps many decades, after places began to acquire urban characteristics. Some towns, moreover, such as Bishop's Stortford, have no charter and never received an official grant to hold a market: in the Middle Ages it was simply said that trade was conducted here on the basis of ancient rights.[15]

It is thus hard to chart with any accuracy the development of urban life in Hertfordshire, or indeed elsewhere, and even at the time of the Conquest there were almost certainly more places with urban or quasi-urban functions than are listed in the pages of Domesday. If St Albans and Ashwell had markets, why was this not also true of a place such as Hitchin, which had for centuries functioned as a major administrative centre and a meeting place for a hundred court? One distinguishing feature of the towns which do appear in Domesday is that they appear, to judge from their plans, to have been provided with particularly spacious market places (Figure 9.1). That outside the gates of St Albans Abbey was huge, originally covering around two and a half hectares; that at Ashwell, now almost completely colonised by houses, was almost as large, at around 1.8 hectares. At Hertford there were two markets, one associated with each of the Edwardian *burhs* and, although their precise form and location remain uncertain, the southern, rectangular one at least appears to have been large.[16] In all cases, moreover, the parish church (or at St Albans, the abbey) stood beside the market place, a reminder of the need for probity in transactions, and perhaps itself a place where business was conducted.

These large market places may, as J.T. Smith has suggested, have originally served the whole of the hundreds in which they lie.[17] Hertford may thus have served Hertford Hundred, Ashwell Odsey and St Albans Albanstowe or Cashio, as it came to be known. This in turn suggests that other hundreds may have had market places around which towns may have been developing by the end of the Saxon period, but, because their inhabitants had not secured particular legal privileges, they appear simply as agricultural settlements in Domesday. Hitchin, for example, had a large population at the time of Domesday, and has a vast market place, comparable in size and shape to that at St Albans, and this may well have been used by those dwelling within its eponymous hundred (Figure 9.1). Braughing, another hundredal centre with, as we have seen, a deep past, still has an open space ('The Square') to the east of the churchyard, with signs that it was once significantly larger. By the time of Domesday this place had lost its ancient importance and was divided between a number of manors, but an ancient market may have continued to function, for in *c.*1148 when King Stephen granted 100s rent in the manor to the Priory of Holy Trinity, London, he also gave six *librates* of land in exchange for a mill and land which the Queen had granted them near the Tower of London: *librates* which, as the charter tells us, included the site of the market.[18] At Bishop's Stortford, where the market was said to have been held by 'ancient prescription', there are likewise signs of a large, now infilled market place in close proximity to the parish church. This may already, at the time of the Conquest, have been serving the south-eastern sections of Braughing and perhaps Edwinstree Hundreds.

Whatever the truth of these speculations, there is no doubt that the number of towns and markets in Hertfordshire increased steadily in the two centuries following the Norman Conquest, as the economy experienced phenomenal growth.[19] Some of these places also had large and, in the case of Ware, Baldock and Royston, double market places.[20] But in many places the market place was little more than a simple widening in a major road, although once again usually located close to the parish church or, in the case of Buntingford, a chapel-of-ease. This was the case at Berkhamsted, established beside the new castle some time after the Conquest; at Hemel Hempstead, where a market was being held by *c.*1300; at Buntingford, where an informal market is recorded in the early thirteenth century; at Watford, where a market was probably established on the abbot of St Albans' manor in the early twelfth century; at Bushey, where there was a market by 1200; and at Barkway, where a market existed by the middle of the thirteenth century.[21] These markets were evidently intended to serve smaller influxes of people than were the great open spaces at Hitchin or St Albans. Most of the places where they were found, moreover, were not ancient centres of power. Hemel, although a large manor, has no very distinguished antecedents; Bushey and Barkway appear as rather minor places in Domesday; Watford, Baldock and Royston are absent altogether.

It is noteworthy that several of these places occupied 'liminal' positions, on administrative boundaries – very different from the earliest towns which, as we have seen, were generally located close to the centres of hundreds and large estates.[22] Buntingford on Ermine Street, for example, which first developed in the early thirteenth century but which received a market grant in 1360, was located on the boundaries of four parishes: Wyddial, Layston, Throcking and Aspenden. Chipping Barnet, which obtained a market charter in 1199, appears to have grown up on the Great North Road as 'a messy and informal development of inn keepers and traders locating on the heath, common land, woods and pastures beside the highway on the parish boundary between East Barnet and South Mimms, which was also the county boundary between Middlesex and Hertfordshire'.[23] Royston, in the north of the county, developed at the point where the Icknield Way crosses Roman Ermine Street – and was likewise bisected by the county boundary, with the northern half being in Cambridgeshire – and at the meeting place of no fewer than five parishes. Baldock, further to the west on the Icknield Way, lay on the boundaries of Clothall, Bygrave, Willian and Weston. It is often assumed that these last two towns were established by local monastic houses: the former by a small Augustinian priory (which obtained a charter in 1189) and the latter by the Knights Templar (receiving a charter in 1185).[24] But in all these cases it is likely that, far from local lords fostering economic development, these towns' liminal locations (coupled with positions on major roads) ensured a degree of freedom from close manorial supervision which initially allowed them to develop as trading places, and that landowners simply cashed in later by formalising commercial activities which were already taking place.

Town formation and the formalisation of markets continued into the later thirteenth and fourteenth centuries, but few of the places concerned actually developed into major urban centres.[25] Stevenage, Sawbridgeworth, Hatfield and Tring, all of which received charters to hold markets in this period, had great futures ahead of them. But settlements such as Codicote, Bygrave, Eastwick, Digswell and King's Langley always remained as villages. The lord of Popes Hall manor in Buckland, on Ermine Street in east Hertfordshire, obtained the right to hold a market and a three-day fair at a place called 'New Chipping' (from the Old English *ciepping*, market) in 1252. The enterprise was a failure and Chipping remains today a minor hamlet on the A10. In 1258 the lord of Buckland manor obtained a grant to establish a market a little further north along Ermine Street, in Buckland village, but this attempt, similarly, met with little success. Neither was well placed to compete with Royston, to the north, or the developing settlement of Buntingford to the south.[26] In this more crowded world most of the prime economic locations – beside major ancient institutions, where major transport routes coincided with administrative boundaries or at key points in the communications network – were already occupied by trading centres. There was also less space in a more general sense: while numerous large tracts of common land still survived in Hertfordshire at this time, areas close to settlements had generally been filled with agricultural land, tenanted by farmers, so that empty space which could be used by traders often lay some distance away. Whereas earlier towns thus generally had market place and church in close proximity, in these places they were often quite divorced from each other. At King's Langley, for example, the present High Street was developed as a market place parallel to the line of the old main road (it was called New Chipping in a document of 1416), but was some way away from the parish church.[27] Stevenage is a more dramatic case: a grant was received in 1281 for a market located on what was becoming the Great North Road, nearly a kilometre away from what was presumably the site of the Domesday manor and the original focus of settlement in the parish, beside the parish church of St Nicholas (Figure 9.1).[28]

Medieval urban morphology

It can thus be argued that an earlier generation of urban historians over-emphasised the active role of kings and magnates – lay and ecclesiastical – in the development of urban life. Many may also still exaggerate the extent to which towns were 'planned' by their supposed creators – that is, with market places and associated burgage plots laid out according to some pre-determined and often highly regular scheme, the character of which is still discernible in the pattern of urban boundaries.[29] While the properties running back

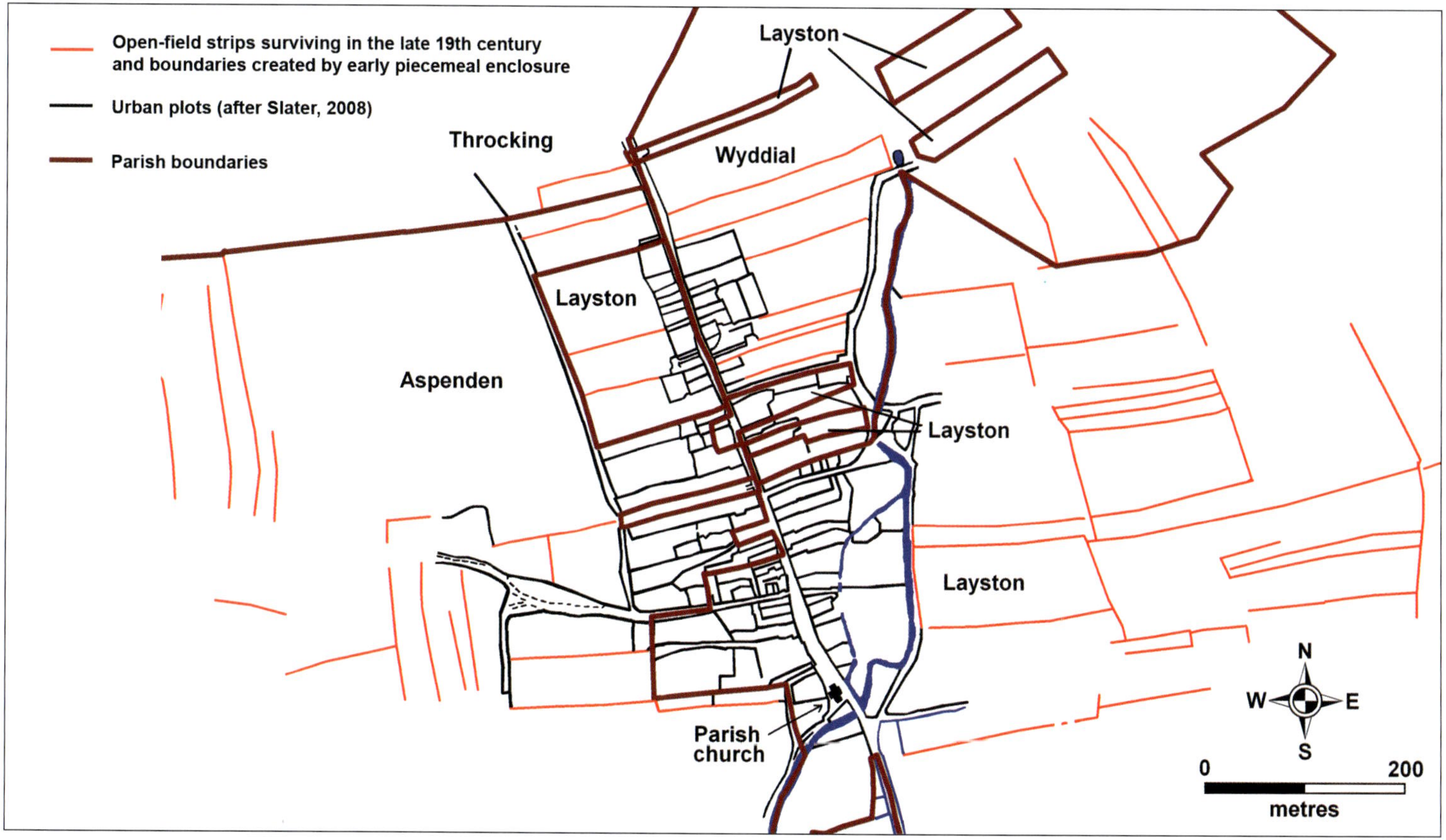

Figure 9.2. The medieval town of Buntingford has a cohesive, 'planned' appearance yet was divided between the parishes and lordships of Aspenden, Layston and Wyddial. The parish boundaries were interdigitated in complex ways and were clearly fixed after the area had become divided into open-field strips. The narrow, parallel plots within the town were probably formed from the consolidation of groups of strips.

from market places do often display a striking appearance of regularity, lying parallel and sharing a common termination – sometimes a back alley or lane – it is noteworthy that the market places themselves do not usually have neat, rectangular shapes, but instead display outlines strongly reminiscent of areas of ordinary common land. Some, like the great spaces at Hitchin or St Albans, take the form of irregular triangles, a typical shape for commons in Hertfordshire as elsewhere, while the 'widened road' form found in many others likewise has close parallels in the morphology of innumerable areas of common land. Most market places, rather than being the planned creations of manorial lords, were probably pre-existing areas of common land on which trading activities developed spontaneously, in the manner already suggested. In the case of the plots running back from market places, and sometimes from the roads leading to them, the situation is more complicated, but it is evident that regular patterns do not necessarily result from unitary planning. The layout of Buntingford is typical of many other medieval towns in the county, with neatly parallel properties running back from Market Hill and Ermine Street, those to the west terminating in part on the lane called the Bowling Green. Yet it is hard to see how this layout could be the consequence of a single planning 'event', given that it embraces the property of at least five separate parishes and lordships (Figure 9.2). Conversely, while parallel plots running back from road and market place are a standard feature of Hertfordshire's towns, these do not always display a high degree of regularity and can even lie at odd angles to the street frontage. Watford, for example, although lying entirely within the jurisdiction of St Albans Abbey when it received its market grant in the early twelfth century, has plots which run back from the market and High Street at all manner of angles (Figure 9.3). Long, narrow plots *per se* were simply the consequence of competition for a place fronting on the market place, or roads leading immediately to it – competition, that is, to secure the most economically advantageous sites. But it may also be that, just as the open spaces for markets were often pre-existing commons, such narrow 'strippy' plots were sometimes adapted from the strips in neighbouring open fields. Many, as on the western side of Royston, are not dead straight, but display the gentle curve characteristic of open-field strips, while at Buntingford the layout of properties seems to fit in neatly with the layout of boundaries in the surrounding open fields, either as still existing at the end of the nineteenth century, or as preserved in the boundaries of hedged fields by earlier piecemeal enclosure. The neat linear terminations of plots like this may thus have originated as the headland at the end of a furlong. The idea that tradesmen gradually acquired small groups of strips beside commons on which markets had developed, and enclosed and then built on them, may be more likely than the accepted model of great lords planning numerous close-packed parallel properties on the assumption that the required numbers of traders would suddenly be attracted there.

This is not to deny that some towns were laid out, in whole or part, by major landowners, if often beside markets which had already come into existence. Royston, studied in detail by Terry Slater, is an interesting case.[30] As already noted, it grew up on the margins of five

Figure 9.3. Watford, as shown on the tithe map of 1842. The narrow urban plots ran back from the High Street and the market place at a variety of angles. Reproduced by kind permission of HALS (DSA4/111/2).

parishes in two counties, strongly suggesting that it was a spontaneous rather than 'planned' market settlement: the town did not become a parish in its own right until 1540. It was focused on a cross roads to the south-east of which – around the present church – a small Augustinian priory was founded in the later twelfth century, on what was then the edge of the parish of Barkway. It obtained a market grant in 1189, 'no doubt to formalise trading already taking place on the site'. The plan of the town suggests that such commercial activity was centred on a long, narrow area of common land lying to either side of Ermine Street: the road, descending the steep escarpment, had evidently grown wide over the years, and also probably occupied a deep hollow way. In Cambridgeshire this wide road continued, in spite of some infilling, to serve as the market place; but on the Hertfordshire side of the boundary, to the south of Icknield Way, it was almost completely infilled with narrow properties by the time that the earliest maps were made. Here, another triangular market place existed to the east, separated from the widened Roman road to the west by an area of regular-looking tofts and bounded on the east by the priory precinct. This was presumably created by the priory in an attempt to attract traders onto a market place on its own land – the parish boundary between Therfield and Barkway ran along the middle of Ermine Street. It was almost certainly an area of open ground, a patch of chalk heath like that which still survives – as Therfield Heath – to the west of the town, but probably formed part of the demesne land of the manor. The priory laid out a block of properties over the western side of this space, backing on to the Roman road. Even where lords took an active, even aggressive role in the development of trading places, in other words, they usually made use of existing structures and spaces in the landscape. Their ability to plan entirely afresh, in the more densely settled parts of the county at least, was usually limited.

Where the margins of cultivation were still expanding, however, especially in parts of the south of the county, it may have been easier to create planned settlements. Bushey, which received a grant of a market and fair in 1141, never developed into a town but – as the historian Grant Longman first pointed out – the village's surprisingly regular herringbone pattern is clearly evident on a parish plan of 1800,[31] especially on the northern side of the High Street, where a series of long tofts runs back, rigidly parallel, as far as Back Lane or Herkomer Road.[32] Although there were some areas of open-field land in the parish, as we have seen, these were located on lower ground, where the soils were lighter, close to the river Colne and its tributaries. It was in these locations, too, that settlement was probably still concentrated at the time of Domesday, on sites later occupied by outlying farms. The village itself, on heavier soils beside the main road to London, may well have been laid out on a virgin site and the regularity of its plan clearly owed nothing to the piecemeal enclosure of open arable. The spacing of the plot boundaries as shown on the parish plan of 1800 – the earliest map of the parish – is no longer obviously regular but this is clearly a consequence of later amalgamations and subdivisions of properties, and it is possible to see that the original pattern consisted of fourteen plots, each with a measured frontage of five rods, laid out to the north of the High Street (Figure 9.4). On the south the overall pattern is less clear, in

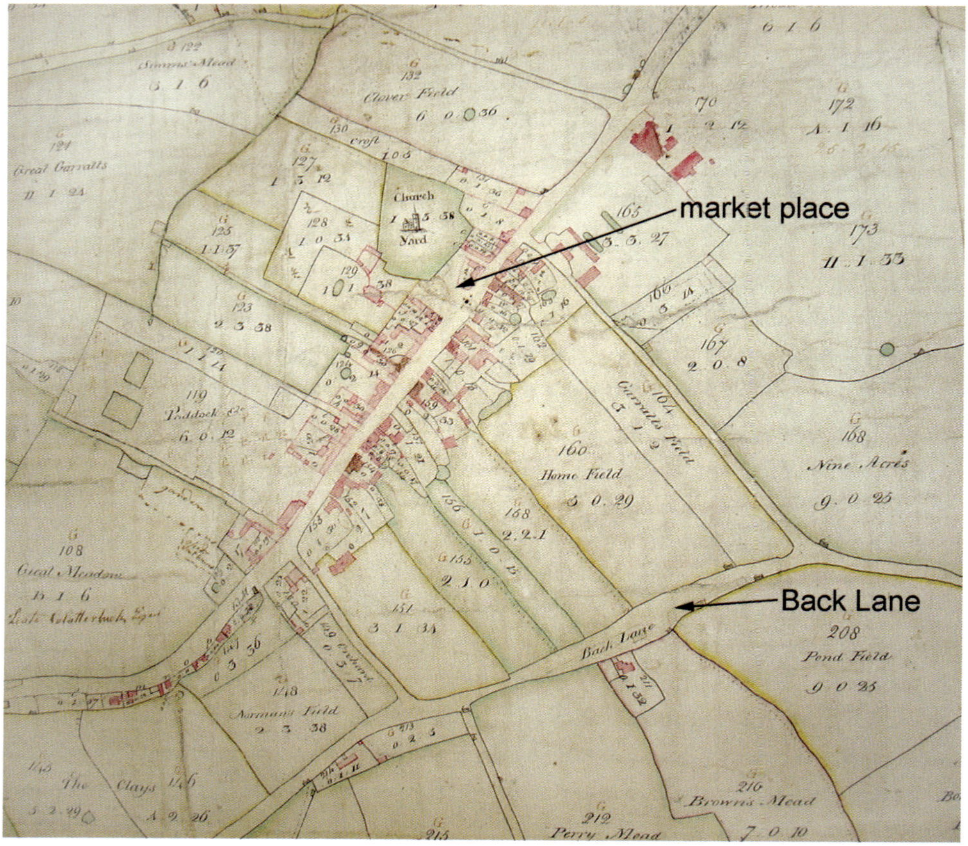

Figure 9.4. Bushey, as shown on the parish plan of 1800. The village has a highly regular, planned appearance, the properties running back from the High Street taking the form of strikingly parallel strips that contrast with the more irregular forms of the surrounding fields. This layout seems to owe nothing to piecemeal expansion over arable strips, and perhaps originated when the settlement was established as a small market centre in the twelfth century. North is towards the bottom right corner. Reproduced by kind permission of HALS (DP/26/29/2).

part because a section of the frontage is taken up by the market place, with the church and churchyard beside it. The church is dedicated to St James, and the fair granted in 1141 was to be held on the feast of St James.[33] The church, originally a chapel within Watford parish, was either already in existence or, more probably, was in the process of being erected. It is noteworthy that, at around the same time as he received his market charter, Geoffrey de Mandeville granted a virgate of land to the chapel at Bushey, suggesting that it was served by a resident priest.[34] A decade or so later, in *c.*1160, an agreement was reached between the abbot of St Albans and the next lord of the manor, Geoffrey de Jarpenville, by which the tithes of Bushey manor were transferred from the church at Watford to the chapel (*capella*) of Bushey, indicating that the latter was becoming an independent parish church.[35] Erection of the church and the laying out of the adjacent village probably occurred at the same time as part of a plan to develop the estate economically as both a market on the road to London and, perhaps, as an agricultural village as the heavier land in the locality was opened up. But all the lordly planning in the world could not ensure economic success, and Bushey was soon eclipsed by the abbot of St Albans' market at Watford, little more than two kilometres to the north-west and on the same road to London.

Late medieval and post-medieval developments

Medieval towns were essentially trading centres, occupied by tradesmen, craftsmen, merchants and their dependants, although some, such as St Albans and Hertford, had administrative or jurisdictional roles. Their plans were thus focused on market places of varying size, and the plots fronting on these were principally occupied by shops and workshops. This basic structure survives, but generally in much-altered form, in most old Hertfordshire towns. As already noted, the market place has usually been filled to varying extents with islands or larger blocks of lanes and buildings on small plots of ground, created as temporary stalls gradually 'fossilised' into more permanent structures, a process which may have begun almost as soon as markets were first held on the sites in question. In some cases, as at Ashwell, this process has gone so far that the original outline of the market place is difficult to reconstruct.[36] More usually, as at Hitchin or St Albans, blocks or islands of complex infilling clearly separate distinct open spaces still used, or used until recently, for market stalls. At the other extreme, some market places have experienced very little infilling. At Baldock the east–west market place, now White Horse Street, has been partly occupied by shops at its western end, close to the church, but the larger High Street market, a fine open space, survives almost intact. In addition to the infilling of and encroachment onto markets, the surrounding plots have become subdivided or amalgamated, the latter the predominant direction of development in the later nineteenth and twentieth centuries. Some of the medieval buildings which occupied the plots still survive, although usually in fragmentary form and embedded within later buildings.[37] More usually, they have been entirely rebuilt on one, two or more occasions. Towns as we see them today are thus, in spite of their medieval origins, largely the consequence of late medieval and post-medieval developments.

Not all medieval towns and markets developed into modern urban centres. Many places with urban aspirations fell back again to purely rural settlements in the course of the fifteenth, sixteenth and seventeenth centuries. Particularly striking was the disappearance of regular markets from a string of villages along Ermine Street, at Standon, Westmill, Corneybury, Chipping and Buckland, allowing the steady growth of Buntingford as the only remaining focus for commercial activity in the district.[38] These were all relatively 'late' markets, developing in the thirteenth or early fourteenth centuries, but some more ancient places also slipped down the scale from urban to rural in this period, including Braughing and, to a large extent, Ashwell. Places which continued to flourish, and in time to expand, were mainly situated on the major roads fanning out from London, especially on the old A411 (Watford, Hemel Hempstead (although slightly offset from the road), Berkhamsted and Tring); on the Great North Road – the modern A1 (Barnet, Hatfield, Stevenage and Baldock); on Ermine Street or its medieval replacements (passing through Cheshunt, Hoddesdon, Ware, Buntingford and Royston); and, in the far east of the county, on the main road from London to Cambridge (Sawbridgeworth and Bishop's Stortford). Barnet was located on a route connecting St Albans and London which developed in early medieval times as an alternative to Watling Street. Only Hertford was a significant exception to this broad rule, although one that highlights the importance in transport terms of the river Lea, which, passing through Ware and Cheshunt, provided from Roman times a navigable route to the Thames and London – although to some extent Hertford's importance, never very great in commercial terms, simply reflected its role as a royal administrative centre.

Not all of these towns were the same size or had the same importance, and their pecking order changed slightly over the centuries. By the time of the 1524–5 lay subsidy St Albans, the great monastic city, was by far the largest place in the county, with a population of over 2,100.[39] Hertford, in contrast, came ninth in terms of size, in spite of its administrative importance reflected physically in the presence of its castle – a standard feature of county towns everywhere. Its neighbour Ware took eleventh place. Each had a population of only around 500 – indeed, there were only 11 places in the county with populations of 500 or more. Interestingly, these were scattered fairly evenly across the county and included places such as Standon, Ashwell and Baldock in the north and north-east of the county, something that is hardly surprising given that this was where the most densely settled rural districts were to be found. But, by 1663, as Nigel Goose has demonstrated, while St Albans was still the largest town in the county by some way, there had been changes in the relative importance of the others.[40] Hertford and Ware now took second and third place respectively, followed by Watford and Hitchin, the latter described a little later, by Sir Henry Chauncy, as 'reputed the second Town in this County for the number of Streets Houses and the Multitude of Inhabitants'.[41] There are clear signs that places nearer to London were rising up the list at the expense of others: Bishop's Stortford was now the only town in the top ten which was located to the north and east of a line drawn from Sawbridgeworth to Hitchin. Throughout the sixteenth, seventeenth and eighteenth centuries there were probably always around 17 true towns in the county – although there are obvious problems of definition, for some places expanded significantly in size without acquiring a market, while others with the right to hold a market, and continuing so to do, declined in status to large villages. Hertfordshire was thus a constellation of small towns without a dominant city, perhaps a consequence of its proximity to the capital. London, while it stimulated urban growth in the county, also militated against the emergence of any major economic centre there, sucking trade and wealth away along the network of roads leading south. Even in 1801 only two towns in Hertfordshire, St Albans and Hertford, were home to more than 3,000 people, and only another four (Ware, Hitchin, Watford and Cheshunt) had populations greater than 2,000.[42]

Towns changed their function in the course of the early modern period, becoming more diversified in their economic and social roles. From the later sixteenth century, and especially from the later seventeenth, many began to attract residents whose wealth did not directly derive from commerce: that is, they became stylish and convenient places to live.[43] From the later seventeenth century many large, elegant houses were erected which might resemble – more than we notice today, because of their settings – small country houses. Inns proliferated as travel increased, for many provided not only accommodation but also stabling for horses – for carriers, and later for coaches – heading to or from London. There were already 27 inns in St Albans and 125 in the rest of the county by 1577,[44] and the inns of Hertford alone had stabling for over 530 horses in 1756.[45] Other kinds of buildings developed: almshouses, often located on the outskirts of towns, such as the Pemberton almshouse of 1627 and the more pretentious Marlborough almshouses of 1736 in St Albans, or Bishop Ward's at Buntingford of 1684; schools, such as Richard Hale Grammar in Hertford or the diminutive Pettits' School of *c.*1562 in Stevenage; and, from the later seventeenth century, nonconformist places of worship such as the Friends Meeting House in Hertford of *c.*1670. From the later eighteenth century a variety of public or semi-public buildings began to appear, including corn exchanges where farmers could sell their harvest to corn dealers, like that at Bishop's Stortford of 1829, or at Hertford of 1859. Hertford's Shire Hall was erected in 1768–9, St Albans Town Hall in 1829–31. Towns, in short, became more complex and diverse places, selling services as much as commodities.

The varied histories of Hertfordshire's towns have created very diverse townscapes, their most recent histories having in many ways the greatest impact, with some ancient urban centres, such as Watford, being almost completely destroyed by twentieth-century redevelopment and road improvement. But visual differences also arise from the fact that the architecture of towns shares in many respects the fashions of their hinterlands, so that, for example, pargeting is more common in the east of the county, in places such as Bishop's Stortford and Hertford, than in the west. Essentially, however, such differences really represent variations on the same essential themes, and in many ways towns have a repetitive form and structure. At the heart of most Hertfordshire towns is the parish church, usually large, flint and embattled, rebuilt and expanded in the later Middle Ages especially, although few were rebuilt completely in this period, Hitchin being a notable exception. The market place, or what remains of it, is usually close at hand. A key visual contrast with most villages is the density of houses, with a continuous building line on the market place itself and usually along the principal roads leading to it. There is generally a closely spaced jumble of buildings of different ages and styles, a sight more common before large-scale amalgamation of plots occurred in the nineteenth and especially the twentieth centuries in order to allow for the erection of larger premises. Timber-framed buildings, often with jetties, thus stand cheek-by-jowl with more elegant Georgian façades, although in many cases the latter conceal similar structures, of seventeenth-century, sixteenth-century or earlier date. Many Hertfordshire towns seem to ooze eighteenth-century wealth, with ranks of neat Queen Anne or Georgian elevations, often with decorative details picked

Figure 9.5. Fore Street, Hatfield: a display of eighteenth-century urban elegance.

out in vitrified brick. Baldock in particular is dominated architecturally by such buildings, but other towns can boast notable streets of eighteenth-century elegance, such as Hatfield's Fore Street (Figure 9.5).

Road, canal and rail

The eighteenth and nineteenth centuries were certainly a period of economic expansion, the periods of both the traditional agricultural and industrial 'revolutions'. This was also, crucially, a time of important improvements in communications and, as well as having an immediate and direct impact on patterns of economic development in the county, the new transport systems also structured its development in later periods. In the early eighteenth century most roads in Hertfordshire, as elsewhere in England, were in a very poor condition, a matter of concern at a time of economic expansion. Defoe famously described how the road from Baldock to Stevenage was so bad that farmers set up voluntary tolls on the gates into the adjacent fields 'which travellers always chose to pay, rather than plunge into sloughs and holes, which no horse could wade through'.[46] Until the seventeenth century even the most important roads in England were maintained by the parishes through which they ran: each parish was solely responsible for its own section, regardless of its population and resources, and regardless of the amount of traffic the road in question had to bear.[47] But from the late seventeenth century 'turnpike trusts' began to be established across England. Created by individual acts of parliament, these comprised boards of trustees which would adopt sections of road, erect toll gates, charge tolls and use the proceeds (after a suitable cut had been taken as profit) to keep the route in adequate repair.[48] As a result of complaints about the state of Ermine Street an act of parliament was passed in 1663 which allowed the Justices of the Peace of Hertfordshire, Cambridgeshire and Huntingdonshire to raise money from tolls for repairs and improvements. The first tollgate in the county was set up at Wadesmill and the southern section of the road (south of Wadesmill) became the Cheshunt Turnpike Trust in 1725, while the northern – as far as Royston – became the Wadesmill Trust in 1733. By the 1770s there were eighteen trusts in the county. Some major roads were administered by a single trust, like the old A41 and A411 from Bushey, through Watford to Tring. Others, like the Great North Road, were managed by several, each responsible for different sections.[49] The traces left in the landscape by turnpikes are minor, being largely restricted to around 200 milestones which still survive beside what are now often major arterial roads.[50]

More impressive than turnpike roads, although much fewer in number, were the new 'navigations' (Figure 9.6). Canals were a major factor in eighteenth-century economic expansion, but Hertfordshire can boast only three examples. The Lea, which had long been navigable, was 'improved' as a result of the Lea Navigation Act of 1739. Locks, replacing earlier 'staunches', were built at Ware, Stanstead and Broxbourne, and in 1765 the engineer John Smeaton was employed to create a number of new cuts and additional locks. Improvements continued well into the twentieth century and at the time of the First World War it was still possible for 100-ton vessels to reach Hertford; horse-drawn lighters were still going as far as Hertford in 1950. The river Stort, which joins the Lea near Hoddesdon, was made navigable as far as Bishop's Stortford in the late 1760s, again with the construction of a number of locks. There were plans for continuing this navigation as far as Norfolk and the engineer John Rennie actually surveyed a possible route. Improvements were still being made into the twentieth century and, like the Lea navigation, it is still much used by pleasure craft.[51]

The most important and impressive of Hertfordshire's canals, however, was the Grand Junction, later the Grand Union, designed to connect London with the Midland canal system and authorised by an act of 1793. It was completed as far as Tring in 1799: from here a branch canal led to a wharf at Tring and, more importantly, to a complex of reservoirs at Wilstone, created to supply water to the flight of locks which carried the navigation over the Chiltern escarpment, and which are now an important nature reserve. Water supply was a problem because the canal reaches a height of 255 metres OD at Cow Roast Lock, Wiggington, and the Chilterns are largely waterless. For the rest of its journey through the county the canal followed closely the course of – and often utilised sections of – the rivers Bulbourne, Gade and Colne. Water levels in the latter were maintained by another reservoir, constructed in Aldenham in the 1790s. The canal's numerous locks and some of the attendant lock-keepers' cottages still survive. It remained a major communications route into the twentieth century.[52]

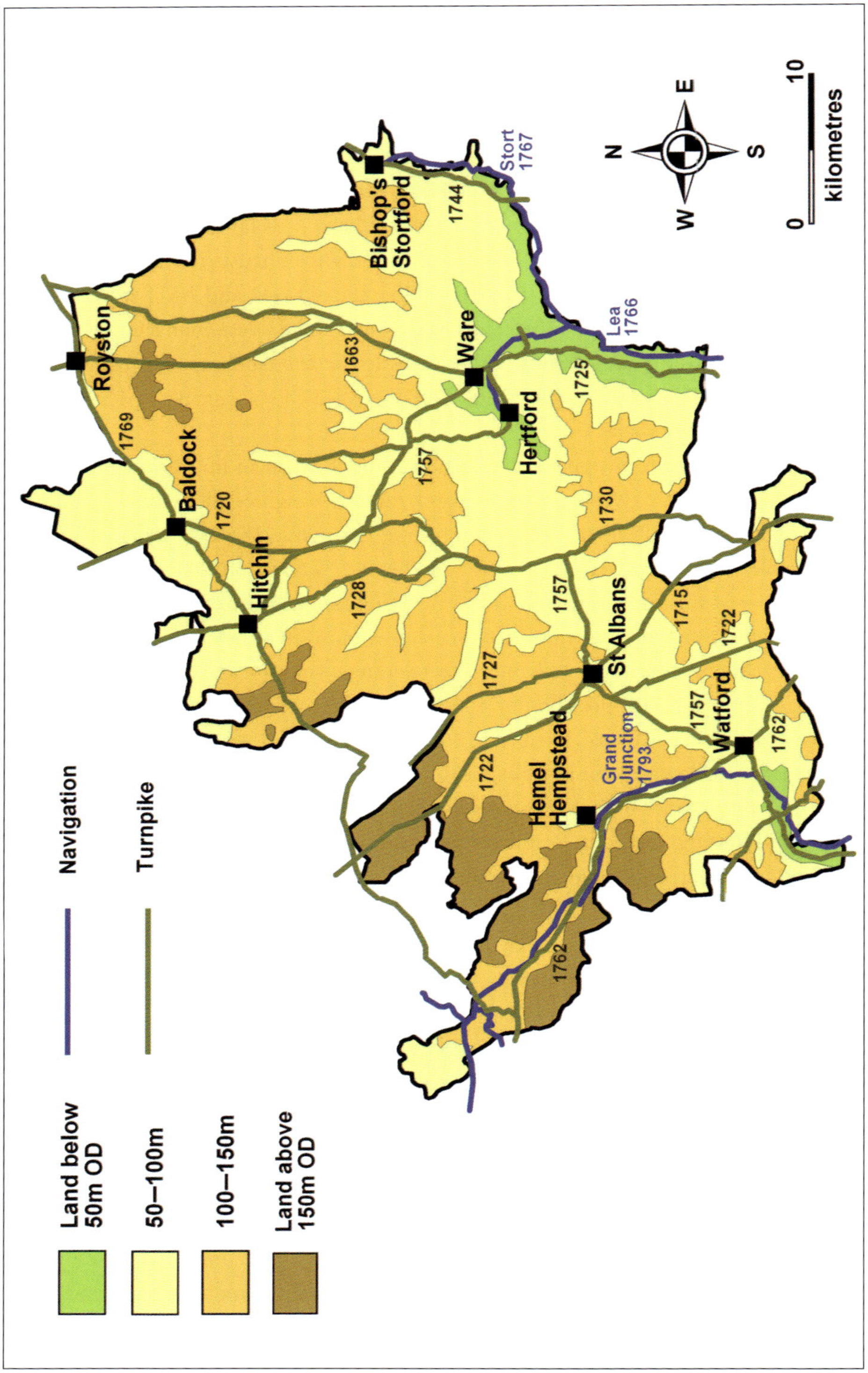

Figure 9.6. Principal canals and turnpikes in Hertfordshire, with dates of establishment.

Significant though canals and turnpike trusts undoubtedly were in stimulating economic growth in the county, the railways made a greater impact on the landscape, both directly – in terms of cuttings, tunnels, embankments and the like – and also indirectly, in that they decided where much subsequent development, industrial and residential, was to take place.[53] The first line to be constructed through the county – George Stephenson's line from London to Birmingham – reached Boxmoor near Hemel Hempstead in 1837 and passed through Tring the following year. This was some 35 years after Trevithick's first locomotive was patented in 1802 and only a dozen after the construction of the first public railway on which steam locomotives were employed, Stephenson's Stockton–Darlington line of 1825. The engineering works required for the new railway were substantial. The line, coming north from London, entered the county near Oxhey and followed the low ground to the west of Carpender's Park before joining the valley of the river Colne at the bottom of Chalk Hill in Bushey, which it followed for two miles before turning west, towards Watford Junction and beyond. This route necessitated a great deal of earth movement, for to maintain the relatively even gradient required by locomotives at this time a cutting had to be made along the first section of the route, while the line had to be raised on a substantial embankment as it crossed the low ground beside the Colne: material removed from the former was employed to construct the latter. In addition, two major bridges had to be built. The first, soon known as 'Bushey Arches', originally comprised five stone-faced arches, each around 30 feet wide and 25 feet high, which carried the two lines of the railway across London Road at the foot of Chalk Hill (although altered and added to on subsequent occasions, the original brick structure of 1837 survives largely intact). A second bridge crossed the river Colne at the point where the railway left the parish of Bushey and entered Watford. Appropriately called 'Colne Bridge', it had five arches, each with a span of around 40 feet, which was likewise widened in 1859 and 1875 but otherwise survives as built. The most important piece of engineering, however, lay further to the north, beyond Watford Junction: the 1.6-kilometre-long Leavesden Tunnel. It would have made more sense to run the line directly to the north-west of Carpender's Park, and thus to the west of the town of Watford, cutting through the fairly narrow (*c.*1.5 kilometre) neck of land separating the valley of the Colne from that of the Gade and from here following the valley of the latter river northwards. But the earls of Essex and Clarendon objected to a route running through their properties, based on Cashiobury House and The Grove, thus necessitating the tunnel's construction, which was only achieved with the loss of eleven lives. Beyond, the line ran close to the Gade, but there were, nevertheless, further embankments and cuttings: at Berkhamsted it smashed through the outer defences of the Norman castle. It left the county through the four-kilometre-long Tring Cutting, itself located within the area of relatively low ground within the Chilterns known as the Tring Gap.[54]

We take all these feats of early railway engineering for granted: we scarcely notice them and, if we do, we would not consider them as part of our 'historic landscape'. Yet, when constructed, they were the subject of engravings produced on a commercial scale

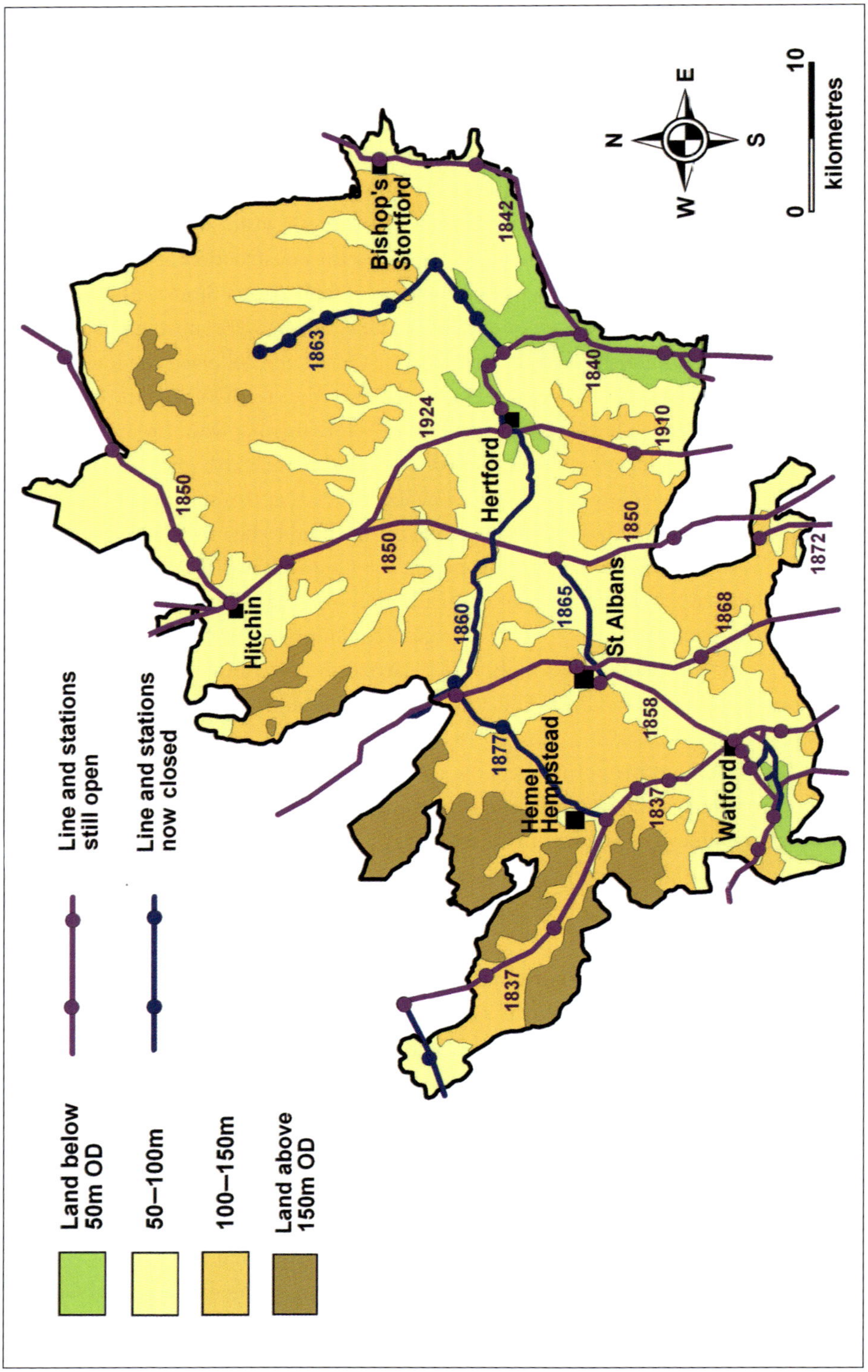

Figure 9.7. Hertfordshire's rail network, with dates of opening. Note the higher density of lines in the south of the county.

and represented eruptions of 'modernity' which, in environmental terms, marked the local advent of an economy based on fossil fuels, a transition whose significance was not lost on contemporaries. When the line first opened crowds from all the parishes around turned up to witness the 'wondrous sight'.[55]

The completion of the London and Birmingham railway (renamed in 1846 the London and North Western) was followed by the construction of the Northern and Eastern, which reached Broxbourne in 1840, Harlow in Essex in 1841 and Bishop's Stortford in 1842, thence continuing to Cambridge. The Great Northern line was constructed through Barnet, Hatfield and Hitchin (and then on to Peterborough and, ultimately, Darlington) in 1850, but it was only in 1868 that the Midland was constructed, southwards from Bedford to Harpenden, St Albans and Elstree. All, perhaps needless to say, had London as their principal destination. Various branch lines and connections between these routes were constructed through the 1840s to 1870s, some of which have since been closed (Figure 9.7). Like the London and Birmingham, the Northern and Eastern followed low ground – the valley of the Lea. The earlier, pre-1850 lines were designed for locomotives with low tractive power, and embankments and cuttings were frequent, to ensure minimal gradient. The later lines tended to cut across more undulating terrain, although still, where possible, following a topographically uneventful course. But embankments and tunnels were inevitably required, especially where major watersheds had to be crossed *en route* for London; these included the Elstree Tunnel on the Midland line and the Potters Bar Tunnel on the Great Northern. And even in Hertfordshire's gently rolling terrain major rivers could necessitate significant bridges, most notably the Welwyn viaduct, also known as the Digswell viaduct, which carries the Great Northern over the river Mimram, with forty arches (the highest 100 feet above the river) made of blue bricks fired on the spot (Figure 9.8). The longest tunnel in the county was the latest to be built – the Ponsbourne Tunnel, to the north of Cuffley, on a loop line to the Great Northern passing through Bayford and Hertford and joining the parent line just to the north of Knebworth, which was completed as late as 1924.

Industrial expansion

While Hertfordshire lacked the raw materials or substantial reserves of water power to fuel significant levels of industrialisation, the period from 1750 to 1870 did see a notable expansion in some manufacturing and processing industries, as well as the growth of towns. Much of the county's industry was concerned with the processing of agricultural produce and was stimulated, above all, by the steady growth of London. Already, by the sixteenth century, Ware was the pre-eminent malting town in England, strategically placed on the edge of the great barley-growing district of north-east Hertfordshire (which extended into Cambridgeshire, north Essex and wider East Anglia) and at the head of the river Lea, providing easy access to the London brewers. In the 1720s a new form of malt – 'brown malt', produced at higher temperatures – was developed here for use in the production of

Figure 9.8. Digswell viaduct, near Welwyn Garden City, is one of the most impressive pieces of industrial archaeology in Hertfordshire.

porter. In the course of the eighteenth century the industry also expanded in Hertford, in part as a consequence of the improvements made to navigation on the Lea in 1767, and also downstream, in Stanstead Abbots. The construction of the Stort navigation also saw the growth of malting in Bishop's Stortford. The industry continued to flourish right through the nineteenth century in spite of increasing competition from firms based in Lincolnshire and Nottinghamshire that were better placed to serve the breweries at Burton-on-Trent which were now rivalling those in London.[56] As David Perman has noted, the first edition Ordnance Survey 6-inch map of 1880 still shows no fewer than 107 malthouses in Ware, 34 in Bishop's Stortford and 20 in Hertford, while smaller concentrations could be found in the other principal towns and some of the larger villages, supplying innumerable local breweries (most towns had at least one). By this time the structure of the industry was changing as larger industrial concerns developed, often located beside railway lines, such as the huge Taylor's works at Sawbridgeworth.

Malthouses have a distinctive appearance and a repetitive layout, comprising an area at one end of the building for storing the grain; a cistern supplying water for steeping the barley so that it swells; one or, in the later examples, several low floors on which the grain was slowly heated and repeatedly turned to germinate it; and, usually at the other end, the tall kiln, with its distinctive cowl, where the germinated barley was dried to malt. The larger businesses of the nineteenth century might have several such complexes. Although by the nineteenth century many Hertfordshire malthouses were fired by coke, hornbeam from the local coppices was the original fuel of choice, and as late as the 1960s Branch Johnson reported that Taylor's of Sawbridgeworth still used 'hornbeam faggots exclusively, of which immense stacks like haystacks are built close by'.[57]

Hertfordshire's malting industry was in steady decline from the later nineteenth century, a reflection in part of the decline in the London brewing industry. Nevertheless, even half a century or so ago it still made an important impact on the county's landscape. In 1964 there were 22 malting works at Ware, of which 7 were still operating; in Bishop's Stortford there were 17, although only 2 were in use; and in Hertford 9, none operational. There were smaller numbers in other towns, most notably at Baldock, where there were 7, of which 4 were still in use. Malting has since largely disappeared from the county, with only one functioning site in Hertford – MacMullen's brewery – and one in Stanstead Abbots – French and Jupps, now one of the only companies producing coloured malts. Many of the buildings have thus been demolished, as the low ceiling height of the earlier examples especially makes conversion to other uses problematic. Nevertheless, a significant number, mainly of nineteenth-century date, have been converted to new uses, like those now used as flats in Watton Road and Hoe Lane in Ware, or the Omega Maltings beside the river in the same town. Some have had complex histories since ceasing to work, such as those in Priory Street in Hertford, built in 1875 for the brewer Benjamin Young, which was used as a council maintenance depot, then as retirement flats and subsequently as a care home.

The eighteenth and nineteenth centuries saw a transformation of grain processing in Hertfordshire, as elsewhere in England, as the growing scale of agricultural production required larger premises than those which had formerly served rural communities. Most surviving water mills in the county were rebuilt on a larger scale after 1750: striking examples include Piccotts End Mill, just to the north of Hemel Hempstead, Moor Mill, Frogmore, and Noke Mill at Great Gaddesden. But smaller mills often survived and those at Mill Green, Hatfield and at Redbournbury are still producing flour.[58] The county's surviving windmills are all likewise of eighteenth- and nineteenth-century date. The earliest is Cromer, built around 1722 and beautifully restored. It is the county's only remaining post mill, although traces of another remain at Great Hormead. In these, the entire body of the mill, perched on a stout post, could be turned to face the wind. Early mills had all been like this, but they were largely replaced by tower mills, taller structures in which only the cap turned, in the course of the eighteenth century. A series of technical improvements made it possible for the cap to be turned and the covering of the sails to be adjusted (to allow for changes in wind conditions) automatically, thus allowing further increases in height. Traces of two 'smock' mills, in which the upper part of the tower was timber framed and only the base of brick, remain in the county, at Brent Pelham (probably 1830s) and Little Hadham (1786–7). The county's seven other windmills, none restored and mainly converted to residential use, are all nineteenth-century brick towers (Colney

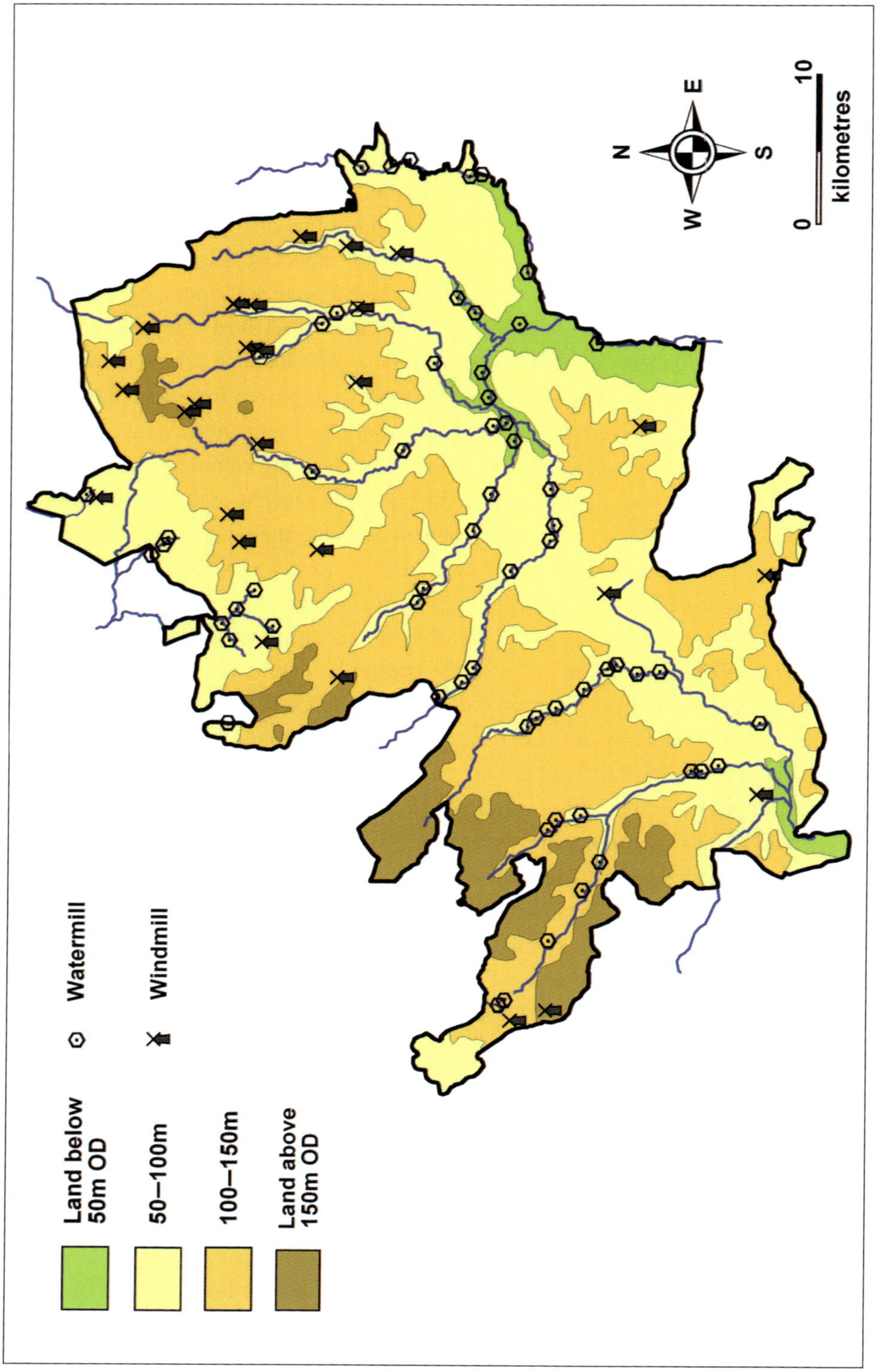

Figure 9.9. The distribution of windmills and watermills in late nineteenth-century Hertfordshire. Source: first edition 6-inch Ordnance Survey maps.

Heath, Croxley Green, Breachwood Green in King's Walden, Great Offley, Reed, Tring and Weston). This is just a fraction of the 46 windmills which were operating in the mid-nineteenth century: mills vanished rapidly from the landscape towards the end of the century, as milling increasingly took place in steam-driven factories located near ports.[59] Both these few survivors, and the lost majority, generally followed the medieval pattern, being located in villages with unreliable water supplies especially on or immediately to the south of the Chalk escarpment and generally on land above *c.*100 metres OD (Figure 9.9).

Papermaking was a speciality of Hertfordshire, although it has left fewer obvious traces in the landscape than malting. Paper was manufactured here – at Sele Mill near Hertford – as early as the late fifteenth century and again in the seventeenth century, at Sopwell near St Albans and at Hatfield. But the eighteenth century saw a significant expansion of the industry, fuelled in part by the London newspaper industry and in part by the requirements of an expanding government bureaucracy. The new mills, initially located in converted corn mills, were concentrated in the south of the county, along the rivers Gade, Colne and Chess.[60] Here they were conveniently located within easy reach of London and could be provided with the ready supply of pure water needed both for the manufacturing process and to power the machinery which pulped rags into paper. In 1801 the rights to use a more rapid system of manufacture, developed a little earlier in France, were acquired by Henry and Sealy Foudrinier, who operated mills at Two Waters and Frogmore in Hemel Hempstead, but they went bankrupt in 1810. By this time, however, John Dickinson had developed his own method of production at nearby Apsley, which he elaborated over the following decades. His business grew. By 1826 he had established a new factory, at Home Park in King's Langley, and in 1830 opened another at Croxley.[61] By the 1870s paper mills were closely spaced along the valleys of the rivers Chess (at Solesbridge in Chorleywood and Mill End and Scots Bridge in Rickmansworth), Colne (at Batchworth in Rickmansworth and Hamper Mill in Watford) and Gade (Frogmore, Apsley, Nash, Home Park and Croxley). Most of the early mills were rebuilt in the course of the twentieth century, where they remained in operation; the rapid decline of the industry in the second half of the century led to many demolitions. Batchworth Mill, near Rickmansworth, operated as a paper mill for much of its life and is partly eighteenth- and partly nineteenth-century in date; while Frogmore Mill, now a heritage centre devoted to the history of papermaking, has buildings largely erected in the 1860s. But elsewhere only fragments, at best, survive.

By the last quarter of the nineteenth century the valley of the Gade downstream of Hemel Hempstead as far as Rickmansworth was one of the most industrialised areas of the county, although with a marked gap around Cashiobury Park and The Grove, where the earl of Essex and Lord Clarendon excluded not only the railway but all other signs of modernity. John Hassell described Apsley Mill in 1819 as 'occupying a large space of ground, and rather resembling a village than a manufactory';[62] all along the valley villages and hamlets expanded with the addition of rows of terraced houses and such things as

Figure 9.10 Apsley Mill, with decorative planting of Lombardy poplar and conifers: undated sketch, early nineteenth century, by Harriet Dickinson (courtesy of Apsley Paper Trail).

nonconformist chapels, while the presence of the near-parallel road (the former Sparrows Herne turnpike), canal and London and Birmingham Railway added to the overall feel of industrialisation (Figure 9.10). The only comparable landscape was perhaps that of the Lea valley, where the Lea navigation, the railway line to Hertford and Cambridge and (in places) the turnpiked Old North Road (Ermine Street) ran parallel, although the malthouses which formed the staple industry of the district were not, for the most part, as closely spaced as the paper mills of the west.

London's growing influence

The population of Hertfordshire, like that of other areas of England, grew rapidly in the first half of the nineteenth century – by 72 per cent, from 97,393 to 167,516, between 1801 and 1851.[63] But its distribution was changing. In the seventeenth century the east and, to some extent, the north of Hertfordshire were, as in the Middle Ages, the most densely settled districts, but by 1801 population was more evenly distributed across the county and by 1851 the earlier situation had been reversed. The south and west were now pulling away, with the growth of, in particular, Hemel Hempstead, Berkhamsted, Watford, St Albans and Hitchin, and also of places in the lower Lea valley, where the Cheshunt area boasted a density of well over 300 people per square mile. In part this pattern reflected the expansion of local industries, including straw plaiting and hat making, as well as paper making in the west and malting in the south-east; but the ease of communication with London was also a factor and, as transport systems improved through the 1850s and 1860s, the southern areas of the county – in and around the London Clay plateau, towards the Middlesex border – experienced rapid growth. New Barnet, for example, was nothing more than a few farms and cottages in 1851 and was not mentioned as a separate place in the census of that year.

By 1871 it had a population of nearly 8,000. The growing size and importance of London was reflected in other ways, too. In 1873 the Colne Valley Water Company was founded in Bushey, partly with the intention of supplying local customers but mainly to provide fresh water to the expanding capital, a pressing requirement given recent outbreaks of cholera.[64] As early as 1832 a scheme had been proposed to take water from the Colne, but it was abandoned owing to opposition from mill owners and others, as was a second scheme in 1840. The new project was likewise opposed, in spite of the fact that it involved only the taking of water from wells sunk in the valley floor, on the grounds that 'the operators of the company would inflict immense injury on the owners and occupiers of wells in the neighbourhood of Watford and Bushey, as well as on the owners and occupiers of mills along the whole course of the river'.[65] It should be noted that the Colne is now a much less substantial watercourse than it was then – in part because the fears of the company's opponents regarding the impact of large-scale extraction from the aquifer on the amount of water flowing in the river proved entirely correct.

Yet it is also important to emphasise that Hertfordshire, even the south of the county, was still largely rural, certainly by modern standards. The built-up area remained small, to judge from the evidence of the first edition Ordnance Survey 6-inch maps, surveyed in the 1870s and 1880s. Admittedly it is hard to quantify this in objective terms, as deciding where to draw the boundaries of towns, given such things as the extensive orchards and diminutive 'villa' parks which were clustered around them, is difficult, but in very general terms – including houses, gardens and industrial sites – there were only four towns in the county which extended over an area of more than a square kilometre: St Albans at *c.*120 hectares was the largest, followed by Hertford at 102 hectares, and Hitchin and Watford at around 100 hectares each. For comparison, this is not much larger than the area currently covered by larger suburbanised villages such as Redbourn or Wheathampstead. It was only after *c.*1870 that the urbanisation, and suburbanisation, of the county proceeded apace.

Notes

1. Niblett and Thompson, *Alban's buried towns*, pp. 178–96.
2. M. Winterbottom (ed.), *Gildas: The Ruin of Britain and other works* (London, 1978), p. 92.
3. M. Biddle and B. Kjolbye-Biddle, 'England's premier abbey: the medieval chapter house of St Albans and its excavation in 1978', *Hertfordshire's Past*, 11 (1981), p. 26.
4. Niblett and Thompson, *Alban's buried towns*, p. 145; Biddle and Kjolbye-Biddle, 'Origins of St Albans Abbey', pp. 45–77.
5. B. Colgrave and R.A.B. Mynors, *Bede's Ecclesiastical history of the English people*, revised edn (Oxford, 1991), p. 35.
6. Crick, *Charters of St Albans*, p. 16.
7. N. Doggett and J. Hunn, 'The origins and development of medieval Berkhamsted', *Hertfordshire's Past*, 18 (1985), pp. 18–36; Williamson, *Origins of Hertfordshire*, pp. 152–3.
8. Kiln and Partridge, *Ware and Hertford*, pp. 72, 142–3; HHER 356143; 356245; C. Partridge, 'Hertford and Ware', in Slater and Goose, *County of small towns*, pp. 140–41.
9. Crick, *Charters of St Albans*, no. 7; HHER 4327; 10902.
10. Wickham Hall: Gower *et al.*, *Place-names of Hertfordshire*, p. 203.
11. D. Hill and R. Cowie (eds), *Wics: the early medieval trading centres of northern Europe* (Sheffield, 2001); R. Hodges, *Dark Age economics* (London, 1982).
12. There is evidence for middle Saxon wharfs, elaborate in character, from the so-called 'Secret Garden Site' (Partridge, 'Hertford and Ware', p. 140).
13. *Ibid.*, pp. 142–5.
14. K.D. Lilley, *Urban life in the Middle Ages, 1000–1450* (Basingstoke, 2002), pp. 42–9.
15. Page (ed.), *VCH Hertfordshire*, vol. 3, p. 292.
16. S. Garfi, 'Appendix A: the Hertford Burghs: a survey of the evidence', in Kiln and Partridge, *Ware and Hertford*, pp. 115–27.
17. Smith, *English houses*, p. 137.
18. Page (ed.), *VCH Hertfordshire*, vol. 3, p. 308.
19. M. Bailey, 'The economy of towns and markets, 1100 to 1500', in Slater and Goose (eds), *County of small towns*, pp. 46–66.
20. D. Plowman, 'Royston: a thirteenth-century planned town?', in Slater and Goose (eds), *County of small towns*, pp. 173–87; Munby, *Hertfordshire landscape*, pp. 99–101.
21. Smith, *English houses*, pp. 138–40.
22. An important point made in T. Slater, 'Roads, commons and boundaries in the topography of Hertfordshire towns', in Slater and Goose (eds), *County of small towns*, pp. 67–95.
23. *Ibid.*, p. 74.
24. Munby, *Hertfordshire landscape*, pp. 99–101.
25. See the excellent summary in I. Thompson, 'Markets and fairs', in Short (ed.), *Historical atlas*, pp. 154–5.
26. Munby, *Hertfordshire landscape*, pp. 99–100.
27. Smith, *English houses*, pp. 138–40.
28. Munby, *Hertfordshire landscape*, pp. 99–100.
29. Lilley, *Urban life*, *passim*.
30. Slater, 'Roads, commons and boundaries', pp. 78–80.
31. HALS DP/26/29/2.
32. G. Longman, *Bushey in the Domesday Book*, Bushey Museum Information Booklet (Bushey, 1998), p. 8.
33. Page (ed.), *VCH Hertfordshire*, vol. 2, p. 183–4.
34. *Ibid.*, p. 185; G. Longman, *Bushey then and now: introduction*, 3rd edn (Bushey, 1986), p. 3.
35. Longman, *Bushey then and now*, p. 3; BL Cotton MS Otho D.3.
36. D. Short, 'Ashwell: an example of Anglo-Saxon town planning', in Slater and Goose (eds), *County of small towns*, pp. 165–7.
37. Adrian Gibson identified six examples of medieval shops in the county: at Abbot's Langley, Ashwell (two), Bishop's Stortford, Hitchin and Ware: A. Gibson, 'Timber buildings', in Short (ed), *Historical atlas*, pp. 24–5.
38. M. Bailey, 'A tale of two towns: Buntingford and Standon in the later Middle Ages', *Journal of Medieval History*, 19/4 (1993), pp. 351–71; Thompson, 'Markets and fairs'.
39. N. Goose, 'Urban growth and economic development in early modern Hertfordshire', in Slater and Goose (eds), *County of small towns*, pp. 105–6.

40. *Ibid.*, pp. 112–14.
41. Quoted in B. Howlett. 'Manorial estate and market town: the early development of Hitchin', in Slater and Goose (eds), *County of small towns*, p. 207.
42. Goose, 'Urban growth', p. 115.
43. Smith, *English houses*, pp. 155–6.
44. P. Plumb, 'Coaching routes, post roads and inns', in Short (ed.), *Historical atlas*, pp. 42–3.
45. Smith, *English houses*, pp. 170–74.
46. Munby, *Hertfordshire landscape*, p. 205 citing D. Defoe, *A Tour through England and Wales, 1724–6*, 1948 Everyman Edition, vol. II, pp. 122–3.
47. B.P. Hindle, *Roads, tracks and their interpretation* (London, 1993), pp. 64–6.
48. W. Albert, *The turnpike road system in England 1663–1840* (Cambridge, 1972); P. Langford, *A polite and commercial people* (Oxford, 1998), pp. 391–408.
49. Branch Johnson, *Industrial archaeology*, pp. 102–6; P. Plumb, 'Turnpike roads', in Short (ed.), *Historical atlas*, pp. 40–41.
50. J. Donovan, 'Milestones', in Short (ed.), *Historical atlas*, pp. 44–5.
51. T. Manning, 'Rivers and navigations', in Short (ed.), *Historical atlas*, pp. 46–7; Branch Johnson, *Industrial archaeology*, p. 116.
52. Branch Johnson, *Industrial archaeology*, pp. 118–22.
53. *Ibid.*, pp. 122–33; G. Boseley, 'Railways', in Short (ed.), *Historical atlas*, pp. 48–9; F.G. Cockman, *Railways of Hertfordshire* (Stevenage, 1983).
54. Branch Johnson, *Industrial archaeology*, pp. 129–30.
55. H. Williams, *History of Watford* (London, 1884), p. 111.
56. D. Perman, 'Malting and brewing', in Short (ed.), *Historical atlas*, pp. 88–9; Branch Johnson, *Industrial archaeology*, pp. 28–54.
57. Branch Johnson, *Industrial archaeology*, p. 33.
58. Only Kingsbury Mill in St Albans and Mill Green mill at Hatfield contain seventeenth-century fabric, concealed behind later façades.
59. Moore recorded 46 working windmills *c.*1862 but numbers then started to fall, with 35 in 1875 and just 9 by 1905 (C. Moore, *Hertfordshire windmills & windmillers* (Sawbridgeworth, 1999), pp. 14, 16).
60. M. Stanyon, 'Papermaking', in Short (ed.), *Historical atlas*, pp. 80–81; Branch Johnson, *Industrial archaeology*, pp. 55–61; A.J. Ward, *The early history of papermaking at Frogmore Mill and Two Waters Mill, Hertfordshire* (Berkhamsted, 2003).
61. J. Evans, *The endless web: John Dickinson and Co. Ltd 1804–54* (London, 1955).
62. J. Hassell, *A tour of the Grand Junction* (London, 1819), p. 30.
63. N. Goose, 'Population, 1801–1901', in Short (ed.), *Historical atlas*, pp. 56–7.
64. Colne Valley Water Company, *The Colne Valley Water Company 1873–1973* (London, 1973).
65. Williams, *History of Watford*, p. 104.

– 10 –

Suburbs and New Towns, 1870–1970

Introduction

In the century after 1870 the landscape of Hertfordshire, and especially the southern parts of the county, became increasingly urbanised and suburbanised. This process was already, in a sense, underway by the start of the nineteenth century, as genteel villas set in diminutive parks were already numerous here, and became more so in the course of the nineteenth century.[1] The enclosure of the great upland heaths towards the county boundary with Middlesex provided opportunities for erecting rather smaller residences. As early as 1810 an advertisement appeared in the *Morning Chronicle* for

> A desirable FREEHOLD ESTATE, delightfully situate near the Windmill Inn, on Bushey Heath, 11 miles from Tyburn Turnpike, in the county of Herts, on an eminence commanding a great extent of rich and interesting scenery, consisting of a large piece or parcel of enclosed ground, together with the handsome Cottage Villa and other buildings now erecting thereon, let on a building lease to Mr Woodward, for seventy years, at a rent of 100l [£100] per annum.[2]

But from around 1850, and increasingly from *c.*1900, much larger numbers of people whose occupations were based in London came to reside in the county, while larger towns experienced significant demographic growth and economic expansion.[3] Equally important was the fact that, as the centre of the capital became increasingly built up, a number of institutions and organisations which required large amounts of space for their activities also migrated out into south Hertfordshire, including mental hospitals (Leavesden near Watford, 1870; Napsbury, 1905; Shenley, 1934) and public schools (Royal Masonic Boys in Bushey, 1902; Royal Masonic Girls in Rickmansworth, 1933; Merchant Taylors, 1933; and numerous others). Such places ensured that there were green interruptions in the spreading streets, as did a number of public parks and private golf courses. Pockets of farmland were, moreover, only slowly eroded (some still remain to this day, even in the far south of the county), as suburbanisation did not spread in neat, concentric circles from London or the county's major towns, but was structured by transport systems and, in the period before *c.*1920, principally by the location of railway stations.

The first phase: 1870–1920

The kinds of landscape created by suburban growth in the period before the First World War are extraordinarily varied. At one extreme are streets composed of rows of terraced cottages – the homes of the lower middle or upper working classes; at the other are the more spacious roads of detached or semi-detached 'villas'. The houses erected on the former are generally of two rather than three storeys, in contrast to the broadly similar dwellings erected in districts closer to London, where land prices were higher: the main exceptions are in parts of Watford.[4] Here, as in parts of Ware and Hertford, development was sometimes on a fairly large scale, leading to streets of near-identical houses reminiscent of the Midlands' industrial towns. Indeed, much of the expansion of Watford in the period between the 1870s and the First World War was the work of William Gough, who came from Aynho in Northamptonshire.[5] In many cases, however – especially where villages such as Bushey or hamlets such as Radlett expanded in this period – what appear at first sight to be serried rows of identical houses soon resolve, on closer inspection, into several short rows of similar but not identical buildings. This is because as fields or groups of fields were sold off for development, and streets laid out across them, the various plots were often acquired by a small number of different builders who sometimes developed them at different rates, so that houses in one street might, in certain cases, have been erected over several decades. Either way, the architectural style of terraced rows in Hertfordshire, as elsewhere, changed over time. Those built in the first phase of suburbanisation, in the 1860s, 1870s and 1880s, often in close proximity to railway stations – such as many of those in 'New Bushey', later renamed Oxhey, to the south of Watford – tend to be rather plain and unadorned, with roofs of slate and walls of local yellow brick. Few boast bow windows or decorative detailing of any kind. By the 1880s more decorative features were being introduced. The houses built by Gough in Watford, for example, have bands of red brick breaking up the façades and sometimes picking out the date of construction.[6] Bow windows at ground-floor level were universal by the 1890s and, by the turn of the century, were often carried up to the second floor and given a separate hipped or gabled roof.[7] At the same time, windows and doors tended to receive more decorative treatment. Simple frames for windows and doors were sometimes replaced by gothic or Norman details, characteristically applied to the windows in the projecting bows. Streets like this continued to be built right up to the beginning of the First World War (Figure 10.1).

The detached or semi-detached dwellings erected for professional and upper-middle-class families, in more spacious streets and often at a greater distance from the railway

Figure 10.1. Typical early twentieth-century terraced development in south Hertfordshire.

stations, display more diversity. Many were individual, architect-designed dwellings. Those erected in the 1860s and 1870s were often neat rectangular brick villas with low roofs, simple detailing and full height bay windows; others were adorned with varying degrees of gothic decoration. But as numbers of new houses increased in the last quarter of the century a high proportion fell into two broad, if ill-defined, groups. Many were of three storeys and semi-detached, with heavy paired gables facing the road and a smaller range, including the entrance, to one side; the top floor provided accommodation for servants. The main range usually had a bay window at ground-floor level and either mock timber framing or tile cladding in the prominent gable above. Roofs were initially of slate, but increasingly tiled. Principal elevations were sometimes rendered but occasionally made use of high-quality red brick brought from the Midlands. Windows and doors usually took simple forms (although often with heavy wooden frames) but the overall outline of the house was often complex, with projecting windows at third-floor level, for example, while the roofs on the non-gabled range were often hipped and massive chimneys were a frequent feature. Many individualistic, idiosyncratic variations on these themes exist, sometimes described as the 'domestic revival' style, which can be traced back to the work of architects such as Phillip Webb, R. Norman Shaw and William Eden Nesfield, who sought their inspiration from the vernacular buildings of the Home Counties (principally Sussex and Kent).[8] Detached, oddly asymmetrical versions of such houses were also erected.

Figure 10.2. Aldenham Road, Bushey *c*.1910. On the right, houses erected in the late 1880s in vaguely 'domestic revival' style; on the left, pseudo-vernacular houses influenced by the designs of H.C. Voysey.

A rather different style, although again inspired by the vernacular architecture of the south-east, was derived from the work of Arts and Crafts architects such as Henry Voysey, who designed a number of houses in south Hertfordshire.[9] In these 'vernacular revival' houses mock timber-framing on the external façades was generally avoided, and long, low elevations were usually rendered in roughcast, although the overall style was of a vernacular cottage or farmhouse, with tiled roofs sweeping low, irregular plans, asymmetrical elevations and massive chimneys. Such houses were invariably detached: some were one-off, architect-designed buildings, using handcrafted materials and skilled builders and workmen, but the style was widely copied in smaller and simpler forms inspired in part by the houses erected, from 1902, at Letchworth Garden City in the north of the county (Figure 10.2).

Houses fusing both of these broad traditions as well as ones built in simpler, plainer styles can also be found, so that suburbs of this type usually display a considerable degree of variety, in part because they were often developed even more gradually, over a number of years or decades, than those comprising terraces. Land was split into numerous plots which were normally developed, in the manner described, by numerous different builders or by individuals who themselves commissioned a builder. Small firms could afford to build only one or two houses at a time, selling them on and using the capital to build more; individuals might wait several years before they had accumulated sufficient to erect the home of their dreams.[10] Even where larger areas of land were developed by single landowners who employed local builders to develop individual plots the process was often slow, in part because the land was purchased with a large mortgage, meaning that insufficient additional capital was available for large-scale construction. In addition, the houses erected by developers were often rented out rather than sold: the financial returns from building, which could be used to fund the erection of further houses, thus came in only slowly. It was relatively hard for middle-class people in the late nineteenth and early twentieth centuries to obtain a mortgage and most rented, rather than purchased, their homes.

All this said, the slow pace of development should not be exaggerated. Towns such as St Albans, Hemel Hempstead and Watford more or less doubled in size between 1860 and 1914, and many of the principal villages and hamlets in the south of the county grew significantly, especially those which lay – like Bushey, Radlett or Borehamwood – on rail lines or major roads leading into London. Most noticeable was the situation in the far south of the county, where Chipping Barnet and East Barnet were, by 1900, effectively connected to London. There were patches of countryside between, but along the principal roads leading to the metropolis the building line was more or less continuous. Yet it should be emphasised that suburbanisation was not simply to accommodate London commuters. Major urban areas around London were themselves expanding economically, and many of the new houses were to house workers in local industries. By the start of the twentieth century a marked contrast was thus developing between the still largely rural north and east of the county and the suburbanising south-west. The *Victoria County History* of Hertfordshire, published in 1904, describes how Barnet had become the terminus of the electric tramway to Highgate and many new houses had lately been erected there: in contrast, at Clothall and Bygrave, in the far north of the county, the medieval open fields still remained unenclosed.[11]

Pre-First World War developments, whether of terraces or of larger detached and semi-detached properties, generally covered relatively small areas of ground: it was rare for any single development to extend across more than *c.*20 hectares and, as they were created by different landowners, their boundaries often perpetuated those of the earlier field patterns. Road layouts were usually simple. Typically, one or more straight streets would be laid out between two more serpentine, pre-existing roads or lanes. Sometimes, to maximise the number of houses and plots which could be fitted into a field or groups of fields bordered by roads running at right angles to each other, a connecting road would run in two straight sections connected by a smooth curve.

Suburbs were not 'planned' in this period, in the sense that there was no 'authority' deciding which areas of the county could or could not be built over. The County Council was established in 1888 and District Councils replaced Sanitary Boards in 1894, but the involvement of local authorities in the planning process was mainly limited, in the period before the First World War and to a large extent after it, to the establishment and enforcement of bylaws which stipulated such things as the width of roads or building plots and the distance back from the street of the building line. Where to build was a matter for private decisions: even the layout of new roads was left entirely to developers. In this matter, as in much else, the involvement of the local authority – generally the Urban District Council – was essentially reactive. Once the properties fronting a road had been sufficiently developed the council, under the terms of the 1892 Private Street Works Act, could resolve that it should be 'sewered, levelled, metalled, kerbed, channelled, lighted and made good', and that the expenses of such work could be apportioned between those owning 'the premises fronting, adjoining or abutting upon the said street'.[12] The necessary work was then put out to tender and, when completed, the road would be formally adopted and henceforth maintained by the council at the expense of ratepayers. Often, given the pace of development, there was a considerable time lag between a street being laid out and the construction of the first houses, and its upgrading in this manner. Local sewers fed into the existing network, which was gradually expanded by the council, often with disastrous effects on local watercourses. The river Colne, for example, was so polluted by the early years of the century that some local councils were paying compensation to angling societies and landowners for damage wrought to fish stocks.[13] The supply of both water and gas was also often haphazard and *ad hoc* in character, with private companies and local authorities carving out their own supply networks in a piecemeal manner. Companies such as the East Barnet Gas and Water Company, established in 1867, the St Albans Gas and Water Company, created in 1852, and the Colne Valley Water Company, set up in 1873, steadily expanded their activities as new housing estates spread through the south of the county. Electricity companies, including the North Metropolitan Electrical Power Distribution Company, followed suit from the turn of the century, although the spread of electricity for domestic use and street lighting was slow and in general it was only in the 1920s that supply was provided to urban and suburban areas, the more remote rural areas of the county sometimes having to wait until after the Second World War.

Although the involvement of public bodies in planning in this period was limited, private developers often considered the size and distribution of building plots carefully, and in one case development was indeed planned on a very large scale: the 'Garden City' of Letchworth, which represents one of the most important parts of Hertfordshire's landscape heritage. In the nineteenth century a number of philanthropic industrialists, appalled by the poor housing and living conditions in the expanding cities of the industrial north, had created planned settlements to house their workers, such as Bourneville, established by the Cadbury family, and Port Sunlight near Liverpool, created by William Hesketh Lever, later Lord Leverhulme.[14] In addition, a number of planned villages had been established in the area immediately around London, such as Ilford, with accommodation for 6,000 people. These ideas were taken further by the writer and social theorist Ebenezer Howard, whose book *Tomorrow: a peaceful path to real reform* was published in 1898 and republished as *Garden Cities of Tomorrow* in 1902.

Howard believed that if large-scale urbanisation was not to lead to social unrest then new, planned settlements, set in the countryside and away from existing urban squalor, would need to be created:

> My proposal is that there should be an earnest attempt made to organise a migratory movement of population from our overcrowded centres to sparsely settled rural districts; that the mind of the public should not be confused, or the efforts of organisers wasted in a premature attempt to accomplish this work on a national scale, but that great thought and attention shall be first concentrated on a single movement yet one sufficiently large to be at once attractive and resourceful.[15]

In 1900 the Garden City Association was formed and in 1902 the Garden City Pioneer Company was established, to be succeeded by the First Garden City Ltd. Some 23,000 acres were purchased in the parishes of Letchworth, Norton and Willian, and the following year construction of the new city began.[16] The area lay in the 'champion' north of the county, and settlement at the time was concentrated in the three villages, which in 1901 had a combined population of only 566 people. These places lay on the periphery of the new city, which was established on a relatively level site – below the Chalk escarpment and to the south of the railway line to London – which was crossed by three roughly parallel roads. The original plan was for a built-up area of 1,250 acres, with 2,500 acres reserved as a protective belt of farmland. The architects were Raymond Unwin and Barry Parker, and the fact that the former continued to be involved in the project until 1914 and the latter until 1943 contributed greatly to the overall coherence of the design. Their plan for the new town was based on wide tree-lined avenues, from which led more intimate, winding, tree-lined roads (many of which were *cul de sacs*) that widened into or terminated at small 'greens'. The most important axis was Broadway, which ran in a straight line south-west from the new railway station and widened to form the Town Square.[17] According to Unwin, the alignment of this axis was established because of the positions of three old oak trees, and it is said that only one tree was felled during the town's construction. The population of the new town began to grow, in part through the publicity generated by the 'Cheap Cottages Exhibitions' (sponsored by the *Daily Mail*), which, from 1905, enabled architects to display their ideas for affordable houses for labourers and which attracted over 60,000 visitors to the town. Initially, the First Garden City company was responsible for almost all aspects of urban life, including the supply of electricity and water, as well as leasing houses, the rents for which were then used, in part, for community improvements. The Garden City was a social experiment as much as a pioneering exercise in town planning and is said to have attracted a variety of idealists – or cranks, depending on your point of view. There was no public house except a teetotal establishment called The Skittles until the ban on the sale of alcohol was lifted in 1958 (existing pubs in the villages of Willian (The Three Horseshoes) and Norton (The Fox) thus flourished).

The houses, designed by Parker and Unwin in a loosely Arts and Crafts, 'cottage' or Tudoresque style, generally have elevations of roughcast, with dormer windows and steep gabled roofs with low eaves, sometimes tile-clad (Figure 10.3). Most were provided with

Figure 10.3. Letchworth Garden City, showing typical cottage-style 'vernacular' houses.

three bedrooms, with a bath in the scullery and a covered way leading to an external toilet. Each had gardens front and rear, and care was taken to prevent houses being overlooked by others, leading to a varied and irregular frontage to many of the streets. A variety of facilities was planned from the beginning – open spaces, libraries and shops – as were factories, to provide employment. Various companies were attracted – after an initially slow start – with cheap rents and subsidies, including the British Tabulating Machine Company and the Spirella corset manufacturing company, with its fine Art Deco style factory, completed in 1920. The modest early growth of the town – there were still only 10,000 people here in 1919, although 14,000 by 1930 – partly explains the rather meagre provision of public buildings, which are mostly concentrated on Broadway and the Town Square.[18] Many of these were erected in the inter-war period, or even after the Second World War, and those that date from the early years are mainly in a Neo-Georgian style rather different from that of the domestic dwellings, although there are exceptions (such as the Jacobean Post Office). The shopping streets are also rather uninspiring, but the design of the residential areas was innovatory. As Pevsner noted, the winding streets, areas of open grass and retention of mature trees all became normal aspects of town planning in the course of the twentieth century: 'to later generations they have thus become a matter of course, but it should not be forgotten that they were for the very first time systematically followed at Letchworth'.[19]

Inter-war expansion

In the twenty years from 1919 to 1939 the pace of suburbanisation in Hertfordshire accelerated, with large areas of the county's farmland lost to housing, industry, parks and playing fields. In this the county's fate was similar to that of other areas in the vicinity of London, or of many large cities in the Midlands: it has been estimated that, in England as a whole, around a third of the modern housing stock was erected in these twenty years.[20] A number of factors influenced this accelerating pace, as well as the character, of urbanisation and suburbanisation. One was the fact that mortgages became much more easily available: houses could therefore be sold quickly, rather than rented out, thus increasing the return on capital and allowing more rapid development on a larger scale. Another was changes in the character of transport. Car ownership rose steadily. Many roads were properly surfaced for the first time and, as traffic problems escalated, a number of major new roads were constructed, including the Watford and Barnet bypasses in the 1920s and the North Orbital in the 1930s.[21] The latter was to be an outer London ring road and was a kind of precursor of the M25, in part constructed by linking lengths of existing road. While it was never fully completed, the modern A405 between the Noke roundabout and the Park Street roundabout south of St Albans, as well as the A414 from there to the A1 junction, were newly built as part of the grand scheme. There were also further improvements in the rail network. A third important factor was the increasing involvement of local government in the provision of houses for the working classes. Legislation to encourage the erection of council houses had begun in 1890, with the Housing of the Working Class Act, which allowed local authorities to apply for central government funding to erect houses. But few Urban or Rural District Councils made use of this and it was only in 1919 that the Addison Act made it obligatory for councils to build houses. Councils were expected to put a penny on the rates to fund housing provision but the shortfall, after account had been taken of income from rents, would be made up by central government. The Housing Acts of 1923 and 1924 resulted in a government subsidy being paid on every house built, whether by local authorities or by private landowners: intended as a way of encouraging the provision of houses for workers, its main effect was to encourage building for the middle classes.[22]

To all this we need to add continued population growth in the south of England, an area largely unaffected by the inter-war depression in heavy industry. Indeed, with the extension of the electric grid the period saw what Hudson has described as a 'Second Industrial Revolution' based on light engineering and technology.[23] In this phase, even more than in the pre-war period, suburbanisation was associated not merely with the practice of commuting to London, but with the growth of manufacturing in the towns close to it. Some of these were businesses which required ample space for their activities and had been forced out of the capital by its continual outward expansion, most notable among them aircraft manufacture. Handley Page moved from Cricklewood to Radlett (Colney Street) in 1930; at the same time, de Havilland moved from Edgware to Hatfield.[24] But many other light industries grew up in the southern towns. It was in this period that printing (especially colour printing) became a major occupation in Watford, with the creation in 1918 of the Sun Engraving Co. through the amalgamation of Andre and Sleigh, the Bushey Colour Press and the Anglo-Engraving Company; in 1934 the Rembrandt Intaglio Printing Company was taken over by them and moved to Watford; while in 1936 Odhams built their factory there.[25] The film industry at Borehamwood, after a slow beginning before the First World War, expanded significantly in the inter-war years, with British National, Imperial, British International, Amalgamated and other companies all developing in the area.[26] New industries needed workers and workers needed houses, and this, as much as direct commuting to the capital, swelled the towns of south-west Hertfordshire.

As in the period before the First World War, development did not take place to the same extent everywhere. Transport and access, now involving road as much as rail, were major influences in its distribution; so too were patterns of land ownership, with some of the most extensive areas of suburbanisation being associated with the break-up of large landed estates, a particular feature of this period in England as a whole but especially in Hertfordshire, where many major landowners, faced with the rising tide of suburbanisation, preferred to sell up and move elsewhere. In addition, spatial planning now began to have some influence on development. In 1909 The Housing and Town Planning Act established that District Councils could prepare planning schemes for any land in the course of development, or likely to be used for building, in order to ensure proper sanitary conditions and to prevent harm to owners of neighbouring properties.[27] The Housing and Town Planning Act of 1919 made such schemes compulsory for places where the population was above 20,000. In the 1920s and 1930s Joint Town Planning Committees were formed by many District Councils and regional plans of a non-statutory nature were drawn up. The Town and Country Planning Act of 1932 extended planning schemes to rural land. However, it was only effective to a limited extent in controlling development. Local authorities were not obliged to draw up such schemes, and while many in Hertfordshire rapidly did so – including Bushey and Watford Rural in 1933, Ware in 1934 and Hertford and District in 1938 – the act instructed that they had to compensate owners for any financial losses incurred by a refusal of permission.[28] The main impact of such schemes was in establishing tighter regulations concerning road layout and width and housing densities, rather than in deciding where development should or should not take place: sale particulars for development land drawn up after 1932 often note that particular parcels had been 'zoned under Town Planning to permit erection of x houses'.[29] There was also an attempt to control low-density sprawl through the Restriction of Ribbon Development Act of 1935, which made all new building within 220 feet of classified roads subject to the control of the local authorities – although, again, its effects were limited.

Inter-war suburbs take a number of distinct forms. The first, perhaps most familiar and, in terms of area covered, the most important, are the ranks of superficially identical 'semis' – semi-detached dwellings of the kind found throughout the outer London suburbs

Figure 10.4. Typical inter-war 'semis' in south Hertfordshire.

(Figure 10.4).[30] These in fact take a variety of forms. Many have simple Tudoresque mock timber-framing in twin gables facing the road above curved or squared full-height bows; a hipped roof to the main block of the house; doors away from each other for added privacy; a central shared chimney stack; and gardens front and back. Variants include simpler, less adorned varieties; examples with twinned swept gables facing the road, a form ultimately derived from Voysey; and, occasionally, houses in the 'moderne' style, with curved 'suntrap' windows, influenced by continental Bauhaus architects such as Le Corbusier or Gropius.[31] All across the south of the county the developments on grids of streets first laid out around the turn of the century, but still not filled at the start of the First World War, were now rapidly completed, in part at least, with such houses. Extensive new areas of fields were similarly built up, often creating larger and more uniform arrays of dwellings than those usually found in the pre-war years, in part because the scale of development – the area being developed by a single builder – was now often larger, but also because, even when a number of developers were responsible, development was often rapid. Large areas of such houses appeared on the outskirts of towns such as Watford (especially to the north) and St Albans, where Beechwood Avenue and the roads around it, on the eastern edge of the town, have an appearance little different from that of the suburbs of north London. Many of the larger villages in the south of the county continued to expand. Roads were still usually laid out on a grid within frameworks provided by existing lanes and field boundaries, but their layout was less rigid and parallel and more complex in character, often featuring, for example, parallel but curving roads. Those built in the 1930s began to feature *cul de sacs* other than, as was the case earlier, those simply dictated by the need to fit a road within a particular block of property. The improvements in road transport – the spread of buses, as well as increased car ownership – ensured that housing could spread far beyond the railway stations or the main roads to London, to which earlier housing had largely been limited. There was much 'ribbon development', in which houses were simply built all along the length of existing roads, economising on construction costs: 1920s and 1930s-era houses thus spread along the long co-axial lanes found in the Gade valley in places such as Abbot's Langley or along the A10 at Cheshunt, while throughout the south of the county small hamlets such as London Colney, Park Street and Chiswell Green grew, partly as a consequence of the opening of the North Orbital. Small clusters of 'semis' appeared in many places within easy reach of major roads, and can still look incongruous facing out onto open countryside. This said, in the areas of the county more distant from London there was still a tendency for new houses to be concentrated close to railway lines, and small developments of this kind often erupted in the midst of fields beside railway stations in more remote parts, as on the hill overlooking Apsley Station near Hemel Hempstead, although this was also stimulated by the continued vitality of the Apsley paper mills. Such developments often include the lower-density sprawl of bungalows – simple boxes with roughcast walls, occasionally embellished with a beamed gable – accompanied by cheap detached versions of the standard 'semi', and interspersed with semi-detached houses of more normal form.

At the other end of the social scale, more up-market developments comprised houses that were generally better built and more individualistic in character than the kinds of dwellings just discussed, although drawing on a similar range of styles. They often boasted 'Jacobethan' details and bow windows, but were also more likely to exhibit features derived from the continental Bauhaus, or elements of Neo-Georgian design. Some were larger versions of the standard 'semi', but many were detached; a few could boast integral garages. Brick tended to be more exposed, the use of roughcast and pebbledash more restrained. Although Tudor details and pebbledash make some of these buildings similar to those of equivalent status erected before the First World War, there are subtle differences. With domestic service declining, few were provided with three full floors; few if any have slate roofs, tile being now universal; features such as tile cladding largely disappeared; and gables facing the road were no longer heavy and overhanging, and in many cases were given hipped roofs.

Houses like these could be found in small numbers on the edges of towns and villages throughout the south of the county, often slotted into undeveloped areas of the grids laid out in the pre-war period. But larger developments, especially in areas more distant from

London, were still closely associated with railway stations. Harpenden, for example, saw extensive development, especially around the margins of the common. Indeed, while in the period before *c.*1920 the immediate vicinity of railway stations on the edges of the larger towns were usually developed as streets of lower-middle and working-class houses, in the inter-war years there was a marked tendency for large up-market developments to be created beside stations in more rural areas – 'halts' and stations serving small villages. Several were laid out on or beside the sites of large mansions that had been demolished or converted to new uses by new owners.

The Gaussen family of Brookmans Park in North Mymms were, like many of their fellows, in increasing financial difficulties in the early years of the century and sold off their estate in stages, disposing of the final 969 acres (for £43,000) in 1923. The purchasers formed a company, Brookmans Park (Hatfield) Ltd., to develop the area for housing, and set about providing two key features which were required to attract middle-class commuters: a golf course was planned in part of the former park and an agreement was reached with the LNER in May 1925 by which the latter agreed to build a new station, with the company agreeing to pay the LNER £500 per annum until 500 houses were built.[32] Brochures made much of the fact that the estate was only 13 miles from Kings Cross, which could be reached by train in less than 40 minutes.[33] By 1927 a dozen new houses had been built, but as interest rates rose sharply at the end of the decade development was slow, and in 1929 the vicar counted only 117 houses on the estate. The following decade saw more rapid development: by 1933, when the company was dissolved and the land divided between two main developers, around 200 homes had been completed. Nevertheless, while building had progressed along much of Brookmans Avenue and Bradmore Way, and on the northern side of Moffats Lane, only the eastern end of Mymms Drive had been developed, and Westland Drive and Oaklands Avenue (to the south of the station) and The Grove had not yet been laid out. The boom years of the mid-1930s, however, saw these roads completed and by the time war was declared in 1939 around 550 houses had been completed or were under construction. This was in spite of conflicts with the council following the introduction of tighter planning controls in 1932, one of which involved the objections of residents in George's Wood Road in 1926 to the erection of bungalows, which they claimed would detract from the value of their homes and generally lower the tone of the place.[34] The war suspended building activity, and only in its aftermath were the existing frontages fully built up and Calder Avenue and the other remaining roads on the eastern side of the estate laid out.

Moor Park in Rickmansworth was in many ways similar. Following the death of Lord Ebury in 1918 his heir, Robert V. Grosvenor, put the estate up for auction and the mansion and its park were acquired by Lord Leverhulme.[35] The house was converted into a country club; part of the park was already in use as a golf course, laid out in 1880 for Lord Ebury, but this was now extended as an attraction for commuters, most of the land being planned as an up-market development.[36] There was already a golf course immediately to the east at

Sandy Lodge, established by James Markes and opened in 1910,[37] and also a small railway station on the Metropolitan Line, Sandy Lodge Halt, which was located between this and the proposed new development. The layout for the new estate was originally designed by the prominent landscape architect T.H. Mawson (Figure 10.5), but little of his plan was actually executed and large areas originally proposed for development were used for the golf course, with others being sold off to build Merchant Taylors, one of the many private schools that moved out from London to more spacious premises in this period.[38] Here, again, golf and fast rail connections to London were the key points emphasised when development began in 1922. An undated brochure described how 'there will be few better inland courses than those now being constructed at Moor Park, under the direction of Mr H.S. Colt, the noted golf course architect'. The brochure also observed that Marylebone and Baker Street Stations could be reached in less than thirty minutes and emphasised that 'there is no more romantic or delightful spot combining at the same time those urban amenities which are rightly regarded as necessities of modern life to be found within equal distance of London'.[39]

Moor Park was not the only up-market development to have been stimulated by the existence of the Metropolitan Line. This had been built as far as Rickmansworth in 1887 and reached Chorleywood in 1889, eventually extending as far as Aylesbury by 1892, but it was the progressive electrification of the line from 1905 which particularly stimulated development, together with the construction of the branch line to Watford in 1925. Some building was undertaken directly by the Metropolitan Railway Company itself, keen to sell not only houses but season tickets, and the charms of living in the countryside were heavily promoted in the annual guide, *Metroland*, published between 1915 and 1932.[40] The term 'Metroland' rapidly came to be used more generally for the expanding suburbs of north-west London and, more especially, for the developments (often involving a lower density of housing) further out, in south-west Hertfordshire and south Buckinghamshire: that mixture of town and country beloved by the poet John Betjeman (who described Chorleywood in 1973 as 'essential metroland', although much of the development here had in fact taken place before the First World War).[41] Development along the line took a number of forms. Large areas of semi-detached houses and bungalows appeared in Croxley Green and, following the completion of the Watford extension, in the neighbouring areas of west Watford. But not far away were more up-market developments, at Chorleywood and Loudwater and at Cashiobury, where the Cashiobury Park Estate in Watford was built on land sold from the Cashiobury estate in 1909, part of which was used to create Cashiobury park, the first public open space in Watford (and further extended in 1930, following the final demolition of the mansion itself in 1927). In such places land was sold subject to restrictive covenants which stipulated that only good-quality detached or semi-detached houses could be erected. By the late 1920s development was spreading far into the Chilterns, with suburban expansion around Berkhamsted and the surrounding villages. Fears that the whole of the

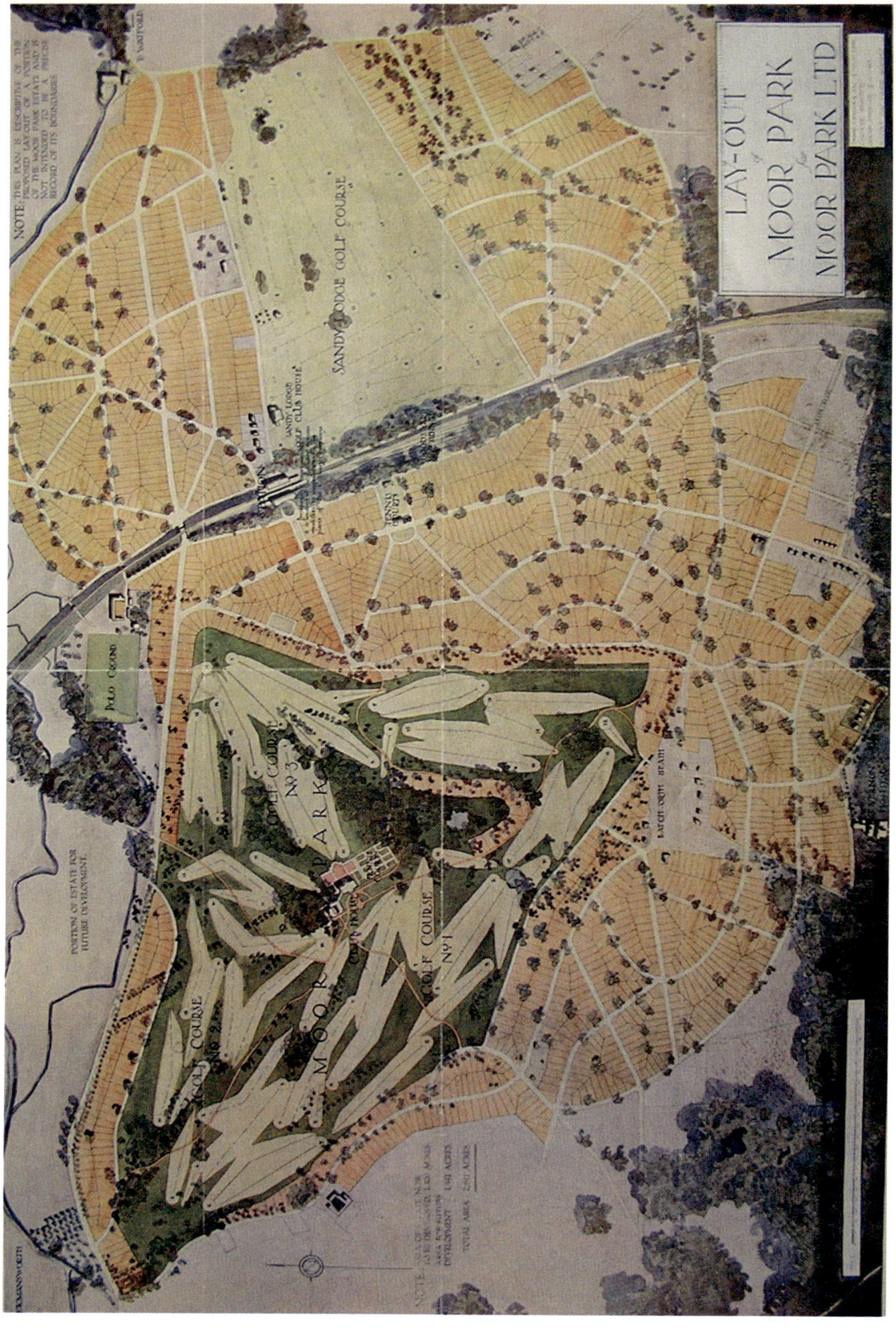

Figure 10.5. T.H. Mawson's design for the upmarket housing estate at Moor Park, c.1921.

Ashridge estate might be developed following its sale led to a fund-raising campaign that eventually acquired around 4,000 acres here for the National Trust.

Council estates are another major feature of the inter-war period. In the immediate aftermath of the Addison Act a number of relatively small developments were erected, often dropped in between the spreading streets of private housing, like the 66 houses erected around Herne, Melbourne and Clapgate roads in Bushey in 1921 and 1922. They usually comprised pebbledash dwellings with hipped roofs arranged in simple rows or as semi-detached pairs – stripped-down versions of the classic inter-war 'semis' – and erected on simple grids of roads. Some council developments, however, were larger and more ambitious in character, such as the Harebreaks estate in Watford, with its complex network of curving and straight roads and small squares and greens that was clearly inspired by the design of Letchworth.

All these developments, while often rapidly built and covering more ground than those of the period before the First World War, were still relatively small-scale and piecemeal in character. Rather different was the county's second Garden City, Welwyn. Ebenezer Howard was seventy when he began this new venture, which was built on fairly level land on either side of the Great Northern Railway to the south of the existing village of Welwyn. Most of the land (1,690 acres), part of the Panshanger estate, was purchased from Lord Desborough in 1919; a further 689 acres was acquired from Lord Salisbury. A number of plans for the new 'city' were drawn up before that by Louis de Soissons, a Canadian-born architect, was selected. It was based around a long formal axis – Parkway – running parallel with the railway, which had a central strip of grass and flanking trees (Figure 10.6).[42] At its northern end was a subsidiary east–west axis, Bridge Road, beyond which lay a formal termination in the form of a semi-circular grass 'exhedra' flanked by what became the civic centre. The shape of this feature was partly determined by the curve of the branch line to Dunstable, just to the north, now closed. Parkway was, and is, a great formal approach, flanked by two churches and rather continental in flavour. The main area of shops developed only slowly through the 1920s owing to the fact that the Welwyn Garden City Company initially allowed only its own Welwyn Stores to operate: it eventually occupied the area to the east of Parkway as far as the railway line. To the west and north of the city centre lay residential districts. To the east of the railway line were areas reserved for industry and business (the Shredded Wheat factory was long a familiar sight, glimpsed from railway carriages) but further east were other areas of housing – something of a planning mistake, as a residence on the 'wrong side of the tracks' implied social inferiority. The houses were less 'vernacular' in character than those at Letchworth, more Neo-Georgian in style, as were the shops and public buildings, many of which were designed by de Soissons or his associates.[43]

The construction of Welwyn Garden City, and Letchworth a little earlier, served to urbanise and suburbanise to a significant extent parts of Hertfordshire lying at some distance from London. Nevertheless, the authors of the 1941 Land Utilisation Survey for Hertfordshire were still able to emphasise the contrast between the north-east and

Figure 10.6. Aerial view of the centre of Welwyn Garden City: Parkway runs from top to bottom of the picture, with the main area of shops to the right (east). Courtesy English Heritage.

the south-west of the county. They described how Watford and Barnet were now 'largely dormitory areas for London in which the dwellers owe little allegiance to the old centres on the fringe of which they now reside', while other market towns in the west had 'entered a new phase: fast and efficient transport to London and to the Midlands combined with a local labour supply have attracted many light industries and much of south-west Hertfordshire has thus become greatly industrialised'. In contrast, factories and dormitory settlement were still 'less marked in the eastern two-thirds of Hertfordshire'.[44]

The expansion of suburbs in the 1920s and 1930s, together with improvements to the road network, had a number of other effects upon the landscape. In particular, they demanded large quantities of concrete, and thus of sand and gravel.[45] Much of this was excavated from low-lying deposits in the valleys of the Colne and the Lea, creating extensive areas of lakes, some of which were subsequently used for boating and swimming – the Rickmansworth 'Aquadrome' – and which more generally became important areas for wildlife in a county in which large areas of open water were in short supply.

The character of suburbanisation

The way in which suburbs developed was endlessly varied, but one vignette – from Bushey, to the south of Watford – may serve to demonstrate some typical patterns. The *c.*800-acre Bushey Hall estate, associated with a grand mansion built as late as 1865, was first placed on the market by its owner, Edward Marjoribanks, as a single lot in 1874. As no buyer was found it was back on sale again in 1877, this time as nine principal lots. This sale likewise proved abortive, and the following year it was up for auction for a third time, now as 17 principal lots (Figure 10.7). Only 60 acres were sold and the remaining 746 were back on the market in 1880.[46] This time the sale was more successful. Although the bulk of the property – the home farm and Bushey Hall Farm to the north of the estate – remained unsold, the hall itself and 240 acres were disposed of as one lot, and 36 acres on the west side of Aldenham Road, bounded to the north by Bushey Hall Road, were acquired by one Frederick Charsley. The following year he auctioned this off as 124 separate building plots, laying out 'Bushey Grove Road' to link the two existing roads: because these ran roughly at right angles to one another the new road followed, in part, a smooth curve. The sales particulars emphasised that 'there are 28 Up Trains daily from Bushey Station, and 35 from Watford Junction; 84 Down Trains to Watford, and 22 to Bushey'. The area was described as 'still quite rural' and much was made of the good-quality water supplied by the nearby Colne Valley Works.[47] The progress of development was slow, however, and when the second edition Ordnance Survey 6-inch map was surveyed in 1897 less than half the plots had been built on. By this time, the area immediately to the south-west had also begun to be developed around a simple grid of roads typical of the period, although, because the site was hemmed in by the adjacent land of the Colne Valley Water Company, there were a number of *cul de sacs*.

In 1881 the remainder of the Bushey Hall estate had been sold and the 146-acre Bushey Grove Farm, the home farm lying to the east of the Aldenham Road, was likewise acquired

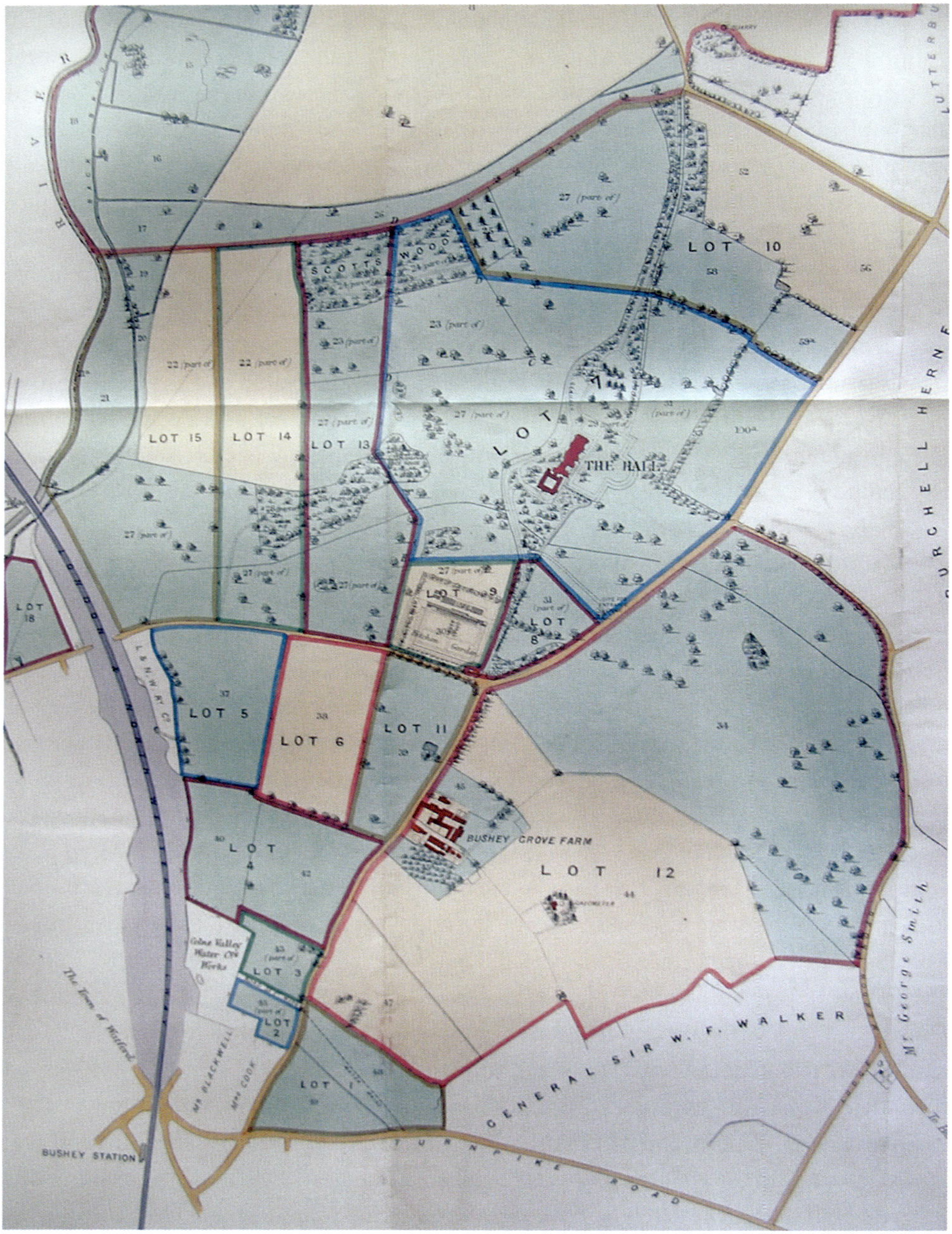

Figure 10.7. Much suburbanisation followed fast on the heels of the break-up of large landed estates: sale catalogue for the Bushey Hall estate, 1878.

Figure 10.8. The complexity of suburban development: the 'Bushey Grove Estate', showing chronology of housing and boundaries in the period up to *c*.1960.
1. Features surviving from the pre-suburban landscape (established by *c*.1880); 2. Houses and boundaries established by *c*.1900; 3. Houses and boundaries established by 1910; 4. Houses and boundaries established by 1920; 5. Houses and boundaries established by 1940; 6. Houses and boundaries established by 1960. Sources: Ordnance Survey 6-inch maps, 1910 Finance Act maps.

by Charsley. Instead of dividing it into building plots, however, he sold it entire in 1883 to Robert Palmer Tebb, who lived in Sussex.[48] The pace of development on what came to be known as the 'Bushey Grove Estate' was much slower and, as a result, the whole texture of the modern landscape is different, more varied (Figure 10.8). This was in part because Tebb initially undertook much of the development himself and, as the estate was heavily mortgaged, different portions could be built on only as they were redeemed, the proceeds paying for further releases and development. By 1885 he had erected five houses along Aldenham Road, all rented rather than sold. In 1887 he attempted to sell a substantial block of the estate, the north-eastern portion of 65 acres, which was put up for auction as 'Bushey Grove Park' (an invented name). Potential buyers were informed that the land was an admirable site for a new mansion but that it would also make a 'highly important freehold building estate' as there was an 'urgent demand for superior class residences by numbers of those who have become acquainted with the attractions of this charming position so easy of access to the Metropolis'.[49] In spite of such attractions the property failed to sell, and was eventually bought by the Masons as a site for a new boarding school, set in spacious playing fields, in 1896.

During the late 1880s and 1890s Tebb erected four further pairs of substantial semi-detached residences along Aldenham Road as well as a single detached house: the latter, and two of the semi-detached dwellings, were sold but the others were again rented out, for £50 per annum.[50] In addition, Tebb made money by selling, at intervals, plots along the road to various individuals and building companies. Yet by the end of the century only two of these had been built on, together with two further plots on 'The Avenue', a fine new road lined with lime trees that was laid out running back from Aldenham Road. Other new roads, running back from Aldenham Road parallel to this and terminating at a new 'Grange Road', were also planned in the 1890s, forming the familiar grid pattern characteristic of the period, although these were only gradually laid out during the first two decades of the next century. Tebb certainly had grand plans for the area, which included the construction of a substantial church, a grand gothic brick edifice, which was never erected.[51] But not all of the land was developed for large middle-class dwellings. In 1891 he secured the release from the mortgage of a block of nine acres in the south-west of the estate, closer to the railway station and to the main road to London, and here small plots were sold off to developers and terraced rows soon began to appear on what became 'Highfield' and 'Hillside' roads. Most of these plots had been built on by 1910, as had a number – this time occupied by semi-detached villas – on Belmont Road.[52] A map drawn up soon after shows that Tebb's whole estate was divided into plots which were 'zoned' in such a way that the smaller ones lay to the south-west, the medium-sized ones in the centre and the most extensive to the north-east.[53] Nevertheless, at this time – some 17 years after Tebb's purchase of the land – a large number of plots remained empty. As we have emphasised, such a slow pace of development, while not universal, was not unusual. Nor was the fact that development tended to occur from the outside of the estate inwards, with the earliest houses being erected on Aldenham Road and Grange Road, and towards the ends of the roads running between them, a pattern probably reflecting the fact that sewers were laid along the principal roads first, greatly facilitating development. On the west side of Aldenham Road, where the land had been divided up by Charsley in 1881 and sold to a number of different individuals, the pace of development was faster, although still relatively slow by modern standards. The street frontages were largely filled by the outbreak of the First World War in 1914, many with houses by the local builder James Rodwell. In 1900 a new church, St Paul's, a chapelry of the parish church, was erected on Bushey Hall Road.

Tebb died in 1907 and his heirs appear to have abandoned the policy of building houses for rent – a practice which was generally falling from favour at this time as the greater availability of mortgages increased the number of people able to purchase houses. In 1909 all the remaining family property was released by the mortgagors and many of the plots were immediately sold: the remainder, 157 in all, was disposed of in 1912, together with a number of the houses which Tebb had erected in the 1880s and 1890s.[54] The sales catalogue emphasised that there was 'a large and increasing demand in the District for good class medium sized residences', but further development was delayed by the advent of war and, subsequently, by the hard economic situation of the early 1920s.[55]

When development resumed, in the mid-1920s, it took two main forms. The medium-sized plots in the centre of the estate were built up mainly with houses of typical inter-war semi-detached form; the larger plots towards the east with detached, more individualistic houses, some already with integral garages. Later developments included, in the post-war period, the subdivision of the largest plots and infilling with 1960s and 1970s housing, in one case involving the construction of a small *cul de sac* in a former spacious garden. The result of this long, slow pattern of development, coupled with the way that Tebb divided up the estate into plots of different sizes, ensures that this small area of suburbia displays a remarkable range of architectural styles. There are cottages built for farmworkers before development commenced; spacious late nineteenth-century houses erected, often to individual architects' plans, in a range of Arts and Crafts and pseudo-vernacular styles; terraces of similar date; inter-war 'semis'; and more affluent inter-war and post-war residences. This is in contrast to the land west of the Aldenham Road, where the streets are largely occupied by semi-detached villas erected between 1885 and 1910. It should be noted that 'mixed' multi-period suburbs like this are an important and under-researched aspect of Hertfordshire's landscape history; many towns, not only in the south of the county, can boast at least one example. The houses set out along Highfield, Morgans and Queens Roads in the south of Hertford are another good example, although there, as in many other cases, there were slight differences in the trajectory of development: the subdivision of large Victorian gardens in the post-war period was a more important factor in generating the diverse chronology of the housing stock.

The other parcels of land which were created by the break-up of the Bushey Hall estate had a variety of fates. Some plots to the west of the Aldenham Road were acquired by the Colne Valley Water Company to allow its premises to expand. Fifteen acres of land in the far south had been purchased by George Bassett in 1878: closer to the railway station and located beside the main London Road, it was developed more rapidly than Tebb's properties, being divided up into narrow plots and rapidly covered in terraced houses and light industrial units. Other sections of the estate, lying to the north of the Bushey Hall Road and close to Watford, had likewise been divided into roads and built up, again mainly with terraces, by 1900. The hall itself, not an old building when sold in 1879, became a 'hydrotherapeutic establishment', later evolving into a hotel of more normal form, and survived until demolition in the 1950s. In 1889 the surrounding parkland became – typically – a golf course; by 1893 a full eighteen-hole course had been laid out and a small club house erected. People staying at the hotel were permitted to use the course, alongside club members, and from the start of the twentieth century other visitors were allowed to play on a daily basis, and offered a special voucher which included both the green fee and a rail ticket to Bushey and Oxhey Station or Watford Junction. In 1907 a new club house was erected (at a cost of £1,100): this still survives, now a listed building. Lastly, the northern section of the estate, comprising Bushey Hall Farm, survived as agricultural land until the 1930s, when development began with the laying out of Mill Way and Park Avenue; interrupted by the Second World War, building resumed in the area in the post-war period, creating a network of semi-detached houses. Subsequently, in the 1950s and 1960s, further development occurred to the east, partly residential but partly a large industrial estate beside the Watford Bypass, built in the 1920s. Parts of the old estate, however, have survived as open land to this day.

This area of Bushey, a small slice of superficially uninspiring landscape, demonstrates well the rich and complex development of many south Hertfordshire suburbs. Of particular interest are the way that the pattern of initial divisions of the property, themselves often following more ancient lines in the landscape, structured subsequent patterns of land use; the variable but often slow pace of development; the use of land for large institutions, such as schools, forced out from inner London; and the way that elements of the old countryside – field boundaries, roads, even some woods – managed to survive within the new landscape.

The post-war years

The Second World War, and the shortage of building materials in its immediate aftermath, slowed down the rate of development in Hertfordshire. All available resources went into defence: airfields were constructed at Bovingdon, Hunsdon, Sawbridgeworth and Nuthampstead, and various 'pillboxes', concrete anti-tank blocks and the like are still to be seen, especially in the south of the county, where one of the outer London defence lines ran from Watford to Cheshunt.[56] But as economic conditions began to recover in the late 1940s the county was poised for a rash of new development. Fortunately, the Labour government was able to capitalise on a new enthusiasm for corporate planning that had been engendered by the war effort and introduce, for the first time, forms of large-scale spatial planning. The 1947 Town and Country Planning Act made land use planning compulsory over the whole country; it also, building on earlier initiatives by the London County Council, made provision for a Metropolitan Green Belt around the capital and obliged local authorities to take account of this in their own development plans.[57] It is difficult to exaggerate the importance of all this for the landscape of England as a whole, and for Hertfordshire in particular, most of which was included within the Green Belt. The 1920s and 1930s had not only seen the piecemeal expansion of towns and villages throughout the county, although especially in the south, but also the appearance of houses and groups of houses even in quite remote spots, throughout the dipslope of the Chilterns especially. Vast tracts of the county would certainly have disappeared under various forms of low-density sprawl without this change in planning law: not least given the fact that in 1935 the Northern Line took over plans formulated by, and permissions granted to, the Watford and Edgware Railway Company in 1898 to extend their line from Edgware as far as Watford, with a general expectation that it would go further. Construction was actually underway when war broke out, but was abandoned after the passing of the act, in part because without a continued expansion in the number of homes in the south of the county it would not have been financially viable.[58]

But the new enthusiasm for state planning also increased the pace of urbanisation elsewhere in the county. New government powers allowed direct state involvement in the creation, not just of small council estates, but of entirely new towns. Faced with the problems of a chronically over-crowded and bomb-damaged metropolis and continuing economic and demographic expansion in the south-east, the government passed the New Towns Act in 1946 which put into practice ideas formulated in the Abercrombie Plan of 1944, for the 'decentralisation of people and industry from London'.[59] The state could now compulsorily purchase tracts of land to create large new urban settlements, the first of which – following Abercrombie – were dotted around London. Hatfield, Hemel Hempstead and Stevenage were all designated as New Towns; Welwyn Garden City was redesignated as one.[60] Hemel Hempstead lay beside the road and rail routes following the Gade valley, later benefiting from the construction of the M1 motorway, but Hatfield, Welwyn Garden City and Stevenage – and also Letchworth Garden City – all lay alongside the A1 (the former Great North Road) and Great Northern rail line, leading to the creation of a largely urban corridor heading north through the middle of the county.

The design of the New Towns was shaped by the latest ideas in town planning, including those emanating from the United States and Europe (especially the Low Countries).[61] As with the earlier Garden Cities, large areas of green space were retained within the new urban areas, while community spirit was encouraged by the creation of distinct 'neighbourhoods': Stoney Hall, Monks Wood, Bedwell, Broadwater, Shephall, Chells and Pin Green in Stevenage; Bennetts End, Chaulden, Warners End and Adeyfield at Hemel.[62] Unlike the earlier Garden Cities, these places were designed with the assumption that most of their inhabitants would have access to cars and, in Stevenage in particular, every effort was made to separate traffic and pedestrians, with numerous footpaths and underpasses. Concerns about the amount of traffic also explain why, more than in any previous developments, the new towns featured numerous *cul de sacs*, so that the minimum number of residents would be troubled by passing vehicles – the complex serpentine plans, with numerous dead ends, are immediately evident on a map, or from the air. Stevenage also boasts the first pedestrianised town centre in the country. Indeed, all the towns had new centres, quite distinct and spatially separate from those of the adjoining

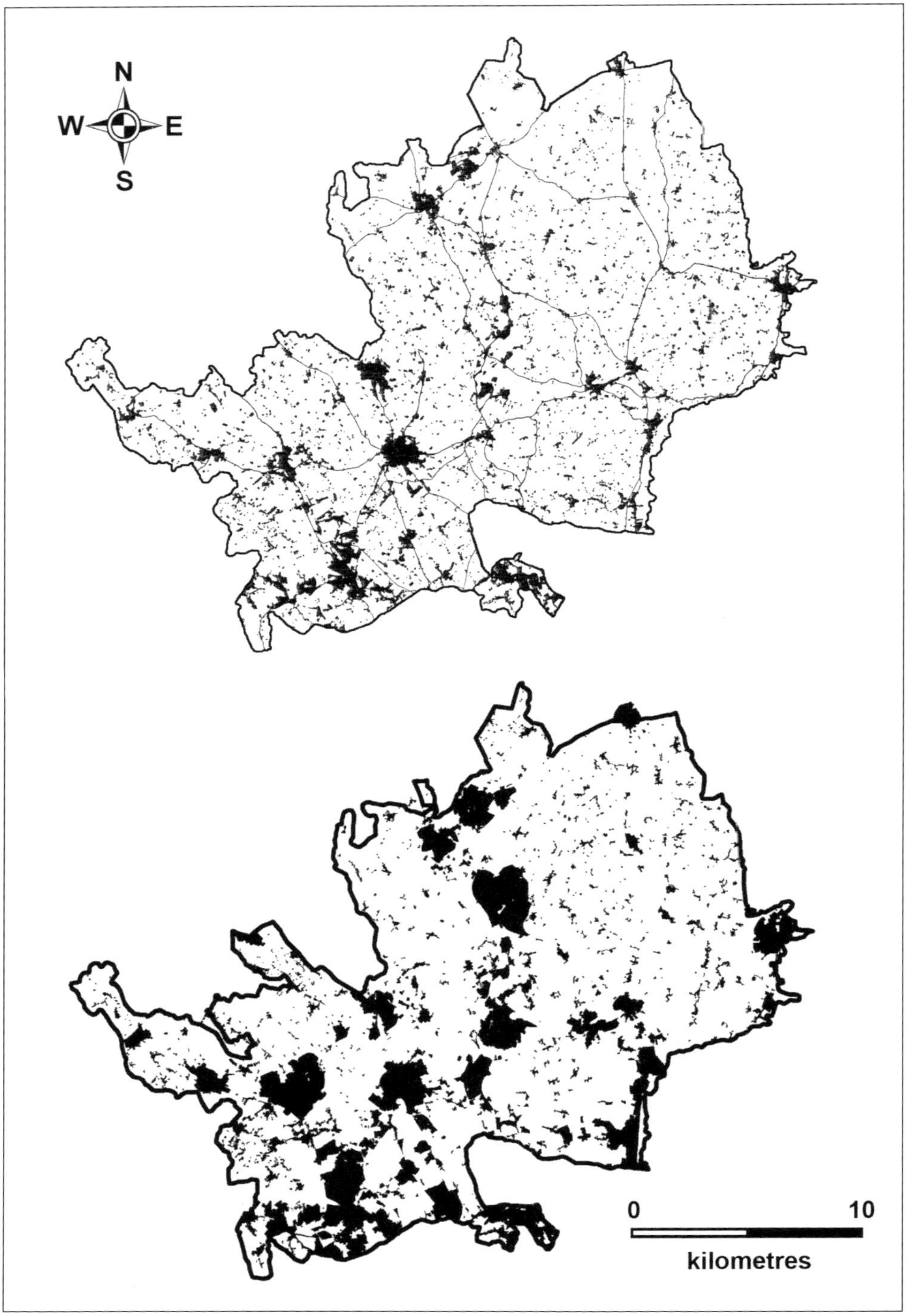

Figure 10.9. The expansion of the built-up area in the second half of the twentieth century. Top: *c.*1940; bottom: *c.*2005. Sources: Cameron, Hertfordshire; HHER, Historic Landscape Characterisation Survey.

'old' towns, considered too antiquated for this new, modern age. The architecture of shops, public buildings and houses was innovative and modernist, and utilised new techniques of prefabrication, some of which have not stood the test of time. Gone were the backward-looking, vaguely vernacular styles of architecture employed at Letchworth, or even the restrained pseudo-Georgian of Welwyn Garden City.

These were not, of course, the only new areas of building to appear in the post-war decades. All the county's towns expanded, with innumerable estates – usually featuring a profusion of *cul de sacs* and sometimes the great areas of mown grass as seen at Stevenage – erected on their peripheries. Smaller groups of houses were forced into any available space, including areas formerly occupied by extensive gardens laid out in a more spacious age, for the inevitable corollary of the institution of 'green belts' was that areas zoned for development often became ever more crowded with houses. These most recent additions to our architectural heritage take a wide range of forms, displaying varying degrees of modernist influence as well as differences related to social status (with more individualistic, architect-designed dwellings at the upper end of the scale); so much so that it is difficult to generalise about their appearance. In broad terms, however, houses became simpler and less cluttered, without bay windows and pseudo-vernacular and Tudoresque detail. Many in the 1950s and 1960s – following the trend set by Welwyn Garden City – display a stripped-down neo-Georgian style, with walls usually of light-coloured brickwork rather than pebbledash or roughcast. Through the 1960s and 1970s shallow pitched roofs became fashionable and houses were provided with few, if any, chimneys; large windows; and façades unadorned or decorated with small areas of applied cladding, often in the form of small sections of weatherboarding between the first and ground floor windows – a style ultimately derived from Scandinavia and certainly displaying no connection or continuity with anything local and traditional. Most, owing to increasing pressure on space, were provided with small gardens and integral garages.[63]

Conclusion

Suburbanisation, and the construction of New Towns and Garden Cities, changed the landscape of Hertfordshire irrevocably (Figure 10.9), but the new landscape was not completely unrelated to what preceded it. Indeed, because of the piecemeal manner in which early suburbs often developed, more of the basic structure and texture of the old rural landscape – the pattern of boundaries and roads – might be preserved in suburbs than in areas which continued to be agricultural, where the intensification of farming in the post-war period often served to create featureless prairies. One under-researched topic is the extent to which continuity went further than this, with the preservation of hedgerows, hedgerow trees and areas of woodland, and the distinctive flora that these contained, within the new urbanised environment. Such features were sometimes consciously retained when Garden Cities and New Towns were planned, but can exhibit a remarkable degree of resilience in other and earlier contexts, especially in the lower-density suburbs.

Conversely, 'where two or more adjacent fields were purchased by the same developer at one time, for example in high density working-class housing, there was less chance of the internal hedges surviving'.[64] Their survival was often best in situations where gardens back onto public parks and recreation grounds. Even when hedges degenerated and disappeared hedgerow trees remained, and often in some numbers. In some areas the patterns of old field boundaries can still be picked out from the air as lines of mature oak trees, as much as a century after development occurred. Such ancient trees provide islands of valuable habitat for wildlife and might continue to be the homes of owls or bats, marooned in the rising tide of housing. The survival of older boundaries was not merely a side-effect of the way that development was structured by existing lines in the landscape, however. In some cases the Town Planning Permits issued by the Urban or Rural District Councils stipulated the retention of established trees as development proceeded. Moreover, those wishing to purchase homes in the suburbs were abandoning the crowded cities, if not for remote countryside, then at least for an environment that was still partly rural in character. Fired up by the example of Letchworth and similar planned towns, developers often went out of their way to emphasise the extent to which such details of the rural landscape would be cherished. Sales particulars for one development in south Hertfordshire, on what was effectively becoming the northern edge of London, emphasised in 1912 that: 'It is desired to preserve the rural characteristics of the locality as much as possible, and with that object in view the natural hedges and as many of the trees will be retained as is consistent with convenient development.'[65] Old trees, and occasionally hedges, also often survived particularly well in the midst of golf courses, sports fields and similar open spaces. Pockets of woodland could demonstrate remarkable resilience, especially on the margins of the more intensively built-up areas, some since becoming nature reserves. Oxhey Woods, close to the county boundary with Middlesex, is now a local nature reserve featuring a mixture of ancient and secondary woodland with areas of remnant heath and wood-pasture, and is noted for its wild service trees and bluebells. Yet it is hemmed in by housing estates and suburbs on all sides except the south, where there is a large golf course. Such survivals are probably more common than we think.

Notes

1. Above, p. 233. See also T. Williamson, 'The character of Hertfordshire's parks and gardens', in Rowe (ed.), *Hertfordshire garden history*, pp. 1–25; Harwood, 'Some Hertfordshire nabobs', pp. 49–77; Munby, *Hertfordshire landscape*, pp. 138–50.
2. *Morning Chronicle*, 3 August 1810.
3. Goose, 'Population, 1801–1901'; E. Lord, 'Population, 1901–2001', in Short (ed.), *Historical atlas*, pp. 58–9.
4. A. Ball, *Street and place-names in Watford* (Watford, 1973), p. 11.
5. *Ibid.*, pp. 10–11.
6. *Ibid.*, p. 11.
7. H. Barrett and J. Phillips, *Suburban style: the British home, 1840–1960* (London, 1993), pp. 86–7.
8. Barrett and Phillips, *Suburban style*, pp. 84–8.
9. *Ibid.*, pp. 88–9. Pevsner and Cherry, *Hertfordshire*, pp. 35–6.
10. J. Burnett, *A social history of housing, 1815–1970* (London, 1980).
11. Page (ed.), *VCH Hertfordshire*, vol. 2, p. 329.
12. J. Scholefield, *The Private Street Works Act, 1892: being a practical guide to the working of the act* (London, 1902).
13. HALS Off Acc 717: Bushey Urban District Council minutes, 16 May 1906; 4 July 1906.
14. G. Darley, *Villages of vision: a study of strange utopias* (London, 1975).
15. E. Howard, *Garden Cities of tomorrow* (London, 1902), p. 112.
16. M. Miller, *English Garden Cities: an introduction* (London, 2010), pp. 18–22; K. Thompson, 'Garden Cities', in Short (ed.), *Historical atlas*, pp. 162–5.
17. M. Miller, *Letchworth: the first Garden City* (Chichester, 2002).
18. Pevsner and Cherry, *Hertfordshire*, pp. 224–5.
19. *Ibid.*, p. 228.
20. T. Rowley, *The English landscape in the twentieth century* (London, 2006), pp. 198.
21. HALS Off Acc 708 contract and plans for Watford bypass, 1924.
22. W. Pedley, *Labour on the land* (London, 1942); S. Wade Martins and T. Williamson, *The countryside of East Anglia: changing landscapes 1870–1950* (Woodbridge, 2008), pp. 159–62.
23. K. Hudson, *The archaeology of the consumer society: the second industrial revolution in Britain* (London, 1983).
24. R. Funnell, 'Handley Page Ltd., Radlett, 1929–1970', *Journal of the Watford and District Industrial History Society*, 18 (1994), pp. 25–9; P.R. Stokes, 'Power for Flight Pt. 1: De Havilland and the Piston Engine', *Journal of the Watford and District Industrial History Society*, 10 (1980), p. 36.
25. C. Archer, 'Printing', in Short (ed.), *Historical atlas*, pp. 98–9; Branch Johnson, *Industrial archaeology*, pp. 146–51.
26. S. White, 'Film and television studios', in Short (ed.), *Historical atlas*, pp. 102–3.
27. E.G. Bentley, *A practical guide in the preparation of town planning schemes* (London, 1911).
28. HALS Off Acc 269 Bushey & Watford Rural Joint Town Planning Committee, Map 2a, 1933; HCP1/1/25 County Development Plan, Bushey map, 1951; HCP2/1/1 East Herts Joint Planning Committee, 1938–46.
29. For example, HALS DE/Le/B13/1 sale particulars for 168 acres on the North Orbital in St Albans, 1934.
30. A. Jackson, *Semi-detached London: suburban development, life and transport 1900–1939* (London, 1973); I. Bentley *et al.*, *Dunroamin: the suburban semi and its enemies* (London, 1981); Rowley, *English landscape in the twentieth century*, pp. 203–5.
31. Barrett and Phillips, *Suburban style*, p. 32.
32. P. Kingsford, *A modern history of Brookmans Park* (North Mymms, 1983).
33. HALS DE/Bt/B34 Brookmans Park sale catalogue, 1926.
34. Kingsford, *Brookmans Park*.
35. M. Pedrick, *Moor Park: the Grosvenor legacy* (Rickmansworth, 1989), p. 60.
36. Hertfordshire Gardens Trust and Williamson, *West Hertfordshire*, p. 89.
37. M. Rowlinson *et al.*, *'In Arena Virtus': 100 years of sand and spirit* (Sandy Lodge, 2010).
38. G. Beard and J. Wardman, *Thomas H. Mawson: the life and work of a northern landscape architect* (Lancaster, 1978), p. 182; HALS DE/Bn(Add)/B177 illustrated brochure advertising development of Moor Park Estate, n.d.
39. HALS DE/Bn(Add)/B177.
40. J.R. Day and J. Reed, *The story of London's Underground* (London, 2008), p. 84.
41. J. Betjeman, 'Metro-land', BBC documentary (1973).

42. Miller, *English Garden Cities*, pp. 31–5.
43. R. Filler, *Welwyn Garden City* (Chichester, 1986).
44. Cameron, *Land of Britain*, p. 336.
45. Wallace, 'Sand and gravel quarrying', pp. 104–5.
46. HALS DE/X37/T6 sales particulars of Bushey Hall and Bushey Grove estates, 1877–81.
47. *Ibid.*
48. Bushey Museum, Abstract of title to land on the Bushey Grove estate, with map.
49. HALS 66423–4 sale particulars of Bushey Grove Park.
50. Leicester and Rutland Record Office, DE 4939/977.
51. Bushey Museum, Sale particulars for freehold building land on Aldenham Road, 1892.
52. HALS IR1/479–80 1910 Finance Act maps, Bushey.
53. Private collection.
54. Bushey Museum, Sale particulars for 157 building plots on Bushey Grove estate, 1912.
55. *Ibid.*
56. M. Osbourne, *Defending Britain: twentieth-century military structures in the landscape* (Stroud, 2004), pp. 63–6.
57. Rowley, *English landscape in the twentieth century*, pp. 112–14.
58. T. Beard, *By Tube beyond Edgware* (London, 2002).
59. P. Abercrombie, *Greater London Plan 1944* (London, 1945).
60. M. Ashby and A. Cudmore, 'New Towns', in Short (ed.), *Historical atlas*, pp. 166–7. A.G. Champion *et al.*, *Facts about the New Towns: a socio-economic digest* (Corbridge, 1977).
61. Rowley, *English landscape in the twentieth century*, pp. 183–5.
62. Pevsner and Cherry, *Hertfordshire*, pp. 180–84, 350–58; S. Hastie, *Hemel Hempstead: the story of New Town development 1947–1997* (Hemel Hempstead, 1997); J. Balchin, *First New Town: an autobiography of the Stevenage Development Corporation, 1946–1980* (Stevenage, 1980).
63. Barrett and Phillips, *Suburban style*, pp. 136–7.
64. L.R. Williams and J. McLaughlin, 'Tracing old hedgerows in suburban areas', *Local Historian* (May 1988), p. 65.
65. Bushey Museum, sale catalogue for 'Bushey Village estate', 1912.

Conclusion

This book is based on much original fieldwork and documentary research, carried out over many years. But it also rests firmly on the endeavours of others. As a glance at the bibliography will make clear, Hertfordshire's history and archaeology have been extensively studied by both 'amateurs' and professionals over many decades. This said, much remains to be researched. Like Lionel Munby's book of 1977, this volume represents not the final word on the subject of the county's landscape history but a provisional statement, a summary of current knowledge and ideas. There remain vast areas of unexplored terrain, both physically and intellectually, in the county, and a number of key topics in particular need of attention have been flagged up over the previous pages.

At the start of this book we emphasised the great diversity and variety of Hertfordshire's landscapes and suggested that this made the county a particularly interesting one to study. So much variety in so small a compass provides a good opportunity for exploring the kinds of factors which generate the particular character of different human landscapes in terms of the patterns of fields and roads, the form and distribution of settlements and the extent and location of woods, commons and meadows. In the course of this study we have explored some of the influences which shaped this diversity and have emphasised the ways in which different phases and modes of social and economic organisation have, over long periods of time, interacted with the environmental characteristics of different regions – aspects of soils, topography, water supply and so on – to create distinct types of landscape. Before industrialisation and the transition to a fossil fuel economy variety was born, above all, of such environmental factors: landscapes were created by the adjustment of farming economies to external demands, but above all to local conditions. Change, moreover, was small-scale and piecemeal, if on occasions relatively rapid: we have emphasised here the organic and gradual development of the medieval landscape, rather than the kinds of sudden change made at the behest of lords and magnates which loom so large in many studies of landscape history, and we have gone so far as to suggest – perhaps controversially

– that even the layout of the county's medieval towns was usually the consequence of gradual growth, rather than of sudden 'planning'.

Yet the relationship between environment and the human landscape is often complex and indirect. The fertility of soils had a profound impact on early medieval population densities, and thus – at one remove – on such things as the character and complexity of parish churches. More importantly, even in medieval times many patterns in the landscape reflected not local environmental factors but networks of contact and communication which shaped the spatial distributions of particular styles and fashions, as with the differences exhibited in the period up to the seventeenth century by styles of vernacular architecture in the east and west of the county. Trade and communication had, arguably since late prehistory, structured patterns of urbanisation in the county, and their influence increased steadily in the course of the post-medieval period. Proximity to London fuelled levels of demographic growth and economic activity above the national average, especially in the period after *c.*1870, when – with the elaboration of the rail network and subsequently the development of the internal combustion engine and the national electricity grid – large areas of the county, particularly in the south and west, became suburbanised. The rules of the game changed further from the middle of the twentieth century, however, with the development of large-scale spatial planning, so that political decisions came to override not only the influence of the local environment but also to some extent the patterns of growth structured by economic and demographic factors, and by the configuration of roads and railways.

Hertfordshire continues to change. With much of Britain's economic and demographic growth concentrated in the Home Counties, pressure on the countryside remains intense even in the current period of recession. The extent to which bricks and mortar have spread across the county's surface during the last 75 years or so has been astonishing, as Figure 10.9 demonstrates, although it arguably underestimates the true extent of suburbanisation, in the sense that many of the blank areas it shows represent not surviving pockets of farmland but such things as playing fields or golf courses. Planning controls have, it is true, ensured that the extensive, low-density sprawl of the inter-war years has been controlled and curtailed, but the corollary has been that the county's urban and suburban areas have become more and more densely built up, with new housing in the south of the county especially being shoe-horned into anything which could reasonably be described as a 'brown field' site.

The long-term fate of Hertfordshire's environment will thus depend not on local environmental and economic influences but on national or even international developments in an increasingly crowded world. Politics, above all, will shape the future landscape, and much may depend upon the outcomes of struggles between local interest groups and the policies of national governments. For the history of Hertfordshire over the last century and a half has been shaped in part by a dream, at the heart of which lies a contradiction. Many people have moved to the county, often from the capital, to find a life in or on the edge of the countryside. But the erection of one house, in one housing development, has led inexorably to the appearance of others, and in many districts the rural has now been entirely destroyed. Even Letchworth Garden City is no longer surrounded on all sides by belts of farmland, as its creators envisaged; instead, it is joined to Baldock, and nearly to Hitchin, by retail parks, playing fields and leisure centres.

Yet we must be careful not to exaggerate the extent to which Hertfordshire has become a built-up county. As the Woodland Trust recently commented:

> Despite its new towns and London commuter-belt tag, this is not a built-up urban or suburban county. On the contrary, half of Hertfordshire is designated as Green Belt, and consequently it is a relatively green county. The landscape is varied, combining picturesque valleys and woodland with majestic hills. Wooded lanes criss-cross a countryside of rolling fields and take you through charming, often ancient, villages. To the north-west and to the west you'll find the Chiltern Hills, an Area of Outstanding Natural Beauty.

Moreover, public pressure and the hard work of a number of conservation organisations have ensured the survival of important pockets of woodland and farmland within the most built-up areas, not for economic reasons but to benefit wildlife and provide opportunities for recreation. Notable examples include the extensive woods in the Broxbourne area and the fields on and around Merry Hill in Bushey, acquired in recent years by the Woodland Trust.

We hope that, in the course of this book, we have avoided any simple assumption that, in the modern landscape, urban and suburban are somehow 'bad', and invariably represent alien intrusions which have destroyed the 'good' things of the countryside. Even in terms of biodiversity, suburban gardens and flooded former gravel pits can be richer than the 'official' countryside where this comprises, as it sometimes does in the county, chemical-soaked arable prairies extending to the far horizon. In terms of the landscape's *historic* heritage, moreover, the situation is even more complex and nuanced. Historical texture may be preserved better in suburban areas than in rural ones. In the arable districts of eastern Hertfordshire ancient hedges and even green lanes were often removed on a large scale as farming modernised in the second half of the twentieth century. Yet, as we have seen, in the suburbs of the south and west of the county the patterns of suburban roads and property boundaries often perpetuate the essential structures of landscapes first formed 1,000 years ago. More than this: built-up areas are themselves 'historic landscapes'. Letchworth, the world's first Garden City, is arguably in international terms the most important landscape in Hertfordshire, and even townscapes with a less prestigious pedigree may soon become hallowed by time. Places which, in the nineteenth century, were regarded as brash intrusions into the countryside are often now protected and celebrated as Conservation Areas. Even Stevenage New Town will, one day, be treasured as an 'historic landscape' in its own right, a testimony to the dreams, aspirations and optimism, as well as the economic and social needs, of the post-war decades.

Bibliography

Primary sources

Apsley Paper Trail
Documents/444 sketch of Apsley Mill by Harriet Dickinson n.d.

Bodleian Library, Oxford
MS Eng 2167 f.37 and f.95 The Grove, Watford
MS Gough Drawings a.3 fol. 27 map of three manors in the south part of Cheshunt, *c.*1575
MS Maps Herts. a.1 plan of Gubbins (Gobions), North Mymms, n.d.
MSGD a.4, fol. 64 design for gardens at Sacombe attributed to Charles Bridgeman, *c.*1715–17

British Library
Add. MS 33,581 Terrier of the manors of Weston, Argentein, Newbury and Lannock, undated but contemporary with a map of 1633 at HALS
Add. MS 42,508 Surveys and Perambulations, with Plans, by John Norden, of the manors of Mincinbury, Abbotsbury and Hoares, in Barley, co. Hertf., made for Sir John Spencer, Lord of the Manors, 1603
Cotton MS Aug.I.i.75 map by J. Thorpe, 1611
Cotton MS Otho D.3
Cotton MS Tiberius E.ii, fos 140–41 survey of manors of the bishopric of Ely, 1222
Egerton MS 3390 'Leeds papers'
Ordnance Survey Drawing 138 pt. 1, Enfield, 1799
Ordnance Survey Drawing 149, Hatfield, 1805
Ordnance Survey Drawing 151, Chipping Barnet, 1805
Phillips MS 82742 holdings of Michael Bellis, 1825

Bushey Museum
Abstract of title to land on the Bushey Grove estate, with map
Sale catalogue for 'Bushey Village estate', 1912
Sale particulars for 157 building plots on Bushey Grove estate, 1912
Sale particulars for freehold building land on Aldenham Road, 1892

Cambridge University Library
EDR G/3/27 Old Coucher Book of Ely, fo. 80

Gloucester Record Office
D1245/FF75 map of an estate belonging to Jeremy Sambrooke Esq

Hampshire Record Office
9M73/958 diary of the Reverend George William Harris, Rector of Egglescliffe, Durham, 1757

Hatfield House Archives
Court Rolls 9/25 bailiff's account for the manor of Little Hadham, 1435/6
Court Rolls 10/16 account for the manor of Bedwell, 1484/5
Court Rolls 10/20 Hoddesdon lands, 1493
Court Rolls 11/4 bailiff's accounts for the manor of Little Hadham, 1322–1507
CP 354 survey of manors in Broxbourne by John Gascoyne, 1692
CPM Supp 69 'Plan of the manor of Holbatches alias Popes … belonging to Mrs Rebecca Asheton surveyed by Taylor and Chilcott in the year 1785'
Hatfield Manor Papers I, 1392/3, p. 253
Hatfield Manor Papers I, 1396 Michaelmas, p. 265
Hatfield Manor Papers: Summaries I, p. 339
Salisbury Papers Sals. General 44/1
Salisbury Papers Sals. Legal 66/7

Hertfordshire Archives and Local Studies, Hertford
XIII/30 map of the manor of Sopwell, seventeenth century
XIII/51 plan of Broadfield, 1775
1985 photograph of a map of 'Barkhampsted Parke', *c.*1638 (original at the National Archives)
6398 Bushey manor court roll, 1600
6424 Bushey manor court roll, 2 October 1630
6442 Bushey manor court roll, 1707
7568 plan of Langley Lodge and Balls Pond Farms, King's Langley, 1797
9607 accounts of the Capel estate at Hadham and Walkern, 1591–1640
10996 A/B notes in case concerning cutting down of trees on Cheshunt common, n.d.
11788 timber survey, Three Houses Farm, St Stephens, St Albans, 1719
15594 plan showing The Golden Parsonage, mid-eighteenth century
21579 map of the demesne of the manor of Bourn Hall, n.d.
21600 Bushey manor, copy of court roll, 1738
26293 map of Bygrave & parts of Clothall, 1808
26872A map of Meesden, 1815
27251 timber survey of Tarrants Farm, Lamer estate, 1734
27252 wood accounts for the Lamer estate, 1760
34137 conveyance of Gobions estate, 1815
37123–33 Bushey manor court roll, 1461–9
43754 plan of estate at Puckeridge, 1778
44737 lease of the manor of Napsbury, 1538
44831 grant of the manors of Champneys and Fosters, 1427
46126 grant of the manors of Champneys and Fosters, 1546
46143 Bourn Hall court roll, 1603–23
46606 lease of Knebworth warren, 1722
49464 survey of the manor of Weston, 1548
54027 plan of farms in Wyddial & Barkway, 1768
54569–72 deeds of Monkenfrith Farm, 1583–4
54804 plan of Tile-house Farm, Haydons Farm, Cressalls Farm, Theobald Street, 1765
54807 manor of Bournhall court roll, 1627–34
54835 plan of the manor of Corneybury in Layston & Wyddial, 1744
56457 plan of Brent Pelham Hall & farms, 1778
56458, 56459 & 56461 plans of estates in Brent Pelham, 1787
57310 map of Little Gaddesden, 1802
57415 map of Anstey Decastrum, mid-eighteenth century
57594–5 Lilley and Offley: list of lands of John Ireland, 1705 and Terrier, 1705–53
58882 'A Plan of the Estates of the Revd Thomas Whitehurst and others; in the Hamlet of Walsworth in the Parish of Hitchin, Inclosed by Act of Parliament, 1767'
63631 map of Barkway, Reed & part of Wallington, 1807
63716 Hormead Dane Farm, Gt Hormead, pre-1815
66423–4 sale particulars of Bushey Grove Park, 1887
71014 map of the parish of Broxbourne, 1829

71182 plan of Thorley Hall, 1673
72036 agreement by parishioners of Pirton, 1692
72351 survey of the manor of Wellbury, Offley, 1553
74423 map of land in Clothall, Wallington & Cottered, 1729
74776A map of Rushden & parts of Clothall, Wallington & Cottered, 1802
76828A–B plan of an estate in Westmill *c.*1670
76830 map of Stonebury in Braughing, Little Hormead and Layston, 1712
76831 plan of Haxten-end, Little Heath and Hill-end Farms, Northchurch and Hemel Hempstead, 1758
79423 lease of the manor of Blackhide alias Corsers, 1547
79699 plan of Furneux Pelham Hall estate, *c.*1800
79703 copy plan of Agnells Manor, Hemel Hempstead, 1645
80226 court rolls manor of North Mymms, 1427
80251 Corner Farm, Hemel Hempstead, 1772
80804 map in Wormleybury estate papers, 1851
80870–87 Bovingdon estate timber accounts, Lane Farm and Little Shantocks, 1800–10
81750 plan of land at Whempstead, Little Munden, 1808
83280–1 plan of Marshalls Farm, Standon, *c.*1730
83353 plan of Marshalls, 1804
A/2832–3 plan of Great Barwick estate, Standon, 1778
A/2834 plan of Fabden Farm, Standon, 1790
A/2837 plan of Colliers End Farm, 1809
A/2838 plan of Standon Green End, 1774
A/2845 Brown's plan for improving Youngsbury, n.d.
Acc 4283 map of the township and parish of Weston by Henry Liley, 1633
AH/680 plan of Boxted Farm, Hemel Hempstead, 1735
AH/681 plan of Fields End Farm, Hemel Hempstead, 1764
AH/1163 re Manor of Frithesden, 25 Sept 1672 reciting Articles of Agreement (to Inclose)
AH/2476 An Abstract of Faggotts Cut in the woods & on Other parts of the Ashridge Estate by William Pratt & Company From December 31st 1834 to [blank]
AH/2790 extracts from Court Rolls of the manors of Aldbury and Berkhamsted, 1764
B/86 account book for manor of Cheshunt, with Broxbourne (mainly timber sales), 1657–94
B/479 map of lands on the Broxbourne Mills Estate, late eighteenth century
CP/107/26/2 Westmill enclosure map, 1819
DAcc 1154 Aston tithe map, 1840
DE/Am/E3 lease of Medburn Farm, Aldenham, 1694
DE/Am/E11 leases of Hartsbourne Manor Farm, Bushey, 1633–90
DE/AS/91 extent and rental of Aston, 1540
DE/B513/E1 farm leases including 320 acres at King's Walden, 1662, Lannock Manor, Weston, 1663, Wilches and Gillies Farms in King's Walden, 1662 and 'the Farm', Weston, 1665
DE/B944/E2 timber survey, Sheeps Farm, London Colney, 1852
DE/B983/E1 survey and valuation of copyhold estates of the manors of Chipping Barnet and East Barnet, 1787
DE/B983/E2 timber survey of farms and lands on the manor of East Barnet, 1786
DE/B1622/E9 timber survey, The Holt Farm, St Stephens, 1817
DE/B1622/P1 map of Albury & Furneux Pelham, 1781
DE/B1622/P3 plan of Kendals Hall, *c.*1740
DE/B1767/M1 survey and custumal of Cheshunt, 1669
DE/B1768/P1 plan of farm at Colliers End, 1767
DE/B1768/P2 plan of estate at Colliers End, 1794
DE/B2067B/E24 a list of woods appertaining to the monastery of St Albans, *c.*1485–1509
DE/B2351/P1 maps of Braughing Bury Farm, Upp Hall Farm, Warren Farm and Gatesbury Mill, Braughing, *c.*1800
DE/Bb/E26 wood accounts for the Broxbournebury estate, 1719–95
DE/Bb/E27 timber survey of farms and tenements on the Broxbournebury estate, 1784
DE/Bc/P6 & P7 plan of an estate in Brent, Furneux & Stocking Pelham, *c.*1800
DE/Bg/4 volume of paintings by J.C. Buckler
DE/Bn(Add)/B177 illustrated brochure advertising development of Moor Park Estate, n.d.
DE/Bt/B34 Brookmans Park sale catalogue, 1926

DE/Bw/28766 probate, Thomas Goodwyn of Baldock, 1665
DE/Cd/E14 lease of Ridge Hill Farm, 1657
DE/Cd/E101 lease of Aspenden Hall Farm, 1745
DE/Cn/E1 Parnell estate, Meesden, 1762
DE/Cn/P4 plan of an estate in Much Hadham, 1803
DE/Cn/P7 plan of Beeches, Brent Pelham, 1769
DE/Cp/L1 dispute between Lord Essex and Falcon re Brickett Wood and Garston manor including deeds to manors of Park and Meriden and various other papers relating to Frythe House, and manors, 1607–16
DE/Cr/55 timber surveys on Beeson End Farm, Redbourn, Westwick Farm, Hemel Hempstead and St Michaels, Market Street Farm, Flamstead, and Folly Farm, Albury, 1838
DE/Cr/125/2 Plan of the Liberty and Manor of Cheshunt together with that part of Theobalds Park within the Parish of Northaw in the County of Hertford, and also that part of Theobalds Park within the Parish of Enfield in the County of Middlesex, 1785
DE/FL/15719 Mortgage 'of land formerly part of a wood called Buckwood', 1631
DE/FL/17075 lease of lands in Studham, 1578
DE/FL/17246 grant of 'Bradwynfelders, Flamstead, formerly the property of Dunstable Monastery', 1545
DE/FL/17248 lease of the 'manor house of Beechwood, late called the Priory of St Giles in the Wood', 1549
DE/FL/18159 mortgage of 'property in Flamstead, near Studham Common, including the two Home Closes, formerly part of Bradwin Fields', 1756
DE/Fn/P1 plan of Redheath estate in Watford, 1760
DE/H/P3 map of Aspenden Hall estate, 1810
DE/H/P5 map of land in Standon & Braughing, 1809
DE/H/P6 plan of Little Hormead Bury & Balons Farm, 1835
DE/H/P14 plan of Briggens in Hunsdon, Stanstead Abbots & Roydon, 1781
DE/H/P15 plan of Tinkers Farm, Furneux Pelham, 1808
DE/H/P16 plan of Stockers and Pearces Farms in Sawbridgeworth and Thorley, 1807
DE/H/P17 plan of Newgates alias Little Briggens Farm, Stanstead Abbots, 1781
DE/Ha/P1 Plan of the Manor and Parish of Ickleford and other Lands, 1771
DE/Hn/Z5 notes about Berkhamsted Common including extracts from the manor court rolls in April 1666 and October 1726
DE/Hx/P2 plan of the Bushey Hall estate, 1685
DE/K/E15 Knebworth estate papers, 1840
DE/Le/B13/1 sale particulars for 168 acres on the North Orbital in St Albans, 1934
DE/Lw/E7–8 Bowers tenement, Harpenden, 1706
DE/Lw/E16 depositions relating to a dispute between Harpenden manor and Rothamsted manor over boundary at Harpenden Common, with Wheathampstead, 1839–43
DE/Lw/E33 account of timber sold from Marshalls and Butlers, 1539
DE/Nc/1 map of Braughing, *c.*1820
DE/Of/6/380 Blakesware entry in volume of topographic paintings and commentary by H.G. Oldfield
DE/P/P1 map of the manor of Digswell, 1599
DE/P/P4 plan of Earl Cowper's estate in Hertingfordbury, 1704
DE/P/P6 plan of Rushingwells Farm, Barkway, 1732
DE/P/P7 plan of Hertingfordbury Farm, 1732
DE/P/P15 plan of Brocket Hall Park, 1798
DE/P/P21 Repton's Red Book for Panshanger, 1799
DE/P/T262 Indenture between George Mym of Woodcott in the parish of Ebbisham, Surrey Esq and William Pryor of Hartingfordberie, warrener, 1637
DE/P/T264 lease of Hertingfordbury warren for nine years, 1634
DE/P/T3313 Release of the manor of Panshanger to Thomas Woodford in trust for the Right Hon. William Earl Cowper, 1719
DE/Pm/72439 Chancery bill in title deeds and estate papers of the Pym family of Norton Bury, Norton, 1661
DE/R/E114/1–10 accounts for Wain Wood, 1739–67
DE/Ry/P1 map of farms in Barkway, 1728
DE/Ry/P3 plan of manor and lands of Newsells in Barkway and Barley, 1788
DE/Ry/P4 map of Nuthampstead Bury Farm, 1789
DE/Sa/P1 estate in Stapleford, *c.*1690
DE/V/P1 Gorhambury estate map, 1634

DE/V/P2 a survey of the estates of The Right Honourable James Lord Viscount Grimston by J.L. Davis, 1768
DE/X3/9 diary of John Carrington, 1806
DE/X10/Z1 W. Minet, *Hadham Hall and the Manor of Bawdes alias Hadham Parva* (Colchester, 1914)
DE/X37/T6 sales particulars of Bushey Hall and Bushey Grove estates, 1877–81
DE/X55/Z2/39 engraving of the ruins of the Elizabethan Gorhambury House, 1808
DE/X58/E1 Lease and tenancy agreement of Newhouse Farm, 1771–93
DE/X58/E9 correspondence to Towse [solicitor], case papers and related papers in a plea of trespass and ejectment, Mills v. William, 1810–11
DE/X195/P1–2 map of Warner's End Estate, 1809
DE/X216/B2 valuation of timber upon lands allotted and exchanged under the Aldenham Inclosure Act 1802–3
DE/X251/P3 map of Cottered, *c.*1806
DE/X437/Z13 Kensworth Inclosure Act, 38 George III c67, 1798
DE/X526/P1 plan of Broadfield Hall estate, 1810
DE/X629/P1 map of the Bayfordbury estate, 1807
DE/X696/P1 map of Anstey, 1730
DE/X905/P1 plan of farm at Flaunden, early eighteenth century
DE/X1010/M1 Tring manor – court book with Tring Rectory, 1691–1708
DE/X1012/M1 Typed transcript of the Steward's account book for North Mymms Manor, 1691
DP/26/10/1–2 Bushey parish rate book (and transcript), 1632–49
DP/26/29/2 plan of the parish of Bushey, 1800
DP/75/3/1 copy of a 1637 terrier of Norton church lands
DP/82/1/1 terrier of glebelands of 1664 inside back cover of register of baptisms, marriages and burials for parish of Radwell, 1590–1718
DP/107/29/1 map of Therfield & Buckland, 1724
DP/111/8/19 Tring vestry minutes 1782–1815
DP/114/3/4 map of land in Walkern, 1810
DP/117/25/7 plan of three fields called Church lands, Watford, 1754
DSA4/25/2 Broxbourne tithe map, 1839
DSA4/27/2 Bushey tithe map, 1840
DSA4/29/1–2 Caldecote tithe award and map, 1838–41
DSA4/44/1–2 Little Hadham tithe award and map, 1843–4
DSA4/56/2 Hunsdon tithe map, 1842
DSA4/66/1–2 Letchworth tithe award and map, 1837
DSA4/72/1–2 Newnham tithe award and map, 1847
DSA4/111/2 Watford tithe map, 1842
D/Z42/Z1 Repton's Red Book for Tewin Water, 1799
D/Z44/P1 plan of Westfield Farm, 1767
D/Z61/P1 map of Bishops Stortford, 1823
D/Z110/P1–2 map of Ardeley, 1744
HCP1/1/25 County Development Plan, Bushey map, 1951
HCP2/1/1 East Herts Joint Planning Committee, 1938–46
IR1/36–7 1910 Finance Act maps, Bygrave
IR1/64 and 79 1910 Finance Act maps, Clothall
IR1/90, 91, 105, 106 1910 Finance Act maps, Hitchin
IR1/479–80 1910 Finance Act maps, Bushey
K100 accounts for the manor of Knebworth, 1370/1
K116 manorial account for Knebworth, 1408/9
K117 bailiff's accounts for the manor of Knebworth, 1410/11
K119 bailiff's accounts for the manor of Knebworth, 1412/13
Off Acc 269 Bushey & Watford Rural Joint Town Planning Committee, Map 2a, 1933
Off Acc 708 contract and plans for Watford bypass, 1924
Off Acc 717 Bushey Urban District Council minutes, 16 May 1906; 4 July 1906.
PC528 Pirton enclosure map, 1818
QS/E3 Aldenham enclosure map, 1803
QS/E9 Ashwell enclosure map, 1863
QS/E54 Norton enclosure map, 1796
QS/E70A–B Watford enclosure award and map, 1854

QS/E80–QS/E81 Great Wymondley enclosure map and award, 1811
QS/E82 Bishops Stortford inclosure map, 1822
RICK/5 drawing of 'More Park on the river Colne' by J.M.W. Turner, 1824
WATF/58 engraving of Cashiobury by J. Kip and L. Knyff, published in *Britannia Illustrata*, 1707

Newspapers:
County Press, p. 3, 5 July 1834
Hertfordshire Mercury, p. 3, 4 May 1861 and p. 5, 15 July 1871
Hertfordshire Mercury, p. 3, 26 May 1888
Morning Chronicle, 3 August 1810

King's College, Cambridge
KCAR/6/2/034 Cheshunt manor accounts, 1435/6

Leicester and Rutland Record Office
DE 4939/977

London School of Economics
Coll Misc 38/3 f.8 journal of Sir John Parnell

The National Archives: Public Record Office (TNA:PRO), Kew
C142/240/40 inquisition post mortem, Henry Hickman, 1593/4
DL43/7/4 survey of the boundaries of the manor of Enfield, 1571–2
E134/26Eliz/Hil7 and E134/MISC/2415 legal depositions concerning a wood called Mayne Wood, in the parish of Tringe, parcel of the manor, 1583–4
E134/43Eliz/East3 legal depositions concerning a warren and ground called the Leith, 1601
E134/4Jas1/Mich19 legal depositions concerning damage to crops and the pasture on the common caused by rabbits from a neighbouring warren, 1606–7
E199/10/2 Extent of manor of Bushey, held by John le Latimer (full details, including stock and grain), 1314–16
E210/407 Ancient Deeds, Series D
E315/391 survey of Hertfordshire manors, lands and possessions of King Philip and Queen Mary, 1556
E315/457 sales of wood in divers counties 35 Hen VIII, 1543–4
E315/458 accounts etc. of sales of woods, 1544–5
E317/Herts/19 parliamentary survey of Cheshunt park, 1650
E317/Herts/24 parliamentary survey of Theobalds manor, 1650
E317/Herts/27 parliamentary survey of Theobalds park, 1650
E317/Herts/29 parliamentary survey of the Manor of Tring, 1650
E326/4213 grant of the manor called 'le Hermebemgate' in Bishop's Hatfield
SC/1145/10 Ministers' and Receivers' Accounts, Bushey, 1321–2
SC2/176/120 Aldbury court roll, 1506
SC2/178/70 Tring estreats 1615–16 and 1621–3
SC6/868/20 bailiff's account for the manor of Standon, 1328/9
SC6/873/21 accounts for the manor of Weston, 1304/5
SC6/873/25 accounts for the manor of Weston, 1396/7
SP14/11/22 valor of the manor of Abbot's Langley, *c.* 1604
SP14/124/65 State Papers Domestic, James I, December 1621
STAC 3/1/49: bill of complaint of Sir William Cavendish concerning resistance to survey of waste land at Northaw for purposes of enclosure, 1547–53

Private collection
Map of the manor and parish of Benington, 1628

Secondary sources

Abercrombie, P., *Greater London Plan 1944* (London, 1945).

Adams, B. (ed.), *Lifestyle and culture in Hertford … Wills and inventories for the parishes of All Saints and St Andrew, 1660–1725*, Hertfordshire Record Publications 13 (Hitchin, 1997).

Adams, S. (ed.), *Household accounts and disbursement books of Robert Dudley, Earl of Leicester, 1558–1561, 1584–1586* (Cambridge, 1995).

Albert, W., *The turnpike road system in England 1663–1840* (Cambridge, 1972).

Archer, C., 'Printing', in Short (ed.), *Historical atlas*, pp. 98–9.

Archer, P.C., *Historic Cheshunt* (Cheshunt, 1923).

Ashby, M. and Cudmore, A., 'New Towns', in Short (ed.), *Historical atlas*, pp. 166–7.

Astle, T., Ayscough, S. and Caley J. (eds), *Taxatio Ecclesiastica Angliae et Walliae Auctoritate P.Nicholai IV circa AD 1291* (London, 1802).

Austin, P., 'Hatfield Great Wood and its inclosure', *Hertfordshire's Past*, 38 (1995), pp. 2–7.

Austin, P., 'The leasing of Lord Burghley's Hoddesdon woods in 1595. An insight into woodmanship', *Hertfordshire's Past*, 41 (1996), pp. 11–21.

Austin, P., 'Traditional woodland management. The woods of the Manor of Digswell 1776–1843', *Hertfordshire's Past*, 51 (2001), pp. 2–12.

Bailey, G. and Hutton, B., *Crown post roofs in Hertfordshire* (Chichester, 1966).

Bailey, H. and Wood, C., 'The Upper Cretaceous Chalk', in Catt (ed.), *Hertfordshire geology and landscape*, pp. 36–60.

Bailey, M., 'A tale of two towns: Buntingford and Standon in the later Middle Ages', *Journal of Medieval History*, 19/4 (1993), pp. 351–71.

Bailey, M., 'Introduction', in J. Brooker and S. Flood (eds), *Hertfordshire lay subsidy rolls, 1307 and 1334*, Hertfordshire Record Society 14 (Hertford, 1998), pp. ix–xxvii.

Bailey, M., 'The economy of towns and markets, 1100 to 1500', in Slater and Goose (eds), *County of small towns*, pp. 46–66.

Balchin, J., *First New Town: an autobiography of the Stevenage Development Corporation, 1946–1980* (Stevenage, 1980).

Ball, A., *Street and place names in Watford* (Watford, 1973).

Banham, D., 'In the sweat of thy brow thou shalt eat bread: cereals and cereal production in the Anglo-Saxon landscape', in N.J. Higham and M.J. Ryan (eds), *The landscape archaeology of Anglo-Saxon England* (Woodbridge, 2010), pp. 175–92.

Banister, K., 'Salads and ornamentals: a short history of the Lea Valley nursery industry', in Spring (ed.), *Hertfordshire garden history II*, pp. 202–29.

Barnes, G. and Williamson, T., *Hedgerow history: ecology, history and landscape character* (Macclesfield, 2006).

Barrett, H. and Phillips, J., *Suburban style: the British home, 1840–1960* (London, 1993).

Barton, P., *Manorial economy and society in Shenley: a Hertfordshire manor in the 13th and 14th centuries* (Hertford, 1981).

Bassett, S.J., 'In search of the origins of Anglo-Saxon kingdoms', in S.J. Bassett (ed.), *The origins of Anglo-Saxon kingdoms* (Leicester, 1989), pp. 3–27.

Beard, G. and Wardman, J., *Thomas H. Mawson: the life and work of a northern landscape architect* (Lancaster, 1978).

Beard, T., *By tube beyond Edgware* (London, 2002).

Bentley, E.G., *A practical guide in the preparation of town planning schemes* (London, 1911).

Bentley, I., Davis I. and Oliver, P., *Dunroamin: the suburban semi and its enemies* (London, 1981).

Beresford, G., 'Excavations at the deserted medieval village of Caldecote, Hertfordshire: an interim report', *Hertfordshire's Past*, 4 (1978), pp. 3–14.

Beresford, G., *Caldecote: the development and desertion of a Hertfordshire village*, Society for Medieval Archaeology Monograph 28 (London, 2009).

Betjeman, Sir J., 'Metro-land', BBC documentary (1973).

Biddle, M. and Kjolbye-Biddle, B., 'England's premier abbey: the medieval chapter house of St Albans and its excavation in 1978', *Hertfordshire's Past*, 11 (1981), pp. 3–29.

Biddle, M. and Kjolbye-Biddle, B., 'The origins of St Albans Abbey: Romano-British cemetery and Anglo-Saxon monastery', in M. Henig and P. Lindley (eds), *Alban and St Albans: Roman and medieval architecture, art and archaeology* (Leeds, 2001), pp. 45–77.

Boseley, G., 'Railways', in Short (ed.), *Historical atlas*, pp. 48–9.

Bowyer, M., *We have to deal with the farmers. Episodes in the history of North Hertfordshire in the 19th and 20th centuries* (Girton, 2010).

Branch Johnson, W., *The industrial archaeology of Hertfordshire* (Newton Abbot, 1970).

Branch Johnson, W. (ed.), *Memorandoms for … the diary between 1798 and 1810 of John Carrington* (London, 1973).

Branigan, K. and Niblett, R., *The Roman Chilterns* (Great Missenden, 2005).

Britten, J. (ed.), *William Turner's The names of herbs* (London, 1802).

Broket Archive, The, <http://www.brockett.info/hertfordshire/Hertsto1599.htm>.

Brooker, J. and Flood, S. (eds), *Hertfordshire lay subsidy rolls 1307 and 1334*, Hertfordshire Record Publications 14 (Hertfordshire Record Society, 1998).

Brown, T. and Foard, G., 'The Saxon landscape: a regional perspective', in P. Everson and T. Williamson (eds), *The archaeology of landscape* (Manchester, 1998), pp. 67–94.

Bryant, S., 'The late Bronze Age to the middle Iron Age of the north Chilterns', in Holgate (ed.), *Chiltern archaeology*, pp. 17–27.

Bryant, S., 'Iron Age settlements', in Short (ed.), *Historical atlas*, pp. 70–71.

Bryant, S. and Burleigh, G., 'Late prehistoric dykes of the eastern Chilterns', in Holgate (ed.), *Chiltern archaeology*, pp. 92–5.

Bryant, S. and Niblett, R., 'The late Iron Age in Hertfordshire and the north Chilterns', in A. Gwilt and C. Haselgrove (eds), *Reconstructing Iron Age societies* (Oxford, 1997), pp. 271–81.

Bryant, S., Perry, B. and Williamson, T., 'A "relict landscape" in south east Hertfordshire: archaeological and topographic investigations in the Wormley area', *Landscape History*, 27 (2005), pp. 5–15.

Burke, J. and Burke, J.B., *A genealogical and heraldic history of the extinct and dormant baronetcies of England, Scotland and Ireland*, 2nd edn (London, 1841).

Burnett, J., *A social history of housing, 1815–1970* (London, 1980).

Burnham, B.C. and Wacher, J., *The 'small towns' of Roman Britain* (London, 1990).

Calendar of Charter Rolls, vol. 2, Henry III, 1257–1300 (London, 1906).

Calendar of Close Rolls, Edward I, 1272–1279 (London, 1900).

Cameron, L.G., *The land of Britain, Part 80: Hertfordshire* (London, 1941).

Campbell, C., *Vitruvius Britannicus*, IV (London, 1739).

Carswell, J., *The South Sea Bubble* (London, 1960).

Catt, J. (ed.), *Hertfordshire geology and landscape* (Welwyn Garden City, 2010).

Catt, J., 'Introduction', in Catt (ed.), *Hertfordshire geology and landscape*, pp. 7–20.

Catt, J., 'Neogene deposits and early landscape development', in Catt (ed.), *Hertfordshire geology and landscape*, pp. 79–117.

Catt, J. and Cheshire, A., 'Quaternary deposits and later landscape development', in Catt (ed.), *Hertfordshire geology and landscape*, pp. 118–82.

Catt, J. and Doyle, J., 'The Palaeogene period', in Catt (ed.), *Hertfordshire geology and landscape*, pp. 61–78.

Catt, J., Edmonds, C., Bailey, H. and Green, C., 'Mineral resources and church building stones', in Catt (ed.), *Hertfordshire geology and landscape*, pp. 299–321.

Champion, A.G., Clegg, K. and Davies, R.L., *Facts about the New Towns: a socio-economic digest* (Corbridge, 1977).

Chauncy, Sir H., *The historical antiquities of Hertfordshire* (London, 1700).

Clutterbuck, R., *The history and antiquities of the county of Hertford*, vol. 1 (London, 1815).

Cockman, F.G., *Railways of Hertfordshire* (Stevenage, 1983).

Colgrave, B. and Mynors, R.A.B., *Bede's Ecclesiastical history of the English people*, revised edn (Oxford, 1991).

Colne Valley Water Company, *The Colne Valley Water Company 1873–1973* (London, 1973).

Colvin, H.M. (ed.), *The history of the king's works, vol. III 1485–1660* (London, 1975).

Cook, M., *The manner of raising, ordering and improving forest trees* (London, 1676).

Countryside Management Service, 'Hertfordshire's woodland strategy', unpublished report (*c.*1997), <http://www.hertsdirect.org/docs/pdf/h/wstrat>, accessed 13 February 2012.

Cowell, B., 'The Commons Preservation Society and the campaign for Berkhamsted Common, 1866–70', *Rural History: economy, society, culture*, 13 (2002), pp. 145–61.

Cowell, F., *Richard Woods (1715–1793): master of the pleasure ground* (Woodbridge, 2009).

Cox, J.C., 'Forestry', in W. Page (ed.), *VCH Hertfordshire*, vol. 4 (London, 1914), pp. 275–80.

Crick, J. (ed.), *Charters of St Albans* (Oxford, 2007).

Cunliffe, B., *Iron Age Britain* (London, 1995).

Curry, D., 'Tertiary', in D. Duff and A. Smith (eds), *The geology of England and Wales* (London, 1992), pp. 389–410.

Cushion, B., 'Braughing: woodland archaeological earthwork surveys', unpublished report for Hertfordshire County Council Environmental Department (2008).

Cussans, J.E., *History of Hertfordshire*, vol. 2 (London, 1876).

Daniell, H.C.N., 'Popes Manor, Essendon', *Transactions of the East Herts Archaeological Society*, 7/2 (1924), pp. 148–60.

Daniels, S., *Humphry Repton: landscape gardening and the geography of Georgian England* (New Haven and London, 1999).

Darley, G., *Villages of vision: a study of strange utopias* (London, 1975).

Davison, A., *The evolution of settlement in three parishes in south east Norfolk*, East Anglian Archaeology 49 (Dereham, 1990).

Dawes, M.C.B., *Register of Edward the Black Prince, part 4, 1351–1365* (London, 1933).

Day, J.R. and Reed, J., *The story of London's Underground* (London, 2008).

De Gray Birch, W., *Cartularium Saxonicum: a collection of charters relating to Anglo-Saxon history*, vol. 1 (London, 1885).

Defoe, D., *A tour thro' the whole island of Great Britain, divided into circuits or journies*, 2 Letter VI (1724–7), <http://www.visionofbritain.org.uk>, accessed 19 September 2010.

Defoe, D., *A Tour through England and Wales, 1724–6*, 1948 Everyman Edition, vol. II, pp. 122–3.

Doggett, N., 'The parish churches of Dacorum: how they reflect social history', *Hertfordshire's Past*, 11 (1981), pp. 28–36.

Doggett, N., 'The medieval chapels of the Chilterns: the evidence from south-west Hertfordshire', in Holgate (ed.), *Chiltern archaeology*, pp. 113–17.

Doggett, N., 'Medieval parish churches', in Short (ed.), *Historical atlas*, pp. 126–7.

Doggett, N. and Hunn, J., 'The origins and development of medieval Berkhamsted', *Hertfordshire's Past*, 18 (1985), pp. 18–36.

Donovan, J., 'Milestones', in Short (ed.), *Historical atlas*, pp. 44–5.

Drury, P. and Rodwell, W., 'Settlement in the later Iron Age and Roman periods', in D.G. Buckley (ed.), *The archaeology of Essex to AD 1500*, Council for British Archaeology Research Report 34 (London, 1980).

Dulley, F., Burleigh, G., Sawford, B., Rowe, S. and Short, D., 'Hedges and local history', *Hertfordshire's Past*, 3 (1977), pp. 15–24.

Dury, A. and Andrews, J., *A topographical map of Hartfordshire* (London, 1766; republished by Hertfordshire Publications (Stevenage, 1980, reprinted 1986; republished again in 2004)).

Elliott, B., *Victorian gardens* (London, 1986).

Ellis, W., *New experiments in husbandry* (London, 1736).

Ellis, W., *The modern husbandman* (London, 1742).

Ely, K.D. and Edwards, K., 'Cheshunt Park, Hertfordshire. An archaeological investigation by the Time Team', unpublished report (Bristol, 2003).

English Heritage, The National Heritage List for England, including listed buildings, scheduled monuments, registered parks and gardens <http://list.english-heritage.org.uk>.

Evans, J., *The endless web: John Dickinson and Co. Ltd 1804–54* (London, 1955).

Evelyn, J., *Silva*, II (York, 1801 edition).

Eyton, R.W., *Court, household and itinerary of King Henry II* (London, 1878).

Falvey, H., 'Crown policy and local economic context in the Berkhamsted Common enclosure dispute, 1618–42', *Rural History*, 12 (2001), pp. 131–42.

Falvey, H., 'The More revisited', *The Ricardian*, 18 (2008), pp. 92–9.

Field, J., *English field names: a dictionary* (Newton Abbot, 1972).

Filler, R., *Welwyn Garden City* (Chichester, 1986).

Fitzherbert, J., *Here begynneth a newe tracte or treatyse mooste profytable for all husbandmen* (London, 1523).

Fleming, A., 'The prehistoric landscape of Dartmoor: Part 1, south Dartmoor', *Proceedings of the Prehistoric Society*, 44 (1978), pp. 97–123.

Fleming, A., *The Dartmoor Reaves* (London, 1988).

Fletcher, A., 'Charles Bridgeman at Tring Park: a reassessment', in Rowe (ed.), *Hertfordshire garden history*, pp. 41–7.

Fletcher, S., 'Watercress growing', in Short (ed.), *Historical atlas*, pp. 94–5.

Flood, S. (ed.), *St Albans Wills 1471–1500*, Hertfordshire Record Publications 9 (Hitchin, 1993).

Fowler, P., *Farming in the first millennium* (Cambridge, 2002).

Friel, I., 'The Hicce – an Anglo-Saxon tribe of the Hitchin area', *Hertfordshire's Past*, 13 (1982), pp. 2–18.

Funnell, R., 'Handley Page Ltd., Radlett, 1929–1970', *Journal of the Watford and District Industrial History Society*, 18 (1994), pp. 25–9.

Gardiner, H.W., *A survey of the agriculture of Hertfordshire*, Royal Agricultural Society of England County Agricultural Surveys 5 (London, 1967).

Garfi, S., 'Appendix A: the Hertford Burghs: a survey of the evidence', in Kiln and Partridge, *Ware and Hertford*, pp. 115–27.

Gatland, E., 'Richard Woods in Hertfordshire', in Rowe (ed.), *Hertfordshire garden history*, pp. 106–33.

Gelling, M., *The early charters of the Thames valley* (Leicester, 1979).

Gelling, M. and Cole, A., *The landscape of place-names* (Stamford, 2000).

Gibson, A., 'A timber-framed smoke hood at Barkway', *Hertfordshire Archaeology*, 2 (1970), pp. 112–13.

Gibson, A., 'A base cruck hall at Offley', *Hertfordshire Archaeology*, 4 (1976), pp. 153–7.

Gibson, A., 'Timber-framed building in Hertfordshire', in D.F. Stenning and D.D. Andrews, *Regional variation in timber-framed buildings in England and Wales down to 1550* (Chelmsford, 1998), pp. 22–30.

Gibson, A., 'Timber buildings', in Short (ed.), *Historical atlas*, pp. 24–5.

Glennie, P., 'Continuity and change in Hertfordshire agriculture, 1550–1700: I – patterns of agricultural production', *Agricultural History Review*, 36 (1988), pp. 55–76.

Goose, N., 'Urban growth and economic development in early modern Hertfordshire', in Slater and Goose (eds), *County of small towns*, pp. 96–126.

Goose, N., 'Population, 1801–1901', in Short (ed.), *Historical atlas*, pp. 56–7.

Gover, J.E.B., Mawer, A. and Stenton, F.M., *The place-names of Hertfordshire* (Cambridge, 1938).

Grieve, M., *A modern herbal* (London, 1931), <http://www.botanical.com/botanical/mgmh/a/ash--073.html>, accessed 9 November 2012.

Hall, D., *The open fields of Northamptonshire* (Northampton, 1995).

Hammond, B. and Hammond, J.L., *The village labourer 1760–1832* (London, 1911).

Hanham, A.A., 'Guy, Henry (*bap.* 1631, *d.* 1711)', *Oxford dictionary of national biography* (Oxford, 2004); online edn, 2008 <http://www.oxforddnb.com/view/article/11798>, accessed 17 October 2011.

Harris, O., 'Nyn Hall, Northaw: an alternative history', *Hertfordshire Archaeology*, 13 (2004), pp. 99–100.

Harrison, S., 'Open fields and earlier landscapes: six parishes in south east Cambridgeshire', *Landscapes*, 3 (2002), pp. 35–54.

Harwood, K., 'A Hertfordshire garden in the eighteenth century', *Herts Past and Present*, 3rd ser., 2 (2003), pp. 12–16.

Harwood, K., 'Some Hertfordshire nabobs', in Rowe (ed.), *Hertfordshire garden history*, pp. 49–77.

Harwood, K., 'Some Arts and Crafts gardens in Hertfordshire', in Spring (ed.), *Hertfordshire garden history II*, pp. 148–76.

Hassell, J., *A tour of the Grand Junction* (London, 1819).

Hastie, S., *Hemel Hempstead: the story of New Town development 1947–1997* (Hemel Hempstead, 1997).

Henderson, P., 'Sir Francis Bacon's water gardens at Gorhambury', *Garden History*, 20 (1992), pp. 116–31.

Hertfordshire Gardens Trust and Williamson, T., *The parks and gardens of west Hertfordshire* (Hertford, 2000).

Hesse, M., 'Fields, tracks and boundaries in the Creakes, south Norfolk', *Norfolk Archaeology*, 41 (1992), pp. 305–24.

Hewett, C.A., 'Aisled timber halls and related buildings, chiefly in Essex', *Transactions of the Ancient Monuments Society*, 16 (1968), pp. 45–99.

Hicks, M., 'Neville, George (1432–1476)', *Oxford Dictionary of National Biography* (Oxford, 2004); online edn, 2008 <http://www.oxforddnb.com/view/article/19934>, accessed 13 October 2011.

Hill, D., '*Sulh* – the Anglo-Saxon plough *c.*1000 AD', *Landscape History*, 22 (2000), pp. 5–19.

Hill, D. and Cowie, R. (eds), *Wics: the early medieval trading centres of northern Europe* (Sheffield, 2001).

Hill, J. (ed.), *Layston*, Hertfordshire Family & Population History Society, Militia Series No. 16 (1990).

Hill, J. (ed.), *Welwyn*, Hertfordshire Family & Population History Society, Militia Series No. 64 (1996).

Hill, J.D., *Ritual and rubbish in the Iron Age of Wessex*, BAR British Series 242 (Oxford, 1995).

Hindle, B.P., *Roads, tracks and their interpretation* (London, 1993).

Hindle, S., 'Persuasion and protest in the Caddington common enclosure dispute 1635–1639', *Past & Present*, 158 (1998), pp. 37–78.

Hinton, D., *Archaeology, economy and society: England from the fifth to the fifteenth century* (London, 1990).

Hodge, C.A.H., Burton, R.G.O., Corbett, W.M., Evans, R. and Searle, R.S., *Soils and their use in eastern England* (Harpenden, 1984).

Hodges, R., *Dark Age economics* (London, 1982).

Holgate, R. (ed.), *Chiltern archaeology: recent work* (Dunstable, 1995).

Holgate, R., 'Early prehistoric settlement of the Chilterns', in Holgate (ed.), *Chiltern archaeology*, pp. 3–16.

Hooke, D., *Landscape of Anglo-Saxon England* (Leicester, 1998).

Hocke, D., *Trees in Anglo-Saxon England* (Woodbridge, 2010).
Hoskins, W.G., *The making of the English landscape* (London, 1955).
Hoskins, W.G. and Dudley Stamp, L., *The common lands of England & Wales* (London, 1963).
Howard, E., *Garden Cities of tomorrow* (London, 1902).
Howlett, B. (ed.), *Survey of the royal manor of Hitchin, 1676*, Hertfordshire Record Publications 16 (Hitchin, 2000).
Howlett, B., *Hitchin Priory Park. The history of a landscape park and gardens* (Hitchin, 2004).
Howlett, B., 'Manorial estate and market town: the early development of Hitchin', in Slater and Goose (eds), *County of small towns*, pp. 188–223.
Hudson, K., *The archaeology of the consumer society: the second industrial revolution in Britain* (London, 1983).
Hunn, J.R., *Reconstruction and measurement of landscape change*, BAR British Series 236 (Oxford, 1994).
Hunn, J.R., *Settlement patterns in Hertfordshire: a review of the typology and function of enclosures in the Iron Age and Roman landscape*, BAR British Series 249 (Oxford, 1996).
Hunn, J.R., *Tyttenhanger: excavation and survey in the parish of Ridge* (Oxford, 2004).
Hunter, J., *Field systems in Essex* (Colchester, 2003).
Jackson, A., *Semi-detached London: suburban development, life and transport 1900–1939* (London, 1973).
James, T., *Flora of Hertfordshire* (Welwyn Garden City, 2009).
Johnson, M., *Housing culture* (London, 1993).
Johnson, W., 'Hedges: a review of some early literature', *Local Historian*, 13 (1978), pp. 195–204.
Jones, A.C., '"Commotion time": The English Risings of 1549', PhD thesis (University of Warwick, 2003).
Journal of the House of Commons, 3, 1643–4 (London, 1802).
Kalm, P., *Kalm's account of his visit to England on his way to America in 1748*, trans. J. Lucas (London, 1892).
Kemp, E., *How to lay out a garden* (London, 1864).
Kiln, R.J. and Partridge, C.R., *Ware and Hertford: the story of two towns from birth to middle age* (Welwyn Garden City, 1995).
Kingsford, P., *A modern history of Brookmans Park* (North Mymms, 1983).
Kip, J. and Knyff, L., *Britannia illustrata* (London, 1707).
Landon, M., 'Making a mint: the archaeology of a late Iron Age industry', *Current Archaeology*, 234 (2009), pp. 20–25.
Langford, P., *A polite and commercial people* (Oxford, 1998).
Le Quesne, C., Capon, L. and Stevens, T., 'Post-excavation assessment of The Grove Estate, Watford', unpublished report, Hertfordshire County Council Environmental Department (2001).
Leiper, H., 'Mr Lancelot Brown and his Hertfordshire clients', in Spring (ed.), *Hertfordshire garden history II*, pp, 92–120.
Levett, A.E., *Studies in manorial history* (Oxford, 1983).
Lewis, C., 'Test pit excavations within currently occupied rural settlements: results of the HEFA CORS project in 2009', *Medieval Settlement Research*, 24 (2009), pp. 43–58.
Lewis, C., Mitchell-Fox, P. and Dyer, C., *Village, hamlet and field: changing medieval settlements in central England* (Manchester, 1997).
Liddiard, R., *The medieval park: new perspectives* (Macclesfield, 2007).
Lilley, K.D., *Urban life in the Middle Ages, 1000–1450* (Basingstoke, 2002).
Loades, D., 'Baeshe, Edward (*c.*1507–1587)', *Oxford Dictionary of National Biography* (Oxford, 2004); online edn, 2008 <http://www.oxforddnb.com/view/article/71698>, accessed 9 October 2011.
Longman, G., *A corner of England's garden: an agrarian history of south west Hertfordshire 1600–1850*, 2 volumes (Bushey, 1977).
Longman, G., *Bushey then and now: introduction*, 3rd edn (Bushey, 1986).
Longman, G., *Bushey in the Domesday Book*, Bushey Museum Information Booklet (Bushey, 1998).
Lord, E., 'Population, 1901–2001', in Short (ed.), *Historical atlas*, pp. 58–9.
Lysons, D., 'County of Hertfordshire: Chipping Barnet', in *The environs of London: volume 4: counties of Herts, Essex & Kent* (1796), pp. 1–8. <http://www.british-history.ac.uk/report.aspx?compid=45461&strquery=hertfordshire waste>, accessed 9 November 2012.
Manning, T., 'Rivers and navigations', in Short (ed.), *Historical atlas*, pp. 46–7.
Mead, W.R., *Pehr Kalm: a Finnish visitor to the Chilterns in 1748* (Aston Clinton, 2003).
Mercer, E., *English vernacular houses* (London, 1975).
Mileson, S., *Parks in medieval England* (Oxford, 2009).
Milledge, J., 'Hadham Hall and the Capel Family', in Spring (ed.), *Hertfordshire garden history II*, pp. 60–91.
Miller, M., *Letchworth: the first Garden City* (Chichester, 2002).

Miller, M., *English Garden Cities: an introduction* (London, 2010).
Millward, D., Ellison, R.A., Lake, R.D. and Loorlock, B.S., *The geology of the country around Epping: memoir for the 1:50,000 Sheet 240 of the British Geological Survey* (London, 1987).
Minet, W., *Hadham Hall and the Manor of Bawdes alias Hadham Parva* (Colchester, 1914).
Mingay, G.E., *Parliamentary enclosure in England: an introduction to its causes, incidence and impact* (London, 1997).
Moody, K., 'The history of the gardens at Holywell House, St Albans', unpublished report for the Hertfordshire Gardens Trust (2003).
Moore, C., *Hertfordshire windmills & windmillers* (Sawbridgeworth, 1999).
Morris, M. and Wainwright, A., 'Iron Age and Romano-British settlement and economy in the Upper Bulbourne Valley, Hertfordshire', in Holgate (ed.), *Chiltern archaeology*, pp. 68–75.
Morris, R., *Churches in the landscape* (London, 1989).
Munby, L., *The Hertfordshire landscape* (London, 1977).
Munby, L.M. (ed.), *The history of King's Langley* (King's Langley, 1963).
Niblett, R., 'A Catuvellaunian chieftain's burial from St Albans', *Antiquity*, 66 (1992), pp. 917–29.
Niblett, R., 'A new site at Verulamium (St Albans)', in Holgate (ed.), *Chiltern archaeology*, pp. 96–102.
Niblett, R., *Roman Hertfordshire* (Wimbourne, 1995).
Niblett, R., *The excavation of a ceremonial site at Folly Lane, Verulamium, Britannia* Monograph 14 (London, 1999).
Niblett, R., 'Roman settlements', in Short (ed.), *Historical atlas*, pp. 72–3.
Niblett, R. and Thompson, I., *Alban's buried towns: an assessment of St Albans' archaeology to AD 1600* (Oxford, 2005).
Norden, J., *Description of Hertfordshire* (Ware, 1598; repr. 1903).
Oosthuizen, S., 'Prehistoric fields into medieval furlongs? Evidence from Caxton, south Cambridgeshire', *Proceedings of the Cambridgeshire Antiquarian Society*, 86 (1998), pp. 145–52.
Oosthuizen, S., 'The roots of common fields: linking prehistoric and medieval field systems in west Cambridgeshire', *Landscapes*, 4 (2003), pp. 40–64.
Osbourne, M., *Defending Britain: twentieth-century military structures in the landscape* (Stroud, 2004).
Page, W. (ed.), *VCH Hertfordshire*, 4 vols (London, 1904–14).
Page, W. (ed.), *VCH Bedfordshire*, vol. 3 (London, 1912).
Page, W. (ed.), *VCH Buckinghamshire*, vol. 3 (London, 1925).
Page, W. and Round, J.H., 'Houses of Austin canons: abbey of Waltham Holy Cross', in W. Page and J.H. Round (eds), *A history of the county of Essex*, vol. 2 (London, 1907), pp. 166–72.
Partridge, C., 'Excavations at Puckeridge and Braughing 1975–79', *Hertfordshire Archaeology*, 7 (1979), pp. 28–97.
Partridge, C., *Skeleton Green: A late Iron Age and Romano-British site, Britannia* Monograph 2 (London, 1981).
Partridge, C., *Foxholes Farm: a multi-period gravel site* (Hertford, 1989).
Partridge, C., 'Hertford and Ware', in Slater and Goose (eds), *County of small towns*, pp. 127–58.
Pedley, W., *Labour on the land* (London, 1942).
Pedrick, M., *Moor Park: the Grosvenor legacy* (Rickmansworth, 1989).
Perman, D., 'Malting and brewing', in Short (ed.), *Historical atlas*, pp. 88–9.
Petit, S. and Watkins, C., 'Pollarding trees: changing attitudes to a traditional land management practice in Britain 1600–1900', *Rural History: Economy, Society, Culture*, 14/2 (2003), pp. 157–76.
Pevsner, N. and Cherry, B., *The buildings of England. Hertfordshire*, 2nd edn (Harmondsworth, 1977).
Pevsner, N. and Wilson, B., *Norfolk* (London, 2002).
Plowman, D., 'Royston: a thirteenth-century planned town?', in Slater and Goose (eds), *County of small towns*, pp. 173–87.
Plumb, P., 'Coaching routes, post roads and inns', in Short (ed.), *Historical atlas*, pp. 42–3.
Plumb, P., 'Turnpike roads', in Short (ed.), *Historical atlas*, pp. 40–41.
Pollard, E., Hooper M.D. and Moore, N.W., *Hedges* (London, 1974).
Potter, J.F., 'The London Basin's gravel churches: indications of geology, medieval history and geographical distribution', *Landscape History*, 23 (2001), pp. 5–28.
Prince, H., 'The changing landscape of Panshanger', *Transactions of the East Herts Archaeological Society*, 14/1 (1955–57), pp. 42–58.
Prince, H., *Parks in Hertfordshire since 1500* (Hatfield, 2008).
Pugh, R.B. (ed.), *VCH Middlesex*, vol. 5 (Oxford, 1976).
Rackham, O., *The history of the countryside* (London, 1986).
Rackham, O., *Ancient woodland: its history, vegetation and uses in England* (Dalbeattie, 2003; first pub. 1980).

Renn, D., *Medieval castles in Hertfordshire* (Chichester, 1971).

Repton, H., *Humphry Repton's Red Books of Panshanger and Tewin Water, Hertfordshire, 1799–1800*, with an introduction by T. Way, Hertfordshire Record Publications 27 (Hitchin, 2011).

Richards, S., *A history of Tring* (Tring UDC, 1974).

Richardson, G., 'Henry VIII and travel', p. 2, <www.hrp.org.uk/Resources/Glen%20RichardsonFINAL.pdf>, accessed 30 December 2009.

Ridgeway, C., 'William Andrews Nesfield: between Uvedale Price and Isambard Kingdom Brunel', *Journal of Garden History*, 13 (1993), pp. 69–89.

Riley, H.T., *Gesta Abbatum Monasterii S. Albani*, Rolls Series (London, 1867).

Riley, H.T., *Gesta Abbatum Monasterii S. Albani*, Vol III (London, 1869).

Roberts, E., 'Totternhoe Stone and flint in Hertfordshire churches', *Medieval Archaeology*, 18 (1974), pp. 66–89.

Roberts, E., 'Parish church studies: an analysis of Hertfordshire plans', *Hertfordshire's Past*, 4 (1977), pp. 22–7.

Roberts, E., *A school of masons in 15th century north Hertfordshire* (Hertford, 1979).

Roberts, E., *The hill of the martyr: an architectural history of St Albans Abbey* (Dunstable, 1993).

Roden, D., 'Woodland and its management in the medieval Chilterns', *Forestry*, 41 (1968), pp. 59–71.

Roden, D., 'Enclosure in the Chiltern Hills', *Geografiska Annaler*, 52B (1969), pp. 115–26.

Roden, D., 'Field systems of the Chiltern Hills and their environs', in A.H.R. Baker and R.A. Butler (eds), *Studies of field systems in the British Isles* (Cambridge, 1973), pp. 325–76.

Rodwell, W., 'Relict landscapes in Essex', in H.C. Bowen and P.J. Fowler (eds), *Early land allotment* (London, 1978), pp. 89–98.

Roe, D., *The Lower and Middle Palaeolithic periods in Britain* (London, 1981).

Rowe, A., 'Country house chameleon. The story of Hamels Mansion', *Hertfordshire's Past*, 43/44 (1998), pp. 44–54.

Rowe, A., *The protected valley: a history of Little Munden* (Buntingford, 1999).

Rowe, A. (ed.), *Garden making and the Freman family: a memoir of Hamels 1713–1733*, Hertfordshire Record Publications 17 (Hertford, 2001).

Rowe, A. (ed.), *Hertfordshire garden history: a miscellany* (Hatfield, 2007).

Rowe, A., *Medieval parks of Hertfordshire* (Hatfield, 2009).

Rowe, A., 'Rabbit warrens', in Short (ed.), *Historical atlas*, pp. 112–13.

Rowe, A., 'Hertfordshire's lost water gardens 1500–1750', in Spring (ed.), *Hertfordshire garden history II*, pp. 31–59.

Rowe, A. and Williamson, T., 'New light on Gobions', *Garden History*, 40 (2012), pp. 82–97.

Rowe, A., Taylor, C. and Williamson, T., 'The earthworks at Benington Park, Hertfordshire: an exercise in dating an archaeological garden', *Landscape History*, 32 (2011), pp. 37–56.

Rowley, T., *The English landscape in the twentieth century* (London, 2006).

Rowlinson M. *et al.*, *'In Arena Virtus': 100 years of sand and spirit* (Sandy Lodge, 2010).

Rushton, N., 'Religious houses', in Short (ed.), *Historical atlas*, pp. 130–31.

Ruston, A., 'Protestant nonconformity', in Short (ed.), *Historical atlas*, pp. 134–5.

Salmon, N., *Hertfordshire* (London, 1728).

Saunders, C., 'A sixteenth-century tannery in St Albans', *Hertfordshire's Past*, 3 (1977), pp. 9–12.

Saunders, C., 'Some thoughts on the *oppida* at Wheathampstead and Verulamium', *Hertfordshire Archaeology*, 8 (1982), pp. 31–9.

Saussure, C. de, *A foreign view of England in the reigns of George I and George II* (London, 1902).

Scholefield, J., *The Private Street Works Act, 1892: being a practical guide to the working of the act* (London, 1902).

Seebohm, F., *The English village community* (London, 1883).

Sheail, J., *Rabbits and their history* (Newton Abbot, 1971).

Sherwood, J., 'Influences on the growth and development of medieval and early modern Berkhamsted', in Slater and Goose (eds), *County of small towns*, pp. 224–48.

Short, D., 'Braughing: a possible Saxon estate?', *Hertfordshire's Past*, 23 (1988), pp. 8–15.

Short, D., 'Ashwell: an example of Anglo-Saxon town planning', in Slater and Goose (eds), *County of small towns*, pp. 159–72.

Short, D. (ed.), *An historical atlas of Hertfordshire* (Hatfield, 2011).

Skelton, A.C., 'The development of the Briggens estate, Hunsdon, since 1720', *Hertfordshire Archaeology*, 12 (1994–96), pp. 111–28.

Slater, G., *The English peasantry and the enclosure of common fields* (London, 1907).

Slater, T., 'Roads, commons and boundaries in the topography of Hertfordshire towns', in Slater and Goose (eds), *County of small towns*, pp. 67–95.

Slater, T. and Goose, N. (eds), *A county of small towns* (Hatfield, 2008).

Smith, A.H., *English place-name elements*, vol. I (Cambridge, 1956).

Smith, J.T., *English houses 1200–1800: the Hertfordshire evidence* (London, 1992).

Smith, J.T., *Hertfordshire houses. Selective inventory* (London, 1993).

Smith, T.P., *The Anglo-Saxon churches of Hertfordshire* (Chichester, 1973).

Spooner, S., '"A prospect two fields' distance": rural landscapes and urban mentalities in the eighteenth century', *Landscapes*, 10/1 (2009), pp. 101–12.

Spooner, S., 'The diversity of designed landscapes: a regional approach *c*.1660–1830', PhD thesis (University of East Anglia, 2010).

Spratt, D., *The linear earthworks of the Tabular Hills, North Yorks*, Department of Archaeology and Prehistory, University of Sheffield (Sheffield, 1989).

Spring, D. (ed.), *Hertfordshire garden history II: gardens pleasant, groves delicious* (Hatfield, 2012).

Spring, D., 'The London connection', in Spring (ed.), *Hertfordshire garden history II*, pp. 1–30.

Stanyon, M., 'Papermaking', in Short (ed.), *Historical atlas*, pp. 80–81.

Stoertz, C., *Ancient landscapes of the Yorkshire Wolds: aerial photographic transcription and analysis*, English Heritage (Swindon, 1997).

Stokes, P.R., 'Power of flight Pt. 1: De Havilland and the piston engine', *Journal of the Watford and District Industrial History Society*, 10 (1980), pp. 36–66.

Summerson, J., 'The building of Theobalds 1564–85', *Archaeologia*, 97 (1959), pp. 107–26.

Tate, W.E., *Domesday of English enclosure acts and awards* (Reading, 1978).

Taylor, H.M. and Taylor, J., *Anglo-Saxon architecture* (Cambridge, 1965).

Taylor, P., 'Boundaries and margins: Barnet, Finchley and Totteridge', in M.J. Franklin and C. Harper-Bill (eds), *Medieval ecclesiastical studies in honour of Dorothy M. Owen* (Woodbridge, 1995), pp. 259–79.

Taylor, P., 'Boundaries, margins and the delineation of the urban: the case of Barnet', in Slater and Goose (eds), *County of small towns*, pp. 249–75.

Thompson, I., 'Braughing: Extensive Urban Survey Project assessment report', unpublished, for English Heritage and Hertfordshire County Council (2002).

Thompson, I., 'Markets and fairs', in Short (ed.), *Historical atlas*, pp. 154–5.

Thompson, K., 'Garden Cities', in Short (ed.), *Historical atlas*, pp. 162–5.

Tittler, R., 'Bacon, Sir Nicholas (1510–1579)', *Oxford Dictionary of National Biography* (Oxford, 2004); online edn, 2011 <http://www.oxforddnb.com/view/article/1002>, accessed 9 October 2011.

Tomkins, M., *So that was Hertfordshire: travellers' jottings 1322–1887* (Hertford, 1998).

Tott, V., *The Braughing of Victor Tott: the visual and written memories of a country gentleman* (Quebec, 2010).

Toulmin Smith, L. (ed.), *The itinerary of John Leland in or about the years 1535–43* (London, 1907).

Turner, M., *English parliamentary enclosure* (Folkestone, 1980).

Vera, F., 'The dynamic European forest', *Arboricultural Journal*, 26 (2002), pp. 179–211.

Vera, F., *Grazing ecology and forest history* (Wallingford, 2002).

Wade Martins, S. and Williamson, T., *The countryside of East Anglia: changing landscapes 1870–1950* (Woodbridge, 2008).

Wallace, J., 'Sand and gravel quarrying', in Short (ed.), *Historical atlas*, pp. 104–5.

Ward, A.J., *The early history of papermaking at Frogmore Mill and Two Waters Mill, Hertfordshire* (Berkhamsted, 2003).

Warner, P., *Greens, commons and clayland colonization* (Leicester, 1987).

Way, T., 'Introduction', in H. Repton, *Humphry Repton's Red Books of Panshanger and Tewin Water, Hertfordshire, 1799–1800*, Hertfordshire Record Publications 27 (Hitchin, 2011).

White, P., *The Arrow Valley, Herefordshire: archaeology, landscape change and conservation* (Hereford, 2003).

White, S., 'Film and television studios', in Short (ed.), *Historical atlas*, pp. 102–3.

Wilkerson, J., *Two ears of barley* (Royston, 1969).

Wilkerson, J.C. (ed.), *John Norden's survey of Barley, Hertfordshire 1593–1603* (Cambridge, 1974).

Williams, H., *History of Watford* (London, 1884).

Williams, L.R. and McLaughlin, J., 'Tracing old hedgerows in suburban areas', *Local Historian* (May 1988), pp. 65–8.

Williamson, T., 'The development of settlement in north-west Essex: the results of a recent field survey', *Essex Archaeology and History*, 17 (1986), pp. 120–32.

Williamson, T., 'Early co-axial field systems on the East Anglian boulder clays', *Proceedings of the Prehistoric Society*, 53 (1987), pp. 419–31.

Williamson, T., 'The "Scole–Dickleburgh Field System" revisited', *Landscape History*, 20 (1998), pp. 19–28.

Williamson, T., *Shaping medieval landscapes: settlement, society, environment* (Macclesfield, 2003).
Williamson, T., 'The character of Hertfordshire's parks and gardens', in Rowe (ed.), *Hertfordshire garden history*, pp. 1–25.
Williamson, T., *Rabbits, warrens and archaeology* (Stroud, 2007).
Williamson, T., *The origins of Hertfordshire* (Hatfield, 2010).
Williamson, T., 'Gardens and industry: the landscape of the Gade Valley in the nineteenth century', in Spring (ed.), *Hertfordshire garden history II*, pp. 121–47.
Willis, P., *Charles Bridgeman and the English landscape garden*, with supplementary plates and catalogue (Newcastle upon Tyne, 2002).
Winterbottom, M. (ed.), *Gildas: The Ruin of Britain and other works* (London, 1978).
Witney, K.P., 'The woodland economy of Kent, 1066–1348', *Agricultural History Review*, 38 (1998), pp. 20–39.
Wooldridge, S.W. and Hutchings, G.F., *London's countryside* (London, 1957).
Young, A., *General view of the agriculture of Hertfordshire* (London, 1804).

Index

Page numbers in *italic* refer to figures and their captions.
Main references to subjects have the relevant page numbers in **bold**.
n denotes footnote.
NB one page may contain multiple references.